A Comprehensive Introduction to DIFFERENTIAL GEOMETRY

VOLUME TWO

MICHAEL SPIVAK

ISBN 0-914098-00-4
Library of Congress Card Catalogue Number 73-76372

PUBLISH OR PERISH, INC.
6 BEACON STREET
BOSTON, MASS. 02108 (U.S.A.)

We sell books direct and do not discount. Books are sold prepaid only. Pro forma invoices can be sent to libraries.

Volume 1 $12.50
Volume 2 $10.50

Add .50 per volume outside the United States, Canada and Mexico.

Please make checks payable to Publish or Perish, Inc.

If you want books sent First Class, Airmail, etc., we will bill you for the additional postage. Please note that the shipping weights are:

Volume 1: 5 lbs.
Volume 2: 3 lbs.

For

Harry

Josh

and

Marc

PREFACE

Somewhat to my surprise, the second volume of these notes came out more or less as I had planned. Actually, that statement is only half true, for only half the material which I had planned to write has come out at all; the rest will have to fit into Volume III, which I pray will be the last. This Volume ends at a suitable stopping point, when we have finally gotten to the fanciest definition of a connection, but hardly begun to start spouting theorems. A glance at the table of Contents will show that the semi-historical path promised in Volume I really has been followed. The most decisive encounters with classical differential geometry occur in Chapters 3 and 4, which are divided into parts; some parts present the classical papers, and other parts explain them. It is possible to get through the notes without reading any of the classical works themselves, but the easy way out misses all the fun. The students in my class found Gauss exciting, and while they found Riemann's own words bewildering, they were amazed when they learned what Riemann had been trying to say. To read Gauss, one should get hold of the Raven Press translation of Gauss' <u>General Investigations of Curved Surfaces</u>. There are two selections from Riemann, both of which have been included in the notes. As far as I know, no translation of the Latin paper in Chapter 4C has previously been available. Since most of the paper had more symbols than words, the fact that I don't know Latin didn't hinder me much; when Riemann got a bit wordier I did need help, which was kindly supplied by Martin Flashman. A translation of Riemann's great paper <u>On the Hypotheses which lie at the Foundations of Geometry</u> appears in Chapter 4A. I am grateful to Robert Rosner for his aid in deciphering the nearly impenetrable German.

The material in this Volume covers about what I would have completed in my class this term, had political events not intervened. There are no Problem sets, which is a shame, but much of the material doesn't lend itself to problems, and even if it did, I would have gone beserk trying to turn out another 100 or so pages in any reasonable length of time. This same petty concern for the remnants of my sanity deterred me from including a Chapter on Classical Mechanics in Volume I, or a Chapter on General Relativity in this volume, or, though I would dearly have loved to write it, a Chapter 4E on Riemann as a prophetic physicist. However, those who know something about General Relativity, and have heard it said again and again how this theory represents a great confirmation of Riemann's ideas, will be very interested to see if this popularized view is borne out by a reading of Riemann.

As compensation for the lack of Problems, I hope to have in Volume III a comprehensive bibliography of the literature of Differential Geometry, including texts where problems may be found. I think I had also better apologize in advance for the misprints; much of this material was completed after the academic year had already suffered its premature death, so that I was the only one to proofread it, which probably means that misprints and outright errors abound. An Errata for Volume I is included at the very end of this Volume; it is preceeded by the table of Contents for Volume I, which several people have told me they missed. I can't think of anything else that needs to be said, except that "Theorem I.3-5" refers, as you could undoubtedly figure out, to Theorem 3-5 of Volume I. Finally, I would like to thank Roger Weissberg, but I can't bring myself to do it.

Michael Spivak
Brandeis University
July 6, 3:30 A.M., 1970

CONTENTS OF VOLUME II

PREFACE

[Although the Chapters are not divided into sections, the list for each Chapter gives some indication which topics are treated, and on which pages.]

CHAPTER 4A. AN INAUGURAL LECTURE (4A-1 to 4A-20)

CHAPTER 4B. WHAT DID RIEMANN SAY? (4B-1 to 4B-38)

CHAPTER 4C. A PRIZE ESSAY (4C-1 to 4C-5)

CHAPTER 4D. THE BIRTH OF THE RIEMANN CURVATURE TENSOR (4D-1 to 4D-26)

CHAPTER 5. THE ABSOLUTE DIFFERENTIAL CALCULUS (THE RICCI CALCULUS); OR, THE DEBAUCH OF INDICES (5-1 to 5-24)

CHAPTER 6. THE ∇ OPERATOR (6-1 to 6-42)

CHAPTER 6. The ∇ OPERATOR (6-1 to 6-42) - Cont.

CHAPTER 7. THE REPERE MOBILE (THE MOVING FRAME)(7-1 to 7-58)

CHAPTER 8. CONNECTIONS IN PRINCIPAL BUNDLES (8-1 to 8-62)

Chapter 1. Curves in the Plane and in Space

Volume I of these notes represents the "differential" part of differential geometry. In this volume we finally get down to some geometry. For the present we are going to study only the simplest geometric objects, curves, and at first our approach will be terribly geometric. Nevertheless, the results of this Chapter span a couple hundred years; we will end with some very modern looking constructions, and prepare the way for a corresponding modernization of the classical surface theory which we shall consider in the succeeding chapters.

We begin by considering only curves in the plane, and we further restrict our attention to curves $c\colon [a,b] \to \mathbb{R}^2$ which are immersions, i.e. which satisfy $c'(t) = dc/dt \neq 0$ for all $t \in [a,b]$. For these curves, the arclength function $s\colon [a,b] \to \mathbb{R}$,

$$s(t) = \int_a^t |c'(u)|\,du \quad ,$$

is a diffeomorphism $s\colon [a,b] \to [0,L]$, where L = length of c. The curve $\gamma = c \circ s^{-1}$ is then a reparameterization of c; clearly γ is parameterized by arclength, $|\gamma'(s)| = 1$. We have just introduced a convention to be used throughout the Chapter; for curves $\gamma\colon [c,d] \to \mathbb{R}^2$ with unit tangent vectors, we will usually denote a typical point in $[c,d]$ by s, even at the risk of confusing it with the arclength function defined above. We also emphasize that throughout this Chapter $\gamma'(s)$ just denotes a vector in $\mathbb{R}^3$, not a tangent vector in $\mathbb{R}^3_{\gamma(s)}$, even though we will often draw it that way.

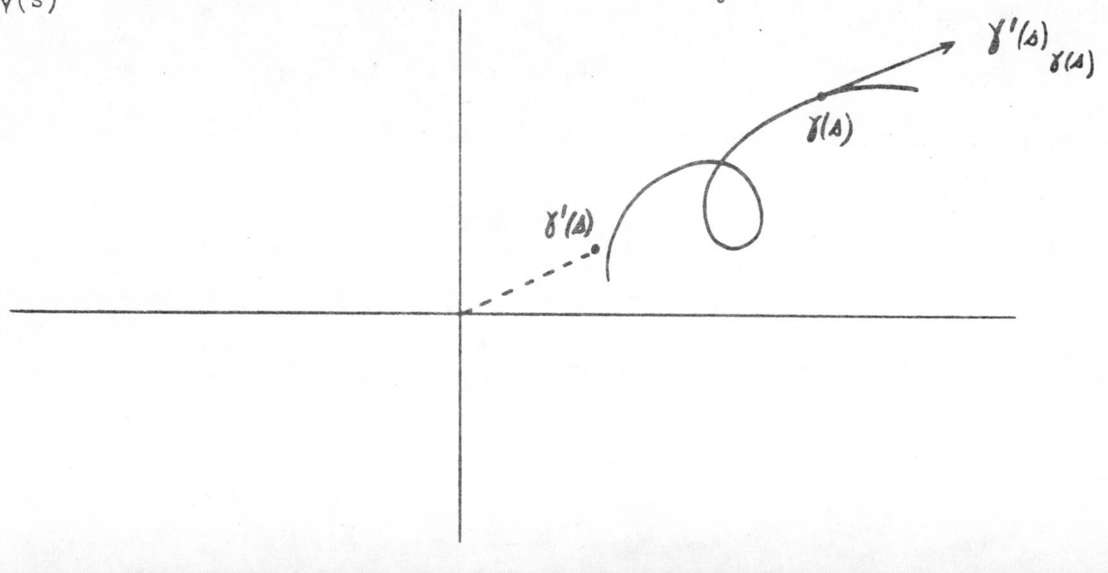

For an immersed curve $c: [a,b] \longrightarrow \mathbf{R}^2$, we would like a way of measuring the amount that c is "curving" at any point. No matter how vague this, as

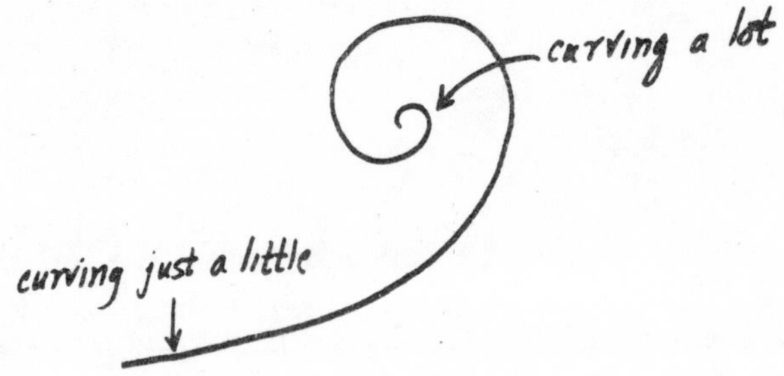

yet intuitive, term may be, we will surely all agree that

a) a straight line is not curving at all

b) a circle of radius $R > r$ is curving less than the circle of radius r.

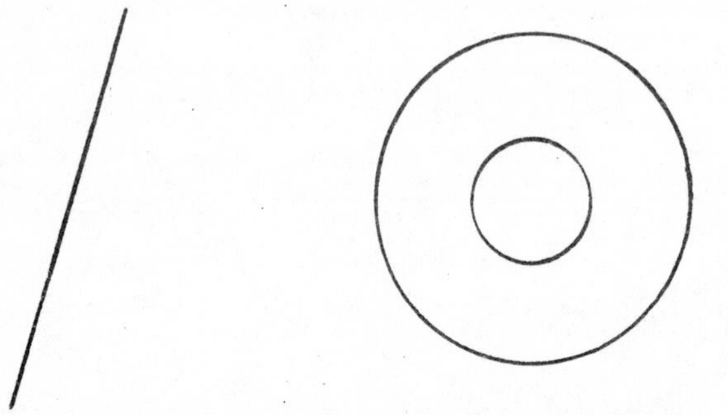

If we have to attach a numerical measure of curvature to these particular curves, it seems reasonable to define the curvature of a straight line at any point to be 0, and to define the curvature of a circle of radius r to be $1/r$ at any point.

From these special cases we want to develop a definition that works for any curve. We take as a clue the precedure which we use in a similar, but simpler, case. We can easily define what we mean by the <u>direction</u> of a curve $c: [a,b] \longrightarrow \mathbf{R}^2$ at any time t. This direction is determined by the tangent line at $c(t)$, which is the limit of lines through $c(t)$ and $c(t')$ as $t' \longrightarrow t$. This

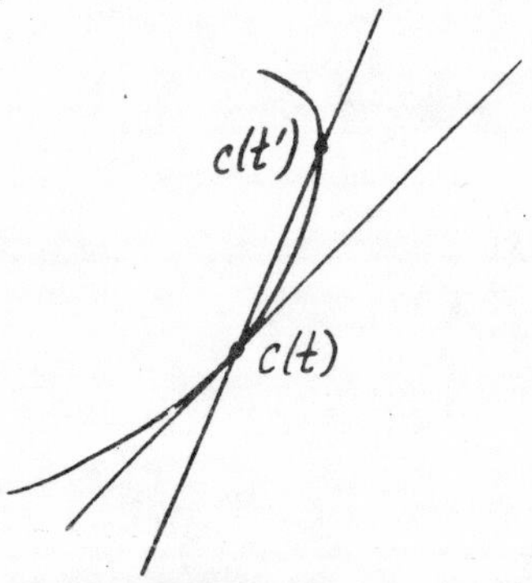

limit exists if $c'(t)$ exists and is non-zero. If, moreover, c' is continuous at t, then this limit line can be described as the limit, as $t_1, t_2 \longrightarrow t$, of the line through $c(t_1)$ and $c(t_2)$; for this line is parallel to the tangent through $c(\xi)$ for some ξ between t_1 and t_2.

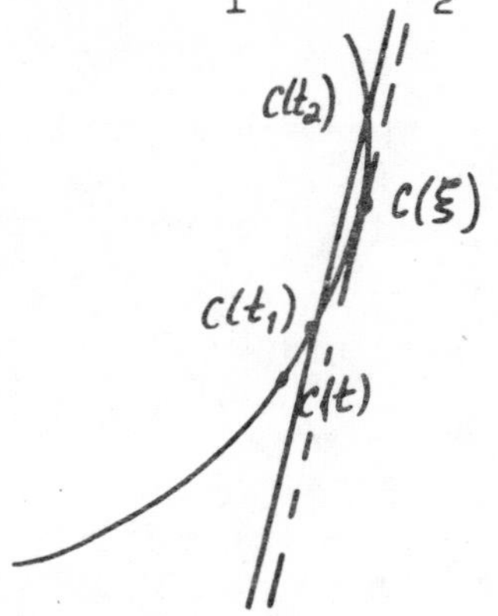

In order to determine the curvature of a curve we follow an analogous procedure. We find the circle which passes through three points $c(t_1), c(t_2), c(t_3)$ and then see if this circle approaches a limiting circle as $t_1, t_2, t_3 \longrightarrow t$. If it does, we can define the curvature of c at t to be the reciprocal of the radius of this circle.

Before proving a precise theorem, we simply try to determine the position of this circle, assuming it does exist. This can be done as follows. First of all, since c is an immersion, it is locally one-one, so for distinct t_1, t_2, t_3 near t, the points $c(t_1), c(t_2), c(t_3)$ are distinct. To be specific, let us say that $t_1 < t_2 < t_3$. Suppose also that $c(t_1), c(t_2), c(t_3)$ do not lie on a straight line, so that there is a unique circle through these three points, with center $C(t_0, t_1, t_2)$. Consider the function

$$t \longmapsto \langle c(t) - C(t_0, t_1, t_2),\ c(t) - C(t_0, t_1 t_2) \rangle \ .$$

At t_1, t_2, t_3 this function has the same value, namely the radius of the circle through $c(t_1), c(t_2), c(t_3)$. So its derivative must be 0 at points $\xi_1 \in (t_1, t_2)$ and $\xi_2 \in (t_2, t_3)$:

$$(1)\quad 0 = \langle c'(\xi_i),\ c(\xi_i) - C(t_0, t_1, t_2) \rangle \qquad \xi_i \in (t_i, t_{i+1}) \qquad i = 1, 2\ .$$

Similarly, the function $t \mapsto \langle c'(t), c(t) - C(t_0,t_1,t_2) \rangle$ must have derivative 0 at some point $\eta \in (\xi_1,\xi_2)$:

$$(2) \quad \langle c''(\eta), c(\eta) - C(t_0,t_1,t_2) \rangle = - \langle c'(\eta), c'(\eta) \rangle .$$

Now, if the points $C(t_0,t_1,t_2)$ approach a point C as $t_0,t_1,t_2 \to t$, and if c'' is continuous, then (1) and (2) clearly imply that

$$(1') \quad \langle c'(t), c(t) - C \rangle = 0$$

$$(2') \quad \langle c''(t), c(t) - C \rangle = - \langle c'(t), c'(t) \rangle .$$

The first of these equations shows that the circle through $c(t)$ with center C must be tangent to c at $c(t)$, which is certainly to be expected.

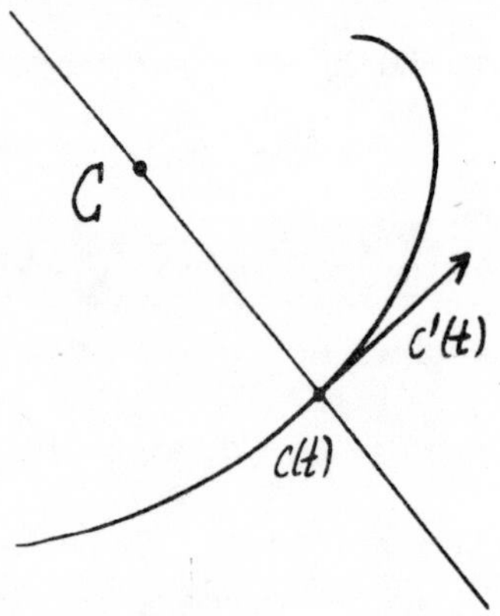

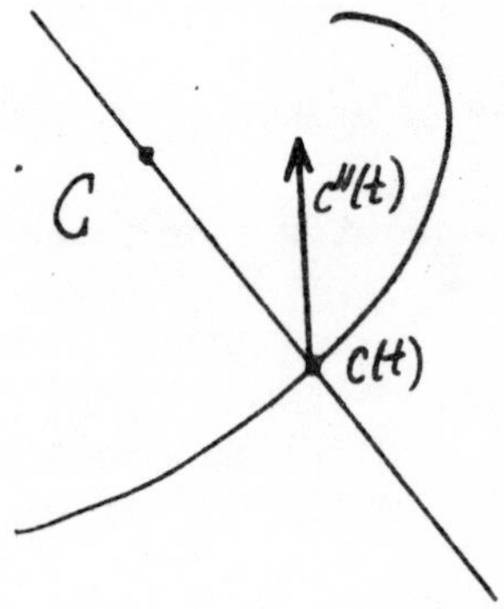

Thus C is already restricted lie along a certain line. If $c''(t)$ is not a multiple of $c'(t)$, the second equation then determines C, since it tells us the inner product of $c(t) - C$ with $c''(t)$. If $c''(t)$ <u>is</u> a multiple of $c'(t)$, then we obtain the contradiction

$$0 \neq - \langle c'(t), c'(t) \rangle = \langle c''(t), c(t) - C \rangle = \text{constant} \cdot \langle c'(t), c(t) - C \rangle = 0.$$

In other words, if $c''(t)$ is a multiple of $c'(t)$, this limiting position cannot exist.

Although equations (1'), (2') could be solved for C, we can make things a lot easier for ourselves by considering a curve c parameterized by arc-length, so that $|c'(s)| = 1$. (This means that $c'(s)$ always lies on the unit circle $S^1 \subset \mathbb{R}^2$, even though we often picture it instead as a tangent vector.) Now the equation

$$< c'(s), c'(s) > = 1$$

can be differentiated to give

$$(*) \qquad < c''(s), c'(s) > = 0 \quad .$$

In other words, $c''(s)$ is always perpendicular to $c'(s)$. In particular, $c''(s)$ is a multiple of $c'(s)$ only when $c''(s) = 0$. Equations 2' and (*) show that $c''(s)$ and $c(s) - C$ are both perpendicular to $c'(s)$. If $c''(s) \neq 0$, then we can write $c(s) - C = a \cdot c''(s)$. Substituting in (2') gives

$$a \cdot < c''(s), c''(s) > = - < c'(s), c'(s) > = -1 \; .$$

Since we also have

$$|c(s) - C| = |a| \cdot |c''(s)| \quad ,$$

we easily deduce that

$$|c(s) - C| = \frac{1}{|c''(s)|} \quad .$$

In other words, our circle is perpendicular to c at $c(s)$, and has radius $1/|c''(s)|$. Its curvature, and hence the curvature of c at s, is thus $|c''(s)|$. We are ready to reverse the order of this reasoning, and take care of details which we have ignored.

<u>1. THEOREM</u> Let $c\colon [a,b] \longrightarrow \mathbb{R}$ be a C^2 curve parameterized by arc-length. If $c''(s) \neq 0$, then for s_1, s_2, s_3 sufficiently close to s, the points $c(s_1), c(s_2), c(s_3)$ do not lie on a straight line. As $s_1, s_2, s_3 \longrightarrow s$ the unique circle through the points $c(s_i)$ approaches a circle passing through $c(s)$, whose radius is $1/|c''(s)|$, and whose center lies on the line through $c(s)$ perpendicular to the tangent line through $c(s)$. If $c''(s) = 0$, then, even if the points $c(s_i)$ do not lie on a line, the circles through them do not approach a limiting circle.

<u>Proof</u>. We first show that if $c''(s) \neq 0$, then the points $c(s_1), c(s_2), c(s_3)$ cannot lie on a line for s_1, s_2, s_3 arbitrarily close to s. Whenever the points $c(s_i)$ lie on a line, for $s_1 < s_2 < s_3$, there are points $\xi_1 \in (s_1, s_2)$ and $\xi_2 \in (s_2, s_3)$ where the tangent lines are parallel to this line (Cauchy mean

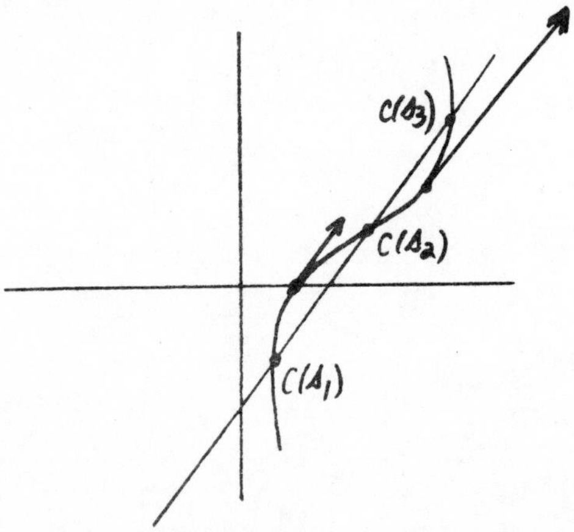

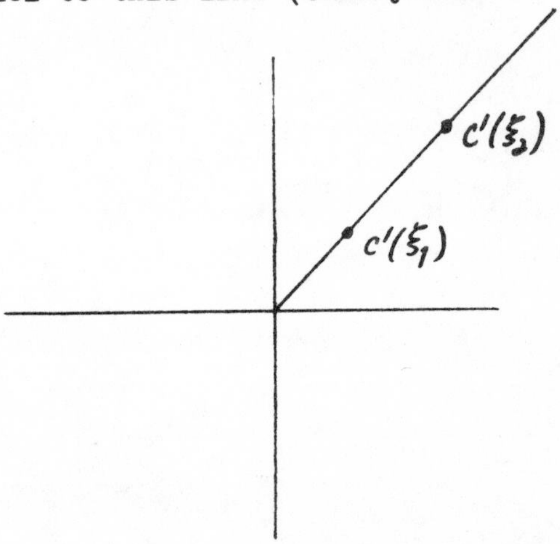

value theorem). This means that the curve c' in $\mathbb{R}^2$ has $c'(\xi_2)$ and $c'(\xi_1)$ on the same line through 0. Consequently, the tangent vector of c' is parallel to this line at some η, which means that $c''(\eta)$ is a multiple of

$c'(\xi_1)$ [and $c'(\xi_2)$]. If this happened arbitrarily close to s, then $c''(s)$ would be a multiple of $c'(s)$, and hence 0, a contradiction.

Now let C be the unique point satisfying

$$(*) \qquad \begin{aligned} &< c'(s), c(s) - C > = 0 \\ &< c''(s), c(s) - C > = - < c'(s), c'(s) > = -1 \;. \end{aligned}$$

For $s_1 < s_2 < s_3$ close to s, let $C(s_0, s_1, s_2)$ be the center of the unique circle through the points $c(s_i)$. We have already seen that

$$< c'(\xi), c(\xi) - C(s_0, s_1, s_2) > = 0 \qquad \xi \in (s_1, s_3)$$

$$< c''(\eta), c(\eta) - C(s_0, s_1, s_2) > = - < c'(\eta), c'(\eta) > = -1 \qquad \eta \in (s_1, s_3)$$

Since $c'(\xi) \rightarrow c(s)$ and $c''(\eta) \rightarrow c(s)$ as $s_i \rightarrow s$, comparison of these equations with $(*)$ shows that $C(s_0, s_1, s_2)$ must approach C.

We have already shown that if $c''(s) = C$, then $C(s_0, s_1, s_2)$ cannot approach a limiting position as $s_i \rightarrow s$. ∎

The circle determined in Theorem 1 is called the <u>osculating circle</u> of c at s ("osculate" means to kiss). It is clearly the circle which best approximates the curve c at $c(s)$. This suggests that we define $|c''(s)|$ to be the curvature

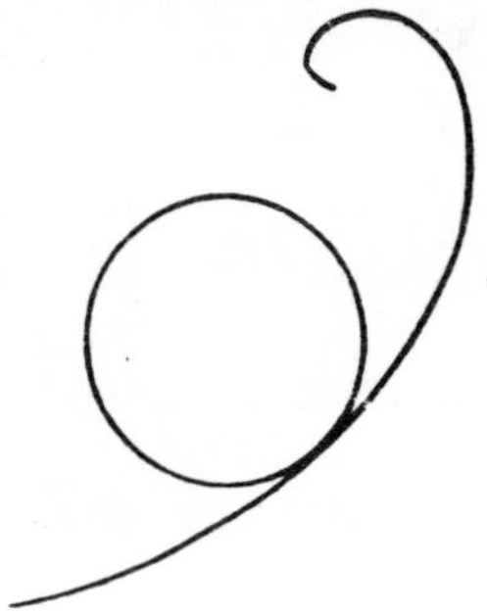

of c at s; even for $c''(s) = 0$, this gives the result we would like. With this definition, curvature would always be non-negative, but we can modify the definition slightly so that we obtain a signed curvature. We will henceforth use the notation $\mathbf{t}(s)$ for $c'(s)$, the unit tangent vector of c at s. We use this notation only for curves parameterized by arclength; recall once again that $\mathbf{t}(s) \in S^1$, even though we usually draw it as an element of $\mathbb{R}^3{}_{c(s)}$.
We then define $\mathbf{n}(s)$, the unit normal at s, to be the unit vector such that $\mathbf{n}(s)$ is perpendicular to $\mathbf{t}(s)$, and $[\mathbf{t}(s),\mathbf{n}(s)]$ is the standard orientation of $\mathbb{R}^2$.

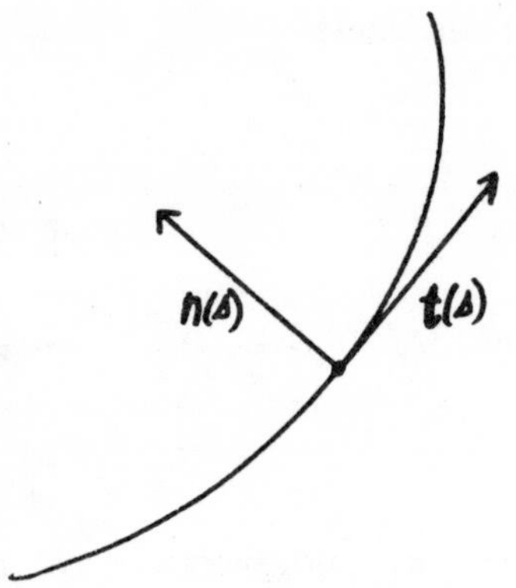

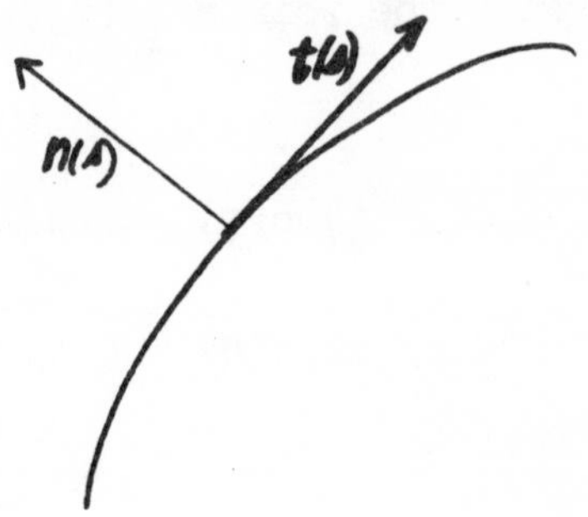

Thus $\mathbf{n}(s) = (-\mathbf{t}^2(s),\mathbf{t}^1(s))$. Now we define the <u>curvature</u> $\kappa(s)$ of c at s by the equation

$$\mathbf{t}'(s) = \kappa(s)\cdot\mathbf{n}(s) \quad .$$

Notice that

$$|\kappa(s)| = |\mathbf{t}'(s)| = |c''(s)| \quad ,$$

so that $1/|\kappa(s)|$ is the radius of the osculating circle at s, for $\kappa(s) \neq 0$.
The significance of the sign of κ is indicated in the figure below.

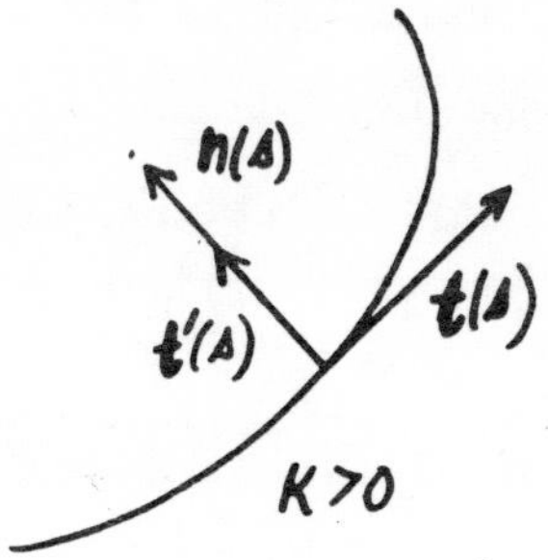

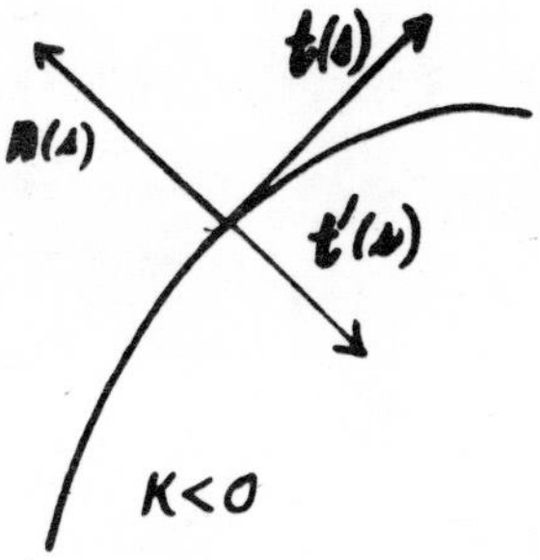

For any curve c we can define a curve $\bar{c}$ "going in the opposite direction" by $\bar{c}(s) = c(-s)$. If we use $\bar{\mathbf{t}}(s)$ for $\bar{c}'(s)$, then clearly

$$\bar{\mathbf{t}}(s) = -\mathbf{t}(-s), \text{ hence } \bar{\mathbf{n}}(s) = -\mathbf{n}(-s)$$
$$\bar{\mathbf{t}}'(s) = \mathbf{t}'(-s) \quad .$$

This shows that the curvature $\bar{\kappa}(s)$ of $\bar{c}$ at s is

$$\bar{\kappa}(s) = -\kappa(-s) \quad .$$

One can see this change of sign in the figure above. Traversing the left curve in the opposite direction, and turning the figure upside down, one obtains the curve on the right.

This relation can also be seen from explicit formulas for κ, which are sometimes useful to have. To write these, we abandon our usual practice of indicating component functions with superscripts, and instead write $c(s) = (c_1(s), c_2(s))$. We have, of course, $|\kappa(s)| = \sqrt{(c_1''(s))^2 + (c_2''(s))^2}$, but we can also develop a formula for $\kappa(s)$ itself. Since $|\kappa(s)|$ is the length of $c''(s)$, which is perpendicular to the unit vector $c'(s)$, clearly $|\kappa(s)|$ is also the area of the rectangle spanned by $c'(s)$ and $c''(s)$. This

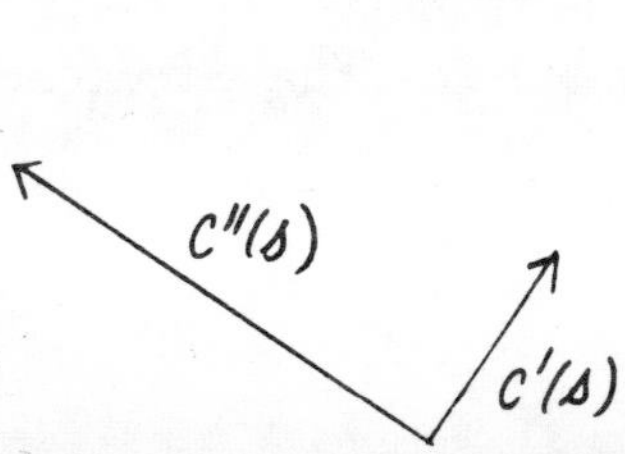

area is given by $\det(c'(s),c''(s))$ which, moreover, clearly has the same sign as $\kappa(s)$. So

$$\kappa(s) = \det\begin{pmatrix} c_1'(s) & c_1''(s) \\ c_2'(s) & c_2''(s) \end{pmatrix} = [c_1'c_2'' - c_2'c_1''](s) \quad .$$

Up til now we have been working exclusively with curves parameterized by arc length. Theoretically this is sufficient, since we consider only curves which can be reparameterized by arc-length. In practice, however, it is usually very inconvenient to actually perform this reparameterization, which involves the inverse of a function defined as an integral. Moreover, since our formula involves only derivatives, only the integrand of this integral should play any crucial role. Consider any (immersed) curve $c\colon [a,b] \longrightarrow \mathbb{R}^2$. Letting $s\colon [a,b] \longrightarrow [0,L]$ be arc-length, and defining $\gamma = c\circ s^{-1}$, we have

$$c = \gamma\circ s$$

$$\frac{dc}{dt} = \gamma'(s)\,\frac{ds}{dt} \qquad \left[\text{i.e. } \frac{dc}{dt}(t) = \gamma'(s(t))\,\frac{ds}{dt}(t)\right]$$

$$\frac{d^2c}{dt^2} = \gamma''(s)\left(\frac{ds}{dt}\right)^2 + \gamma'(s)\,\frac{d^2s}{dt^2} \quad .$$

Thus

$$\gamma'(s) = \frac{dc}{dt}\Big/\frac{ds}{dt}$$

$$\gamma''(s) = \frac{\dfrac{d^2c}{dt^2} - \dfrac{d^2s}{dt^2}\,\dfrac{dc}{dt}\Big/\dfrac{ds}{dt}}{\left(\dfrac{ds}{dt}\right)^2} = \frac{\dfrac{d^2c}{dt^2} - \alpha\,\dfrac{dc}{dt}}{\left(\dfrac{ds}{dt}\right)^2} \quad , \quad \text{say.}$$

Denoting dc_i/dt by $\dot{c}_i(t)$, we have for the curvature $\kappa\colon [a,b] \longrightarrow \mathbb{R}^2$ of c ,

$$\begin{aligned}\kappa(t) &= \text{curvature of } \gamma \text{ at } s = s(t) \\ &= \det(\gamma'(s),\gamma''(s)) \\ &= (ds/dt)^{-3} \det(\dot{c}(t),\ddot{c}(t) - \alpha\,\dot{c}(t)) \\ &= (ds/dt)^{-3} \det(\dot{c}(t),\ddot{c}(t)) \quad .\end{aligned}$$

Thus

$$\boxed{\kappa = \frac{\dot{c}_1\ddot{c}_2 - \dot{c}_2\ddot{c}_1}{(\dot{c}_1{}^2 + \dot{c}_2{}^2)^{3/2}}\ .}$$

Naturally, this formula becomes meaningless for a non-immersed curve, where $\dot{c}_1(t) = \dot{c}_2(t) = 0$. In such cases, we should not generally expect to have a meaningful notion of curvature, for there is nothing to stop the curve from "changing its curvature" at this point.

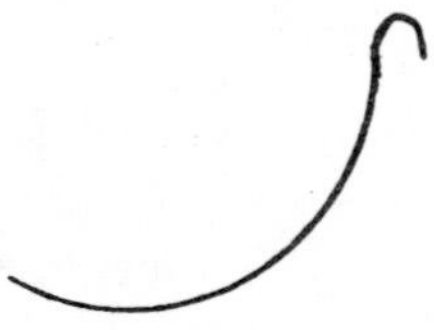

Before examining the significance of the curvature function κ further, we pause for an observation. Since the osculating circle is the limiting position of the circle through $c(s_1), c(s_2), c(s_3)$ no matter how $s_1, s_2, s_3 \longrightarrow s$, it is clearly also the limiting position as $s' \longrightarrow s$ of the circle which is tangent to c at $c(s)$ and which passes through $c(s')$. To see this, we just choose

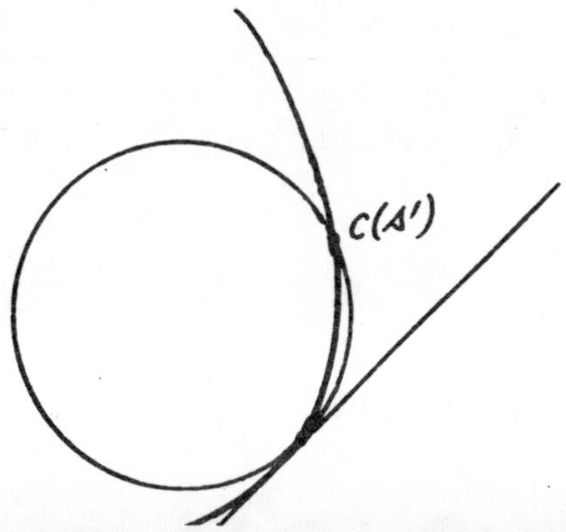

s_1, s_2 much closer to s than s' is. The center of the osculating circle can thus be described as the limiting position of the points S in the figure below

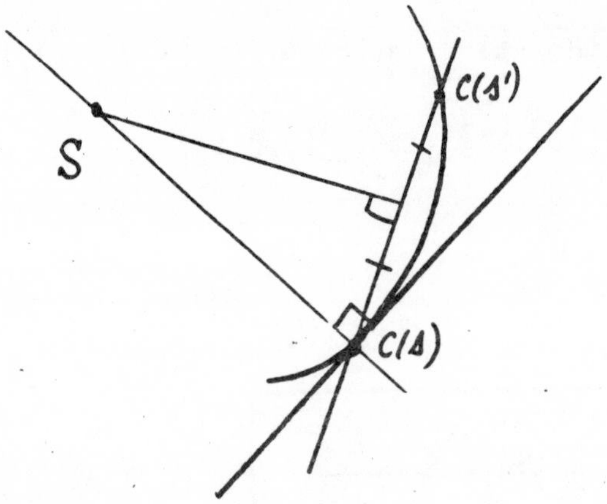

as $s' \longrightarrow s$. These descriptions of curvature go back to Huygens, Liebniz, and Newton.

We began our discussion of curvature by considering straight lines and circles, which were our original models of curves which ought to have constant curvature. It is certainly clear that our definitions do assign curvature 0 to a straight line c, for $c'' = 0$ if c is parameterized by arc-length, and absolute curvature $|\kappa| = R$ to a circle of radius R, since the circle is its own osculating circle. A really satisfactory measure of curvature ought to assign constant curvature to these curves alone. It is clear that if c, parameterized by arc-length, has curvature 0 everywhere, so that $c'' = 0$, then c' is constant, so c is a straight line. The analysis of a curve c with non-zero constant curvature κ (which we might as well assume positive) becomes frustratingly complicated if approached in too straightforward a way. We assume, as usual, that c is parameterized by arclength. Let us introduce the components of the unit tangent vector curve,

$$\mathbf{t}(s) = (\alpha(s), \beta(s)) \quad .$$

Then

$$\alpha'^2 + \beta'^2 = \kappa^2 \quad ,$$

so

$$(1) \qquad \alpha'\alpha'' + \beta'\beta'' = 0 \quad .$$

Recall also that $< \mathbf{t}(s), \mathbf{t}'(s) > = < c'(s), c''(s) > = 0$, so

$$(2) \quad \alpha\alpha' + \beta\beta' = 0 \quad .$$

Equations (1) and (2) show that (α,β) and (α'',β'') are always perpendicular to (α',β'), so one is a multiple of the other,

$$(3) \quad \begin{aligned} \alpha''(s) &= \mu(s)\alpha(s) \\ \beta''(s) &= \mu(s)\beta(s) \quad . \end{aligned}$$

Moreover, differentiating (2) gives

$$(4) \quad \begin{aligned} \alpha'^2 + \beta'^2 + \alpha\alpha'' + \beta\beta'' &= 0 \\ \kappa^2 + \alpha\alpha'' + \beta\beta'' &= 0 \quad . \end{aligned}$$

Substituting from (3), and using $\alpha^2 + \beta^2 = 1$, gives $\mu(s) = -\kappa^2$. Thus,

$$(5) \quad \begin{aligned} \alpha'' + \kappa^2\alpha &= 0 \\ \beta'' + \kappa^2\beta &= 0 \quad . \end{aligned}$$

The solutions of the differential equation in (5) are $s \longmapsto a \sin \kappa s + b \cos \kappa s$, which can also be written as $s \longmapsto A \sin(\kappa s + B)$. In order to have $\alpha^2 + \beta^2 = 1$, we clearly need $A = 1$ and

$$\alpha(s) = \sin(\kappa s + B)$$
$$\beta(s) = \cos(\kappa s + B) \quad .$$

Thus $\mathbf{t}$ traverses a circle, which implies that c itself is a circle of radius $1/\kappa$.

The complicated calculations in the preceding paragraph conceal a basic principle which is much simpler. Suppose that we are given an arbitrary continuous function $\kappa: [a,b] \to \mathbb{R}$. We can ask how many arc-length parameterized curves $c: [a,b] \to \mathbb{R}^2$ there are whose curvature function is κ, without necessarily trying to find a specific formula for these curves. If there is one such curve, then there are automatically others, for translating or rotating a curve will not change its curvature. However, this is the only extent to which the curve is not determined.

2. THEOREM Let $\kappa: [a,b] \to \mathbb{R}$ be continuous. Then there is a curve $c: [a,b] \to \mathbb{R}^2$, parameterized by arc-length, whose curvature at s is $\kappa(s)$ for all $s \in [a,b]$. Moreover, if c and $\bar{c}$ are two such curves, then $\bar{c} = A \circ c$ where A is some proper Euclidean motion (a translation followed by a rotation [an element of SO(2)]).

Proof. Theorem I.5-17 implies that there is a function $\mathbf{t}: [a,b] \to \mathbb{R}^2$ with

$$(*) \quad \mathbf{t}'(s) = \kappa(s) \cdot (-\mathbf{t}_2(s), \mathbf{t}_1(s)) \quad .$$

We can choose $\mathbf{t}(a)$ arbitrarily; choose it to be a unit vector. Now

$$\begin{aligned}({\mathbf{t}_1}^2 + {\mathbf{t}_2}^2)' &= 2\mathbf{t}_1\mathbf{t}_1' + 2\mathbf{t}_2\mathbf{t}_2' \\ &= 2 < (\mathbf{t}_1, \mathbf{t}_2), (\mathbf{t}_1', \mathbf{t}_2') > \\ &= 2 < (\mathbf{t}_1, \mathbf{t}_2), \ \kappa(-\mathbf{t}_2, \mathbf{t}_1) > \\ &= 0 \quad .\end{aligned}$$

So $\mathbf{t}(s)$ is a unit vector for all s. There is, again by Theorem I.5-17, a curve $c\colon [a,b] \to \mathbb{R}^2$ with $c'(s) = \mathbf{t}(s)$. Since $\mathbf{t}(s)$ is always a unit vector, c is parameterized by arc-length. Equation (*) then says that $\mathbf{t}'(s) = \kappa(s)\cdot\mathbf{n}(s)$, so that $\kappa(s)$ is the curvature of c at s.

If c and $\bar{c}$ have the same curvature fundtions κ, then their unit tangent vectors $\mathbf{t}$ and $\bar{\mathbf{t}}$ both satisfy (*). Now if $\mathbf{t}$ is any solution of (*), clearly $B\circ\mathbf{t}$ is also, for any rotation B. Choosing B so that $B(\mathbf{t}(a)) = \bar{\mathbf{t}}(a)$, and using the uniqueness of solutions of (*) with a given initial condition, we see that $\bar{\mathbf{t}} = B\circ\mathbf{t}$. This implies that $\bar{c}$ differs from $B\circ c$ by a translation. ∎

Notice that this theorem makes the previous calculation unnecessary: Since a circle of radius R has constant curvature $1/R$, any curve with constant curvature $1/R$ differs from this circle by a Euclidean motion, and is consequently another circle of radius $1/R$. More generally, Theorem 2 seems to make further study of curves almost pointless. Although one may still study properties of curvature, there is clearly no point in introducing any similar concept; we would only be interested in concepts that remained the same for c and $A\circ c$, and all of these are already determined by the curvature.

Despite these remarks, we are by no means ready to write off the study of plane curves. Many interesting results remain, of which we will be able to mention only a few. However, these results were all proved many years after the study of curves had been initiated, and are all global results, rather than local ones. To begin, we define certain kinds of curve, with which we will be almost exclusively concerned.

A C^1 curve $c\colon [a,b] \to \mathbb{R}^2$ is called <u>closed</u> if $c(a) = c(b)$ and $c'(a) = c'(b)$.

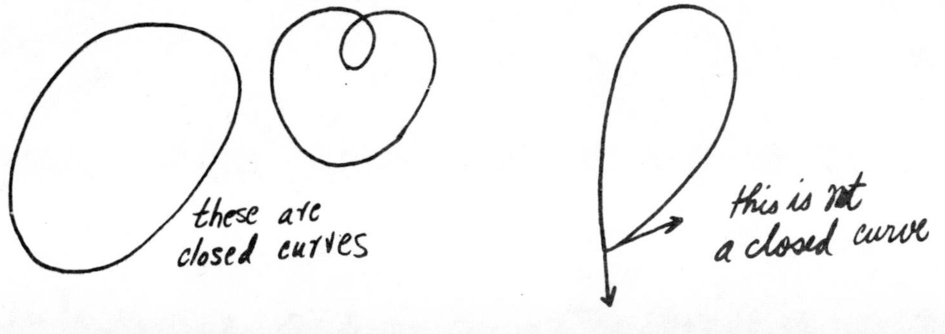

One can also regard a closed curve as an immersion of S^1 in $\mathbb{R}^2$. A curve is called simple if it is one-one. Finally, among the simple closed curves we distinguish a special class of curves called convex. These are defined to be the simple closed curves which always lie on one side of their tangent lines.

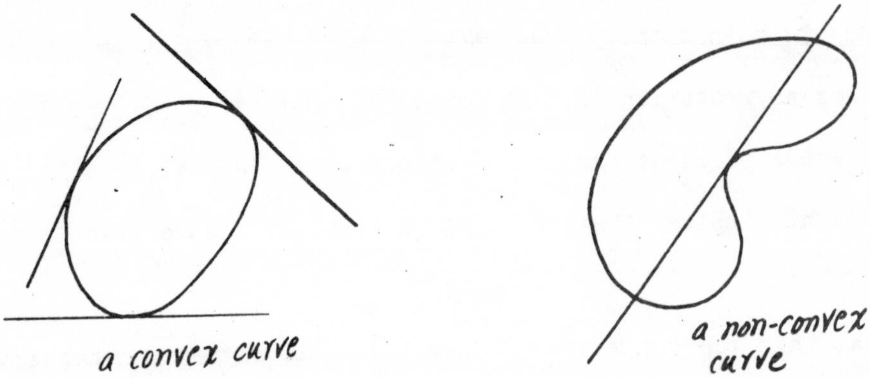

Although the property enunicated in this definition is precisely the one which is used in all proofs about convex curves, we will nevertheless take time out to equate this definition with a more common one.

Any subset A of $\mathbb{R}^2$ is called convex if the line segment $\overline{pq}$ from p to q is contained in A whenever $p,q \in A$. Suppose A is convex and p is

a point in the boundary of A. A line L through p is called a support line of A if A lies completely in one of the closed half spaces into which L divides $\mathbb{R}^2$ (see the figure below).

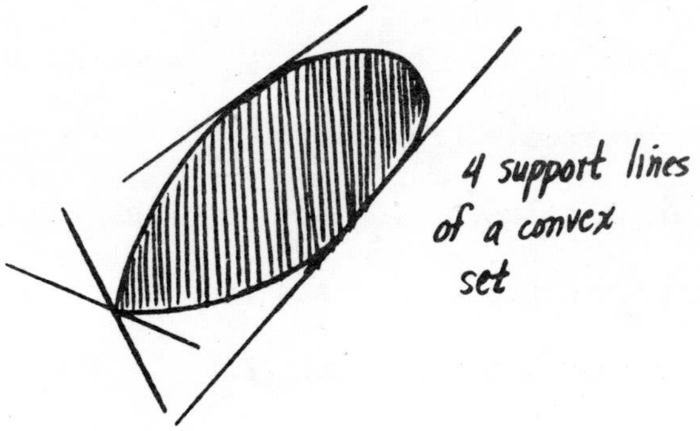

3. PROPOSITION. If A is convex, and p is in the boundary of A, then there is at least one support line L through p.

Proof. If A has no interior points it lies on a line, and the proof is trivial. If A has an interior point q, let ℓ be the ray from q through p, and let ℓ' be the part starting at p. Clearly ℓ' intersects A only at p,

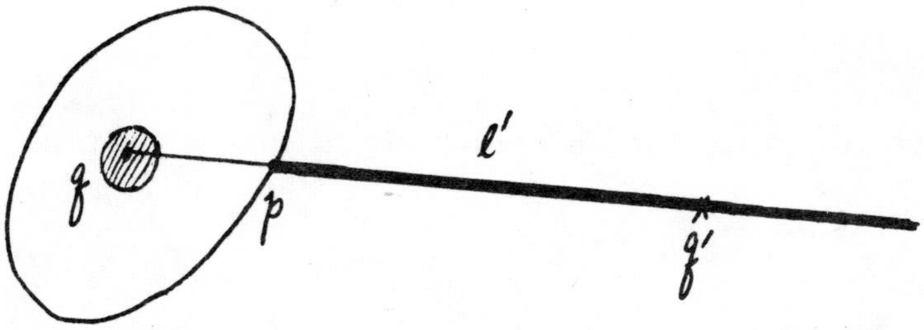

for if a point q' on ℓ' were in A, then all points between q' and the points in a neighborhood of q would be in A, so p would be an interior point of A.

Choose one side of ℓ', and consider angles θ such that rays from p making an angle of θ with ℓ' on this side do not intersect A except at p (it may be that $\theta = 0$ is the only possibility). Let θ_1 be the least upper

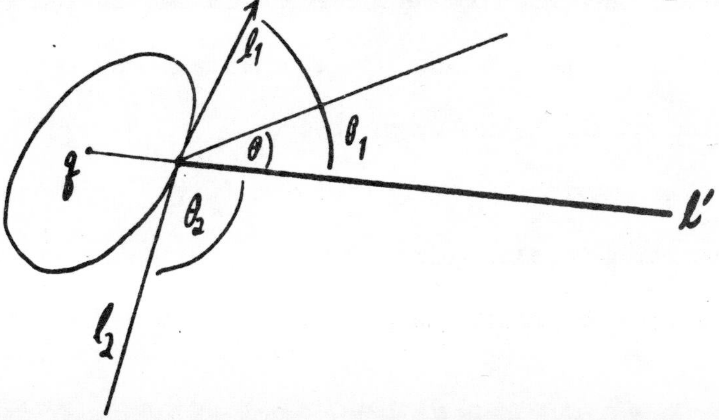

bound of all such θ, and let ℓ_1 be the ray through p making an angle of θ_1. Let ℓ_2 be the corresponding ray on the other side of ℓ'.

I claim that the angle between ℓ_1 and ℓ_2 is $\geq \pi$, which will surely prove the theorem. To prove the claim, note that there are points of A arbitrarily close to (or perhaps even on) both ℓ_1 and ℓ_2. If the angle between ℓ_1 and ℓ_2 were $< \pi$, the triangle containing a suitable pair of such points, and

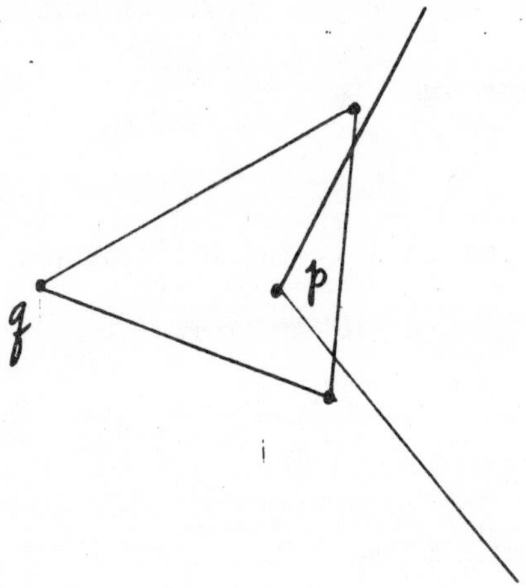

q, would contain p in its interior, which cannot happen, since p is not an interior point of A.∎

<u>Note (for those who are familiar with Banach spaces).</u> This theorem is essentially the Hahn-Banach Theorem. If A were symmetric about the origin, then $\bar{A}$ would be the unit ball in $\mathbb{R}^2$ for a Banach space norm $\| \ \|$. If $W \subset \mathbb{R}^2$ is the subspace spanned by p, and $\lambda: W \to \mathbb{R}$ is the linear functional with $\lambda(p) = 1$, then the desired support line is just a translate of the kernel of an extension $\bar{\lambda}: \mathbb{R}^2 \to \mathbb{R}$ of λ with $\|\bar{\lambda}\| = 1$. Symmetry of A is really unimportant, for the Hahn-Banach theorem only requires a norm satisfying $\|av\| = a\|v\|$ for $a > 0$. The proof given here is just a geometrical translation of the main step in the usual proof of the Hahn-Banach Theorem.

We now want to show that a simple closed curve c is convex if and only if the set A consisting of all points on c or inside c is a convex subset of $\mathbb{R}^2$. This is going to be pretty hard, since we have never defined the inside of a simple closed curve, and are just assuming that the content of Corollary I.11-15 is intuitively obvious. There is really no need to go through the proof of all this right now; it is only necessary to accept the following fact:

> Suppose c is a simple closed curve, and ℓ is a ray from p which intersects c at just one point $q \neq p$. Suppose, moreover, that the tangent line of c at q does <u>not</u> lie along ℓ. Then p is inside c.

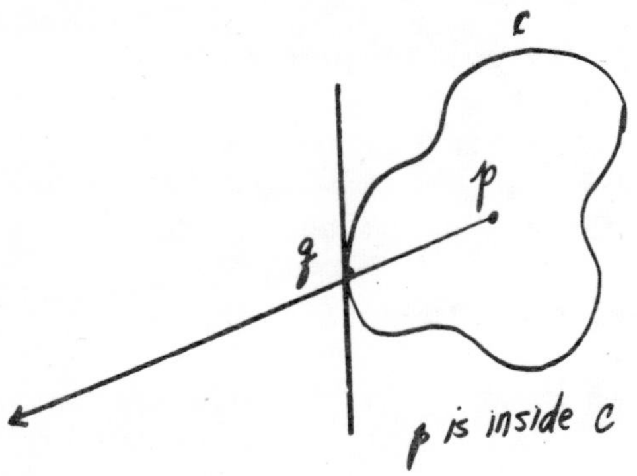

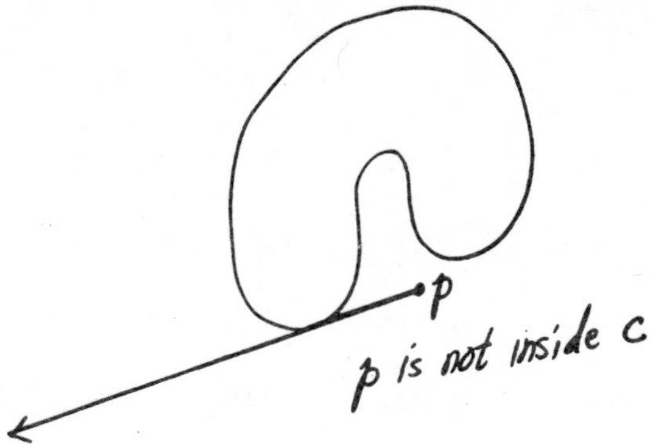

<u>4. PROPOSITION.</u> Let c be a simple closed curve, and let A be the set of all points on or inside c. Then c is convex (that is, c lies on one side of each of its tangent lines) if and only if A is convex.

<u>Proof</u>. Suppose first that A is convex. Any point p on c is a boundary point of A, so there is a support line L of A through p. This line is clearly the tangent line of c at p, so c lies on one side of the tangent line through p.

Now suppose c is convex. For each p on c, let H_p be the closed half plane, bounded by the tangent line through p, in which c lies. Clearly $A \subset \bigcap H_p$. Since the intersection $\bigcap H_p$ of all H_p is convex (any intersection of convex sets is convex), it suffices to show that we actually have $A = \bigcap H_p$. So consider a point q which is outside A. Let p be the point on c closest to q. Clearly the tangent line L of c at p is perpendicular to the ray ℓ

from q through p. It suffices to show that c lies on the opposite side

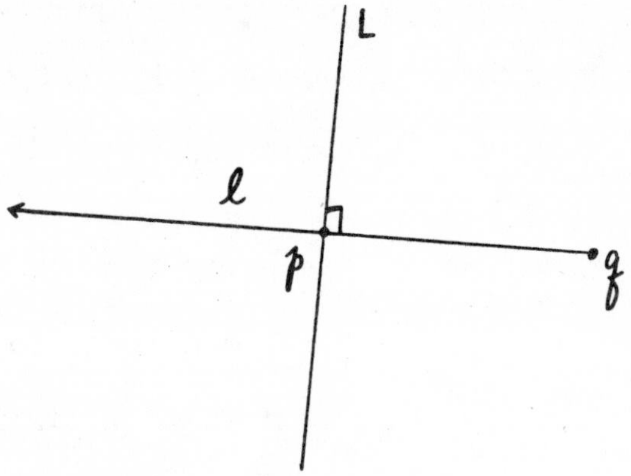

of L from q, for then $q \notin H_p$. Now if c lay entirely on the same side of L as q, then c could not intersect ray ℓ at any other point, for it cannot intersect the open segment $\overline{pq}$, since p is the closest point on c to q; and it certainly could not intersect ℓ at points on the other side of L. By the remark preceding the Proposition, this would mean that q is inside c, a contradiction. ∎

Our first global results about curves depend upon a corresponding global formulation of the curvature function for a curve $c\colon [a,b] \to \mathbb{R}^2$. As usual, we assume c is parameterized by arc-length, and consider its associated unit tangent vector curve $\mathbf{t}\colon [a,b] \to S^1$. Any point in S^1 can be described as $(\cos\theta, \sin\theta)$ for a real number θ. Of course, it is not possible to do this continuously. More precisely, if we define $\mu\colon \mathbb{R} \to S^1$ by

$$\mu(\theta) = (\cos\theta, \sin\theta) \in S^1 ,$$

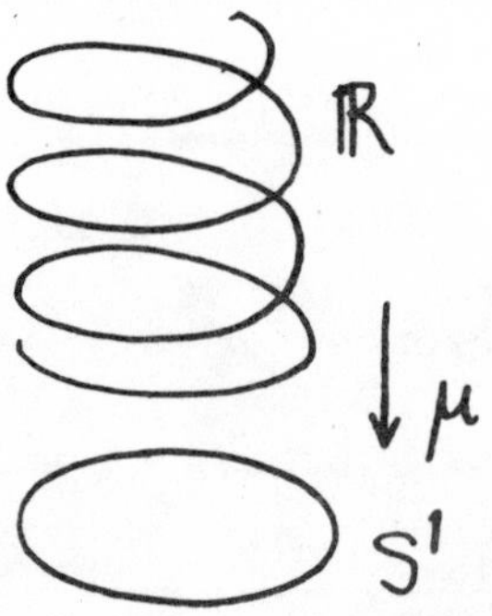

then there is no continuous function $f\colon S^1 \to \mathbb{R}$ with $\mu \circ f$ = identity. On the other hand, for our curve $c\colon [a,b] \to \mathbb{R}^2$ there <u>is</u> a continuous function

$f\colon [a,b] \longrightarrow \mathbb{R}$ with

$$\mu(f(s)) = \mathbf{t}(s).$$

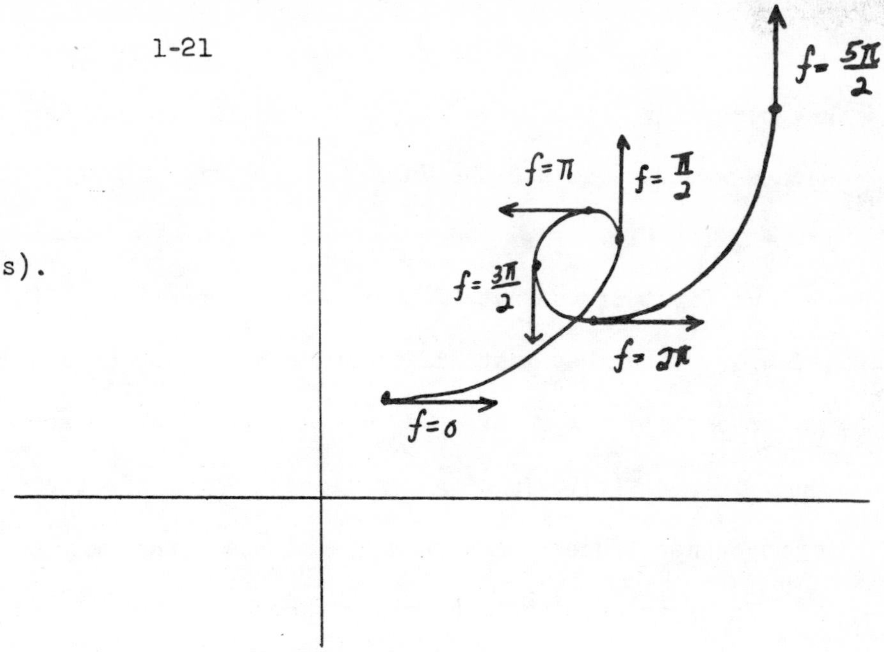

More generally,

5. PROPOSITION. Let $\gamma\colon [a,b] \longrightarrow S^1$ be continuous. Then there is a continuous function $f\colon [a,b] \longrightarrow \mathbb{R}$ with $\mu \circ f = \gamma$. Moreover, if f and $\overline{f}$ are any two such functions, then $f - \overline{f} = 2\pi k$ for some k.

Proof. For any $t \in [a,b]$ there is an open connected subset I_t of $[a,b]$ containing t such that $\gamma(I_t)$ is a proper connected subset of S^1. Clearly $\mu^{-1}(\gamma(I_t))$ is then a disjoint union of connected sets and μ restricted to each of them is a homeomorphism onto $\gamma(I_t)$. This shows that if we choose

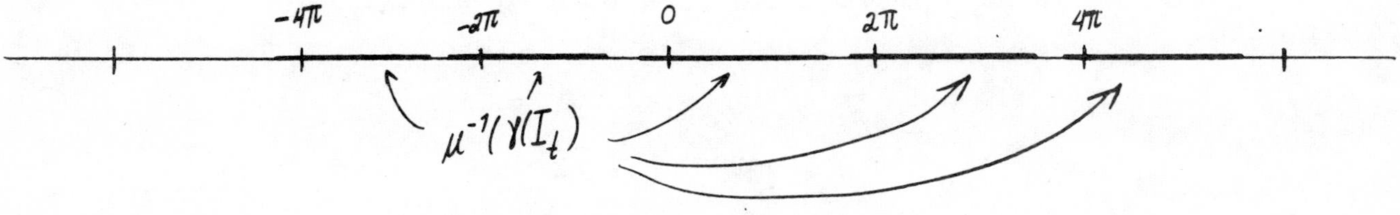

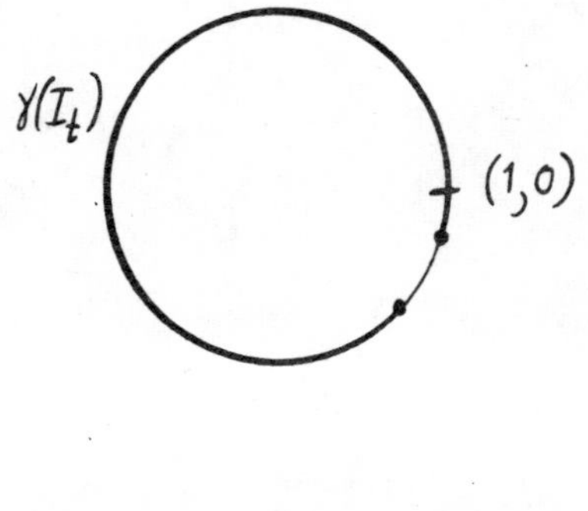

any number v with $\mu(v) = \gamma(t)$, then f can be defined uniquely on I_t in such a way that it has the value v at t, is continuous, and satisfies $\mu \circ f = \gamma$ on I_t.

We first prove that if there are two continuous functions f and $\bar{f}$ with $\mu \circ f = \gamma$, then they must differ by $2\pi k$ for some k. It obviously suffices to prove that $f = \bar{f}$ if $f(a) = \bar{f}(a)$. Let A be set of all $t \in [a,b]$ such that $f(t) = \bar{f}(t)$. Then A is closed, while the previous paragraph shows that A is open. Since $a \in A$ and $[a,b]$ is connected, this shows that $A = [a,b]$.

To prove existence, consider the set B of all $t \in [a,b]$ such that a continuous f can be defined on $[a,t]$, and let t_0 be the least upper bound of B. There is $t_1 \in A$ with $t_1 < t$ and $t_1 \in I_t$. Define a continuous $\bar{f}$ on I_{t_0} (with any value v at t_0 satisfying $\mu(v) = \gamma(t_0)$). Then $\bar{f}(t_1) - f(t_1) = 2k\pi$ for some k. Now $\bar{f} - 2k\pi$ must equal f on $[t_1,t_0]$, by the uniqueness proved previously. So we can extend f to be $\bar{f} - 2k\pi$ on $[0,t_0]$. This shows that $t_0 \in B$. If $t_0 < b$, then we obtain an immediate contradiction by extending f to $[a,t_0] \cup I_{t_0}$. ∎

If f is the function given by Proposition 5 for the curve $\mathbf{t}\colon [a,b] \to S^1$, so that

$$c'(s) = \mathbf{t}(s) = \mu(f(s)) = (\cos f(s), \sin f(s)) \quad ,$$

then

$$c''(s) = (-f'(s)\sin f(s), f'(s) \cos f(s)) \quad .$$

So

$$\begin{aligned} \kappa(s) &= c_1'(s)c_2''(s) - c_2'(s)c_1''(s) \\ &= \cos f(s)\cdot(f'(s) \cos f(s)) - \sin f(s)\cdot(-f'(s) \sin f(s)). \end{aligned}$$

We thus have

$$\kappa(s) = f'(s).$$

Notice that this gives us an easy way to reconstruct the curve from its curvature function: We first reconstruct f as $f(s) = \int_0^s \kappa(\theta)d\theta$; this give us $\mathbf{t}(s) = \mu(f(s))$, so one more integration gives us the curve. Notice also that

$$\begin{aligned} f(b) - f(a) &= \int_a^b f'(s)ds \\ &= \int_a^b \kappa(s)ds \quad ; \end{aligned}$$

this quantity is called the <u>total curvature</u> of c. If c is a closed curve, the function $\mathbf{t}: [a,b] \to S^1$ satisfies $\mathbf{t}(a) = \mathbf{t}(b)$, so we may regard it as a map $\mathbf{t}: S^1 \to S^1$. The total curvature then has a special interpretation.

6. PROPOSITION. The total curvature of a closed curve $c: [a,b] \to \mathbb{R}^2$ is 2π times the degree of the map $\mathbf{t}: S^1 \to S^1$. (The degree of a map is defined on page I.8-50.)

Proof. Since the form "$d\theta$" on S^1 has integral 2π, the degree of $\mathbf{t}$ is

$$\begin{aligned} \frac{1}{2\pi}\int_a^b \mathbf{t}^*(d\theta) &= \frac{1}{2\pi}\int_a^b (\mu\circ f)^*(d\theta) \\ &= \frac{1}{2\pi}\int_a^b f^*(\mu^*(d\theta)) \quad . \end{aligned}$$

Now

$$\mu_*\left(\left.\frac{d}{dt}\right|_t\right) = (-\sin\theta, \cos\theta)_{\mu(\theta)} \in S^1_{\mu(\theta)} \quad ;$$

this is a unit tangent vector of S^1, on which $d\theta$ has the value 1. So

$$\mu^*(d\theta) = dt.$$

Thus

$$\text{degree of } \mathfrak{t} = \frac{1}{2\pi}\int_a^b f^*(dt)$$

$$= \frac{1}{2\pi}\int_a^b f'dt$$

$$= \frac{1}{2\pi}[f(b) - f(a)] \quad . \blacksquare$$

The degree of $\mathfrak{t}$ is also called the rotation index of c. The left most figure below illustrates the first of our global theorems.

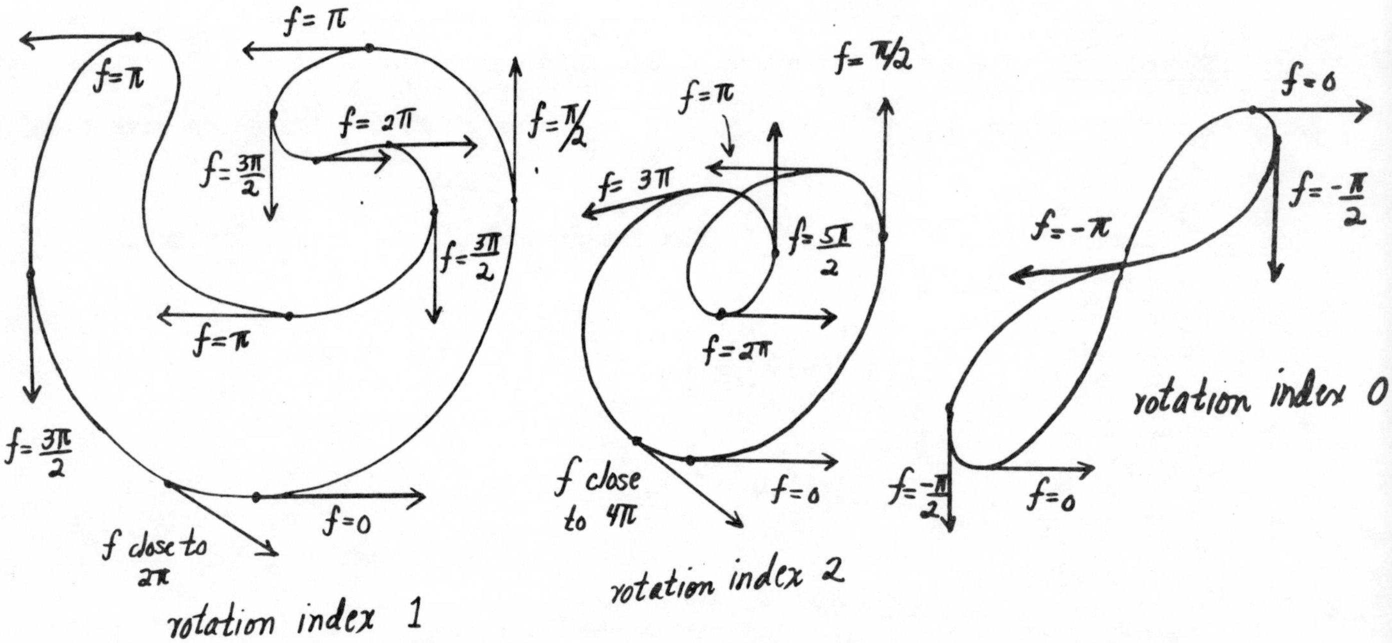

7. THEOREM. The rotation index of a simple closed curve is ± 1 (depending on the direction in which it is traversed).

Proof. Let $c: [0,L] \longrightarrow \mathbb{R}^2$ be the curve, parameterized by arc-length, and let $\Delta \subset \mathbb{R}^2$ be

$$\Delta = \{(s_1, s_2): 0 \le s_1 \le s_2 \le L\} .$$

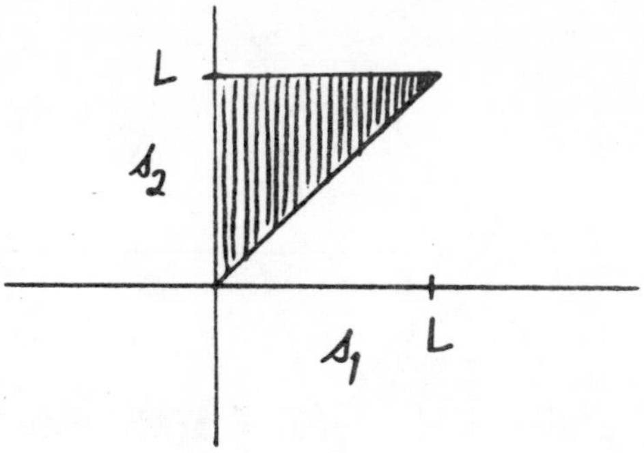

We define $\varphi: \Delta \longrightarrow S^1$ by

$$\varphi(s_1, s_2) = \frac{c(s_2) - c(s_1)}{|c(s_2) - c(s_1)|} \qquad s_1 < s_2 \text{ and } (s_1, s_2) \neq (0,L)$$

$$\varphi(s,s) = c'(s) = \mathbf{t}(s)$$

$$\varphi(0,L) = -c'(0) = -\mathbf{t}(0) .$$

It is easy to see that φ is continuous. Now the map $s \mapsto \varphi(s,s)$ is just $\mathbf{t}: [0,L] \longrightarrow S^1$. On the other hand, this map is homotopic to the map γ obtained by applying φ to the curve which goes along the other two sides of the triangle, from $(0,0)$ to (L,L). So it suffices to compute the degree of γ, which we

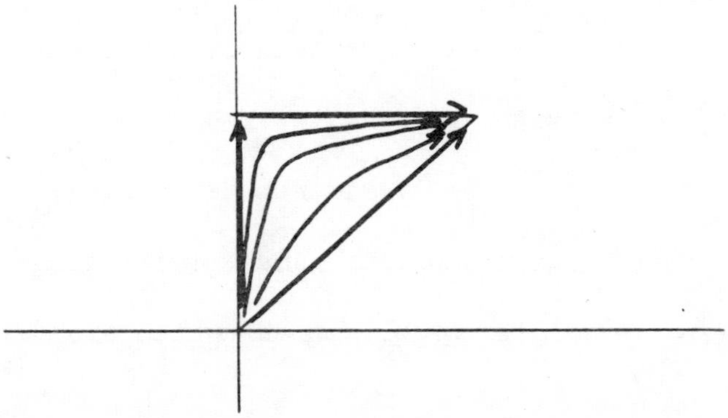

break into two pieces γ_1 and γ_2, defined on $[0,L]$ and $[L,2L]$, say.

The rotation index of c clearly does not change if we rotate or translate c, so we can assume that $c(0) = (0,0)$ and that the tangent line at $c(0)$ is the x-axis. We assume that c is traversed in such a way that $\mathbf{t}(0) = (1,0)$. Now, $\gamma_1(s) = \varphi(0,s)$ clearly always lies in the semi-circle in the upper half

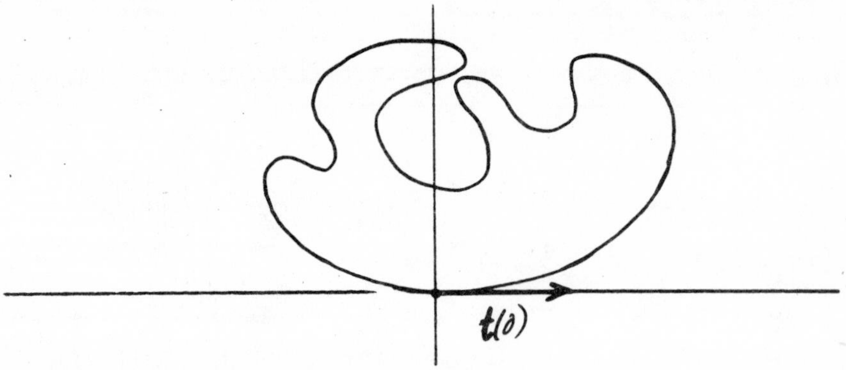

plane, and $\gamma_1(0) = \mathbf{t}(0)$, while $\gamma_1(L) = -\mathbf{t}(0)$. Consequently, the function f given by Proposition 5 clearly has its image in $[0,\pi]$, with $f(0) = 0$ and $f(L) = \pi$. Similarly, γ_2 lies in the lower half plane, so the f for γ_2 satisfies $f(L) = \pi$ and $f(2L) = 2\pi$. Thus the degree of $\mathbf{t}$ is $1/2\pi \cdot [2\pi - 0] = 1$. ∎

Using Theorem 7, we can now relate convexity and curvature.

8. THEOREM. A simple closed curve c is convex if and only if its curvature κ satisfies $\kappa \geq 0$ or $\kappa \leq 0$ (depending on the direction in which c is traversed).

Proof. Let $c\colon [0,L] \to \mathbb{R}^2$ be parameterized by arc-length, and choose a continuous $f\colon [0,L] \to \mathbb{R}$ with $\mu \circ f = \mathbf{t}$.

If $\kappa \geq 0$, then $f' \geq 0$, so f is non-decreasing. Suppose that c were not convex, so that c lies on both sides of the tangent line ℓ through some point p. There are points q_1, q_2 on c which are furthest away from ℓ on both sides of ℓ.

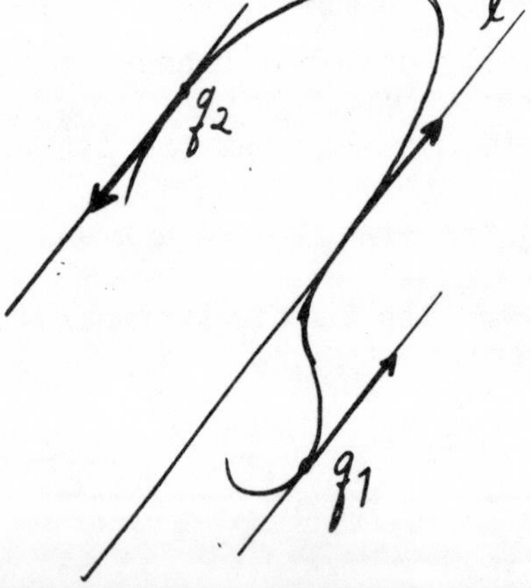

The tangent lines at q_1 and q_2 are clearly parallel to ℓ, so of the three unit tangents at these points, at least two are identical, say $\mathbf{t}(s_1) = \mathbf{t}(s_2)$ for $s_1 < s_2$, with $(s_1, s_2) \neq (0, L)$. Thus $f(s_2) - f(s_1)$ is a multiple of 2π. But f is non-decreasing, and $f(L) = f(0) + 2\pi$, by Theorem 7. So either $f(s_2) = f(s_1)$ or $f(s_2) = f(s_1) + 2\pi$. In the first case it follows that f is constant on $[s_1, s_2]$, so $\mathbf{t}$ is constant on $[s_1, s_2]$. This implies that c is a straight line on $[s_1, s_2]$, contradicting the fact that none of the points p, q_1, q_2 lies on the tangent lines through the others. If $f(s_2) = f(s_1) + 2\pi$, the other arc of c must similarly be a straight line, which is again a contradiction. Thus c must be convex.

Now suppose c is convex. If there is $s_1 < s_2$ with $f(s_1) = f(s_2)$, so that $\mathbf{t}(s_1) = \mathbf{t}(s_2)$, then there is also s with $\mathbf{t}(s) = -\mathbf{t}(s_1)$, since $\mathbf{t}$ has

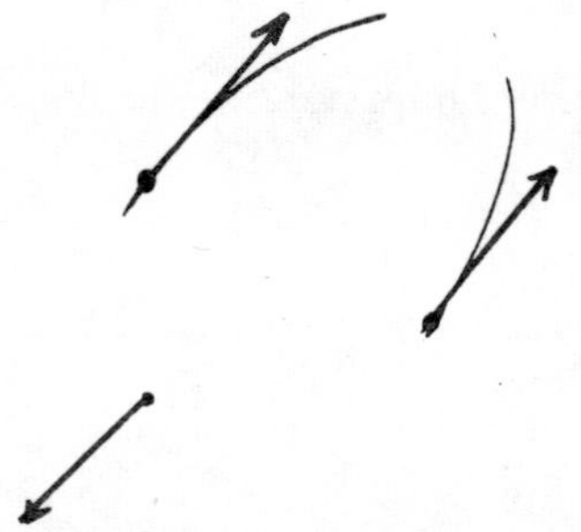

degree 1, and is consequently onto S^1. The tangent lines through two of the three points $c(s)$, $c(s_1)$, $c(s_2)$ must coincide (otherwise c would cross one of them). Thus c is tangent to the same line ℓ at two points, $c(t_1)$ and $c(t_2)$, say. These two points divide c into two arcs, α and β. If ℓ'

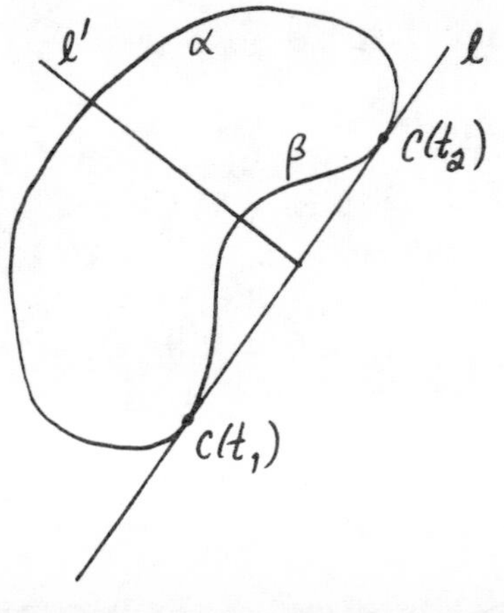

is perpendicular to ℓ, and intersects ℓ between $c(t_1)$ and $c(t_2)$, then α and β must intersect ℓ'; using convexity of c, it is easy to see that α and β intersect ℓ' exactly once. Clearly one arc, say α, always intersects ℓ' at a point further away from ℓ than the other, β. We claim that β lies along ℓ. If not, consider the tangent line through the point P of β furthest from ℓ. This tangent line is parallel to ℓ, and c would lie on both sides of it, a contradiction.

Now, since β lies along ℓ, we have $\mathbf{t}(t_1) = \mathbf{t}(t_2)$. Thus $t_1 = s_1$ and $t_2 = s_2$. Since the curve lies along a line on $[s_1, s_2]$, we have $f(s) = f(s_1)$ for all $s \in [s_1, s_2]$. Thus, f is non-decreasing. ∎

Our final global result about curves in the plane involves another local concept. A vertex of a curve c is a point where $\kappa'(s) = 0$. It is easily seen that an ellipse which is not a circle has exactly four vertices, at the ends of the major and minor axes; these are the points where κ has a local maximum or minimum (though a vertex need not generally be of this type). Before proceeding with the next theorem we need a preliminary observation. The curvature κ is defined by

$$\mathbf{t}'(s) = \kappa(s)\cdot\mathbf{n}(s) \quad ;$$

the existence of such a number $\kappa(s)$ followed from the equation $\langle \mathbf{t}, \mathbf{t} \rangle = 1$, by differentiation. Similarly, from $\langle \mathbf{n}, \mathbf{n} \rangle = 1$ we obtain

$$\langle \mathbf{n}'(s), \mathbf{n} \rangle = 0,$$

which implies that $\mathbf{n}'(s)$ is a multiple of $\mathbf{t}(s)$, say $\mathbf{n}'(s) = \alpha\cdot\mathbf{t}(s)$. On the other hand, from $\langle \mathbf{t}, \mathbf{n} \rangle = 0$ we obtain

$$\langle \mathbf{t}'(s), \mathbf{n}(s) \rangle + \langle \mathbf{t}(s), \mathbf{n}'(s) \rangle = 0 \quad ,$$

$$\kappa(s) + \alpha = 0 \quad ;$$

hence,

$$(*) \qquad \mathbf{n}'(s) = -\kappa(s)\cdot\mathbf{t} \quad .$$

9. THEOREM (THE FOUR VERTEX THEOREM). Every simple closed convex curve has at least four vertices.

Proof. If $c: [0,L] \longrightarrow \mathbb{R}^2$ is the simple closed curve, parameterized by arc-length, then c has at least two vertices — namely, the maximum and minimum points for the curvature. Choose the coordinate system so that the x-axis passes through these two points. Now integration by parts gives the following equation,

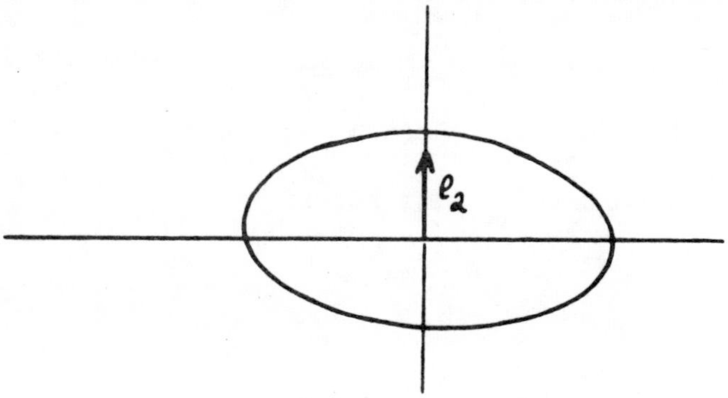

in which we are taking integrals of $\mathbb{R}^2$-valued functions by integrating each component separately:

$$\int_0^L \kappa'(s)\cdot c(s) = -\int_0^L \kappa(s)\cdot\mathbf{t}(s)ds$$

$$= \int_0^L \mathbf{n}'(s)ds \qquad \text{by } (*)$$

$$= \mathbf{n}(L) - \mathbf{n}(0) = 0 \quad .$$

Consequently, we certainly have

$$(**) \qquad \int_0^L \kappa'(s) < c(s), e_2 > ds = 0 \quad .$$

If there are no other vertices, then $\kappa' > 0$ on one half of c and $\kappa' < 0$ on the other. So $\kappa'(s) < c(s), e_2 >$ has the same sign on both halves, contradicting (**). Thus c must have at least one more vertex.

The argument just given actually shows that c cannot be formed of two arcs with $\kappa' > 0$ on one and $\kappa' < 0$ on the other; the same conclusion clearly holds even if $\kappa' \geq 0$ on one and $\kappa' \leq 0$ on the other, since $\kappa' \neq 0$ somewhere. This shows that c must have a fourth vertex: if it had only three, then some pair would divide c into two arcs with $\kappa \geq 0$ on one and $\kappa' \leq 0$ on the other. ∎

We now turn our attention to curves in space, and ask, once again, how to measure the curvature of $c: [a,b] \longrightarrow \mathbb{R}^3$. We can still look for the limiting position of circles through $c(s_1), c(s_2), c(s_3)$ as $s_1, s_2, s_3 \longrightarrow s$. If this limiting circle exists, its center C must satisfy

$$< c'(s), c(s) - C > = 0$$

$$< c''(s), c(s) - C > = - < c'(s), c'(s) > \quad ;$$

the derivation of this necessary condition still works. However, for space curves these equations do not even determine C; the first equation merely restricts C to lie on a certain plane, not on a certain line. We must first see whether the planes through the points $c(s_i)$ approach a limiting position.

10. PROPOSITION. Let $c: [a,b] \longrightarrow \mathbb{R}^3$ be a C^2 curve parameterized by arclength, with $c''(s) \neq 0$. For s_1, s_2, s_3 sufficiently close to s, the points $c(s_1), c(s_2), c(s_3)$ do not lie on a line. As $s_i \longrightarrow s$, the unique plane through the points $c(s_i)$ approaches the plane P spanned by $c'(s)$ and $c''(s)$.

Remark. The plane P should really be described as the plane through $c(s)$ which is parallel to the plane spanned by $c'(s)$ and $c''(s)$, but we will allow ourselves the elliptical terminology suggested by the picture.

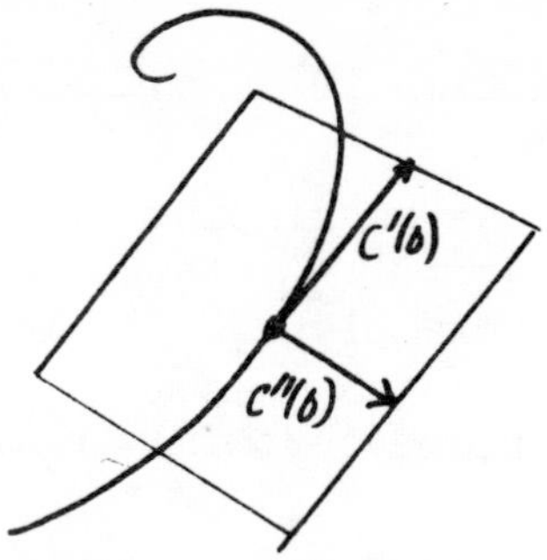

<u>Proof</u>. Assuming for the moment that the points $c(s_i)$ do not lie on a straight line, let $P(s_1,s_2,s_3)$ be the plane spanned by these points, and let $a(s_1,s_2,s_3)$ be a unit vector perpendicular to $P(s_1,s_2,s_3)$. Then the function

$$(*)\quad s \mapsto < c(s) - c(0),\ a(s_1,s_2,s_3) >$$

is 0 for $s = s_i$. So we have

$$(1)\quad < c'(\xi_i), a(s_1,s_2,s_3) > = 0 \qquad \xi_i \in (s_i,s_{i+1}) \quad i = 1, 2 \ .$$

It follows that

$$(2)\quad < c''(\eta), a(s_1,s_2,s_3) > = 0 \qquad \eta \in (\xi_1,\xi_2) \ .$$

Equations (1) and (2), together with continuity of c' and c'', clearly show that $a(s_1,s_2,s_3)$ approaches a unit vector perpendicular to $c'(s)$ and $c''(s)$, so that $P(s_1,s_2,s_3)$ approaches P.

It the points $c(s_i)$ do lie on a straight line, then we can choose a whole circle of unit vectors $a(s_0,s_1,s_2)$ for which the function (*) vanishes at $s = s_i$. If this were true for s_i arbitrarily close to s, the remaining part of the argument would imply that all of these vectors are nearly perpendicular to P, which is absurd. ∎

The plane described in Proposition 10 is called the <u>osculating plane</u> of the curve at s. Notice that, unlike the osculating circle, the osculating plane may exist even if $c''(s) = 0$. For example, if c is a plane curve, which is not straight, then the osculating plane certainly exists. The exact conditions for the existence of an osculating plane are not very important for us, but we will pause to indicate the actual state of affairs.

The worst possible situation occurs for the curve

$$c(t) = \begin{cases} 0 & t = 0 \\ (t, e^{-1/t^2}, 0) & t > 0 \\ (t, 0, e^{-1/t^2}) & t < 0 \end{cases},$$

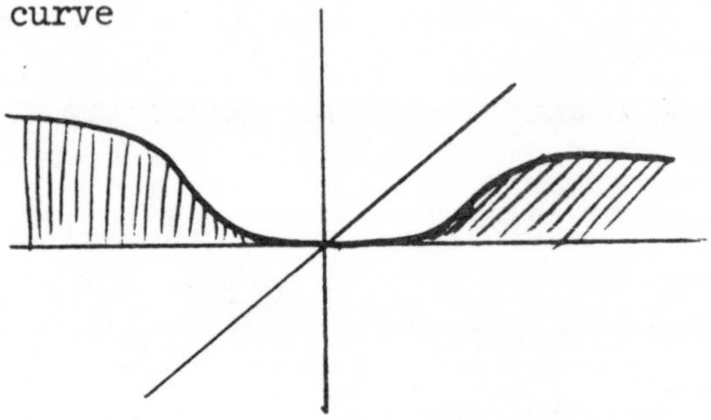

which can't make up its mind whether to osculate in the x-y plane or in the x-z plane. For this curve we have $c^{(k)}(0) = 0$ for all $k \geq 2$, which suggests that we consider only curves with $c^{(k)}(s) \neq 0$ for some $k \geq 2$. Notice that for curves parameterized by arclength, $< c'(s), c''(s) > = 0$ implies

$$< c'(s), c'''(s) > + < c''(s), c''(s) > = 0 \ .$$

So if $c''(s) = 0$, we have $c'''(s)$ perpendicular to $c'(s)$. Similarly, $c'''(s) = 0$ implies that $c^{(4)}(s)$ is perpendicular to $c'(s)$, etc. So the first non-zero higher derivative $c^{(k)}(s)$ is the same as the first derivative which is linearly independent of $c'(s)$. Now suppose P is the plane spanned by $c'(s)$ and this first nonzero $c^{(k)}(s)$. Let s_1, s_2, s_3 be parameter values <u>on the same side</u> of s, i.e. $s \leq s_1 < s_2 < s_3$ (or $s_1 < s_2 < s_3 \leq s$). If $a(s_1, s_2, s_3)$ is a unit vector perpendicular to all $c(s_i) - c(s)$, then, as before, we have

$$< c'(\xi_i), a(s_1,s_2,s_3) > = 0$$
$$< c''(\eta), a(s_1,s_2,s_3) > = 0 \ .$$

Now $\eta \in (s_1,s_3)$ so $\eta \neq s$. Hence, if $c''(s) = 0$ we have $\theta \in (s,\eta)$ with

$$< c'''(\theta), a(s_1,s_2,s_3) > = 0 \ .$$

Continuing in this way, we finally obtain $\lambda \in (s,s_3)$ with

$$< c^{(k)}(\lambda), a(s_1,s_2,s_3) > = 0 \ .$$

As before, this shows that the plane through the points $c(s_i)$ approaches P, and that the points $c(s_i)$ cannot lie on a line for s_i arbitrarily close to 0.

If the s_i are allowed to lie on both sides of s, this result is no longer true. For example, consider the curve

$$c(t) = (t,t^4,t^3) \ .$$

We have

$$c'(0) = (1,0,0)$$
$$c''(0) = (0,0,0)$$
$$c'''(0) = (0,0,6) \ ,$$

so $(0,1,0)$ is perpendicular to the plane spanned by $c'(0)$ and $c'''(0)$. On the other hand, for a vector perpendicular to the plane spanned by $c(0),c(t),c(-t)$ we can choose

$$\text{normalized } [c(t) - c(0)] \times [c(-t) - c(0)] = \text{normalized } (t,t^4,t^3) \times (-t,t^4,-t^3)$$
$$= \text{normalized } (-2t^7,0,2t^5)$$
$$\longrightarrow (0,0,1) \quad .$$

This strange behavior is clarified by a look at the Taylor expansion

$$c(s) = c(0) + sc'(0) + 0 + \frac{s^3}{6} c'''(0) + o(s^3) \quad ,$$

which shows that

$$[c(s) - c(0)] \times [c(t) - c(0)] = [\frac{st^3}{6} - \frac{ts^3}{6}] \; c'(0) \times c'''(0) + \text{higher order terms.}$$

When s and t have the same sign, the dominant term is the first. But when s and t have opposite signs this is no longer true — this term may even be 0. I suspect, but have not checked, that the parameter values s_i may be picked on both sides of 0 if and only if the first $k \geq 2$ with $c^{(k)}(0) \neq 0$ is even.

For space curves $c: [a,b] \longrightarrow \mathbb{R}^3$ with $c''(s) \neq 0$, we now clearly have an osculating circle, the limit as $s_1, s_2, s_3 \longrightarrow s$ of the circle through the points $c(s_i)$; it lies in the osculating plane. We define the curvature κ to be the reciprocal of the radius of this circle, so that

$$\kappa(s) = |\mathbf{t}'(s)| \quad .$$

This definition, once again, determines a curvature even when $c''(s) = \mathbf{t}'(s) = 0$. Unlike the case of plane curves, we cannot obtain a signed curvature, for there is no natural way to pick a vector orthogonal to $\mathbf{t}(s)$. However, when $\kappa(s) \neq 0$, we can now <u>define</u> $\mathbf{n}(s)$ by the equation

$$\mathbf{t}'(s) = \kappa(s)\,\mathbf{n}(s) \;, \qquad \mathbf{n}(s) = \text{normalized } \mathbf{t}'(s).$$

The vector $\mathbf{n}(s)$ is called the principal normal of c at s. We also define the binormal $\mathbf{b}(s)$ by

$$\mathbf{b}(s) = \mathbf{t}(s) \times \mathbf{n}(s) \;,$$

so that $(\mathbf{t},\mathbf{n},\mathbf{b})$ is always a positively oriented orthonormal basis.

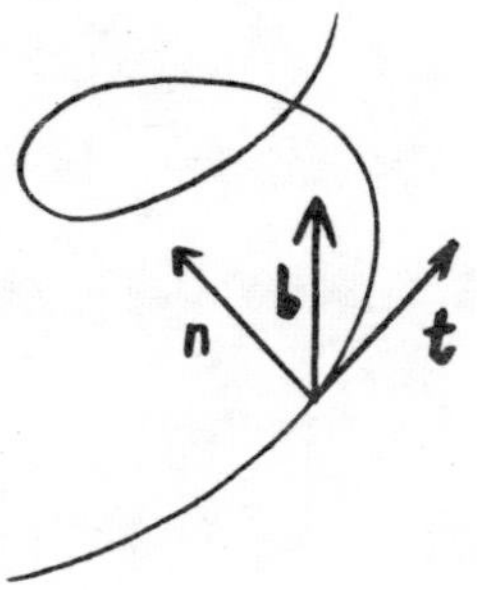

Note that $\langle \mathbf{b},\mathbf{b} \rangle = 1$ implies that $\langle \mathbf{b}',\mathbf{b} \rangle = 0$, so that $\mathbf{b}'$ is a linear combination of $\mathbf{t}$ and $\mathbf{n}$. We also have $\langle \mathbf{b},\mathbf{t} \rangle = 0$, which implies that

$$\langle \mathbf{b}',\mathbf{t} \rangle = -\langle \mathbf{b},\mathbf{t}' \rangle = -\langle \mathbf{b},\mathbf{n} \rangle = 0 \;.$$

Thus $\mathbf{b}'$ is actually a multiple of $\mathbf{n}$, and we can define a new function τ, the torsion, by

$$\mathbf{b}' = -\tau\,\mathbf{n} \;.$$

Of course, we can define τ only at points where $\mathbf{n}$ exists, i.e., where $c'' \neq 0$. This is analagous to the fact that κ can be defined only at points where $c' \neq 0$. The reason for choosing the negative sign in this equation will be explained in a moment; we first interpret the absolute value $|\tau|$. The function

$\mathbf{b}: [a,b] \longrightarrow S^2$ has an arclength function

$$\text{length of } \mathbf{b} \text{ on } [a,s] = \int_a^s |\mathbf{b}'(u)|\,du$$

$$= \int_a^s |\tau(u)|\,du \quad .$$

Consequently, $|\tau(s)|$ is the derivative of this arclength function. Since $\mathbf{b}$ is the perpendicular to the osculating plane, this derivative of the length of $\mathbf{b}$ can be thought of as the rate at which the osculating plane is changing. Thus $|\tau|$ measures in some sense, the rate at which the curve deviates from being a plane curve. Classically, curvature and torsion were also known as first and second curvature, and space curves were called curves of double curvature.

To develop a formula for the torsion, we first recall that the cross product $v \times w$ of v and w is defined by the equation

$$(*) \qquad \langle z, v \times w \rangle = \det\begin{pmatrix} z \\ v \\ w \end{pmatrix} \quad .$$

(Since the ith component of $v \times w$ is $\langle e_i, v \times w \rangle$, this shows that $v \times w$ can be obtained by formally computing the determinant

$$v \times w = \det\begin{pmatrix} e_1 & e_2 & e_3 \\ v_1 & v_2 & v_3 \\ w_1 & w_2 & w_3 \end{pmatrix} = (v_2w_3 - w_2v_3)e_1 + \ldots \quad .)$$

Using the fact that det is alternating, the left side of $(*)$, classically known as the "scalar triple product" (v,w,z), is seen to satisfy

$$(**) \quad \langle z, v \times w \rangle = -\langle w, v \times z \rangle = \langle w, z \times v \rangle \quad .$$

Now for the torsion τ of a curve c parameterized by arclength we clearly have

$$\begin{aligned}\tau = \langle -\mathbf{n}, \mathbf{b}' \rangle &= \langle -\mathbf{n}, (\mathbf{t} \times \mathbf{n})' \rangle \\ &= \langle -\mathbf{n}, \mathbf{t} \times \mathbf{n}' \rangle + \underset{\overset{\parallel}{0}}{\langle -\mathbf{n}, \mathbf{t}' \times \mathbf{n} \rangle} \\ &= \left\langle -\frac{c''}{\kappa}, c' \times \left(\frac{c''}{\kappa}\right)' \right\rangle \\ &= \left\langle -\frac{c''}{\kappa}, c' \times \left(\frac{\kappa c''' - \kappa' c''}{\kappa^2}\right) \right\rangle \\ &= -\frac{1}{\kappa^2} \langle c'', c' \times c''' \rangle ,\end{aligned}$$

$$\boxed{\tau = \frac{1}{\kappa^2} \langle c' \times c'', c''' \rangle}$$

This shows that $\tau > 0$ when $\mathbf{t}, \mathbf{n}, c'''$ form a positively oriented basis for $\mathbb{R}^3$, which is the same as saying that c''' is on the same side of the osculating plane as $\mathbf{b}$. Now in Taylor's formula,

$$c(s + h) = c(s) + hc'(s) + \frac{h^2}{2} c''(s) + \frac{h^3}{6} c'''(s) + o(h^3) \quad ,$$

the term $c(s) + hc'(s) + h^2c''(s)/2$ is in the osculating plane at s, so

$$c(s + h) - [c(s) + hc'(s) + \frac{h^2}{2} c''(s)] = \frac{h^3}{6} c'''(s) + o(h^3)$$

points from the osculating plane to $c(s + h)$. Consequently, if $c'''(s) \neq 0$, then the curve pierces the osculating plane at s, and the points $c(s + h)$ for small $h > 0$ are on the same side as $c'''(s)$, while points $c(s + h)$ for $h < 0$ are on the other side. Together with our previous remarks, this shows that

> if $\tau(s) > 0$, then points $c(s + h)$ for small $h > 0$ lie on the same side of the osculating plane as $\mathbf{b}(s)$, while points $c(s + h)$ for small $h < 0$ lie on the opposite side.

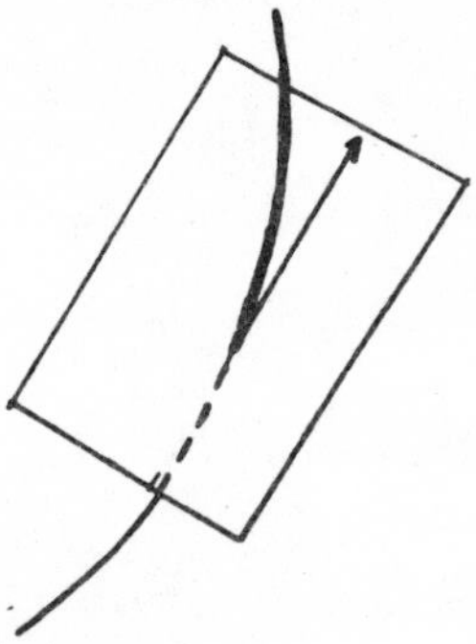

Our formula for τ shows that the torsion for the curve $\bar{c} = s \mapsto c(-s)$ has the same sign as that of c. This is because reversing directions also reverses the binormal. For a curve with an arbitrary parameterization we obtain, preceding just as in the case of curvature, the formula

$$\tau = \frac{\langle \dot{c} \times \ddot{c}, \dddot{c} \rangle}{\langle \dot{c} \times \ddot{c}, \dot{c} \times \ddot{c} \rangle} .$$

A standard and possibly illuminating way of examining the geometrical significance of κ and τ is to examine the projections of c on the planes spanned by any two of **t**, **n**, **b**. The plane spanned by **t** and **n** is, of course, just the osculating

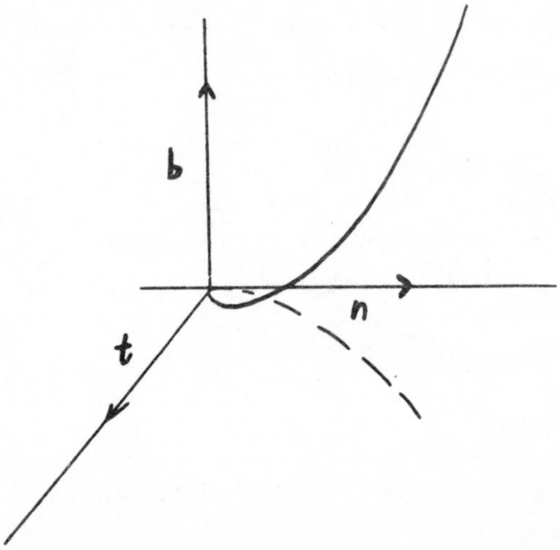

plane. The plane spanned by the principal normal **n** and binomial **b** is called, naturally enough, the <u>normal plane</u>, and the plane spanned by **t** and **b** is called the <u>rectifying plane</u>. We can choose a coordinate system for $\mathbb{R}^3$ so that $c(0) = 0$ and so that the osculating plane of c at 0 is the x-y plane. Further choosing $c'(0) = (1,0,0)$ we obtain

$$c(0) = (0,0,0)$$
$$c'(0) = (1,0,0)$$
$$c''(0) = (0,\kappa,0)$$
$$c'''(0) = (-,-,\kappa\tau) \quad ,$$

the last equation following from the fact that $\tau\kappa^2 = < c' \times c'', c''' >$, as shown above. The three components of the Taylor expansion

$$c(s) = c(0) + sc'(0) + \frac{s^2}{2} c''(0) + \frac{s^3}{6} c'''(0) + \ldots$$

give

$$c_1(s) = s + \text{terms of order 3 or more}$$
$$c_2(s) = \frac{\kappa}{2} s^2 + \text{"}$$
$$c_3(s) = \frac{\kappa\tau}{6} s^3 + \text{terms of order 4 or more} \quad .$$

So the projections look like

$y = \frac{\kappa}{2} x^2$ up to order 2 on osculating plane

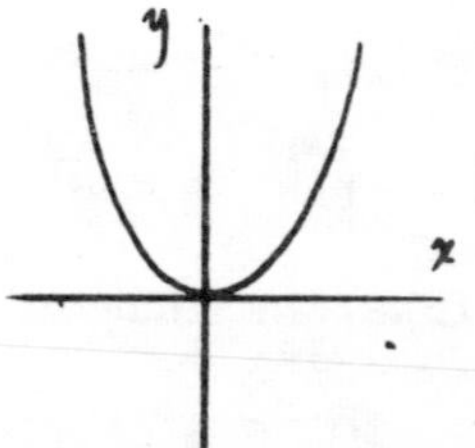

$z = \frac{\kappa\tau}{6} x^3$ up to order 3 on rectifying plane

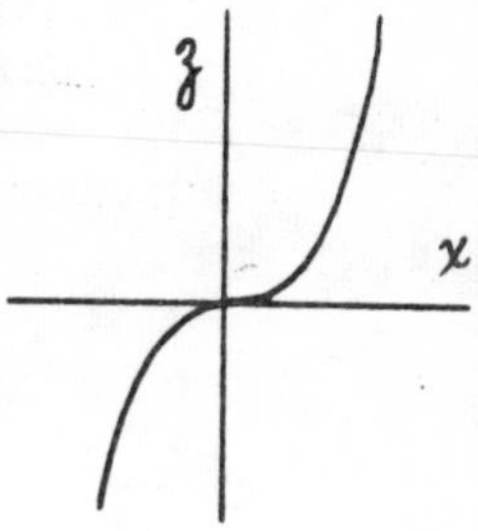

$z^2 = \frac{2}{9} \frac{\tau^2}{\kappa} y^3$ up to order 3 on normal plane.

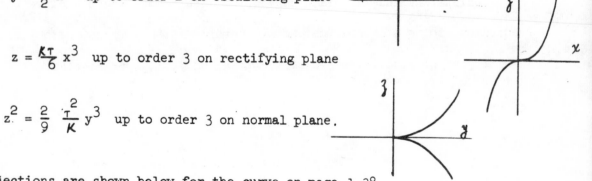

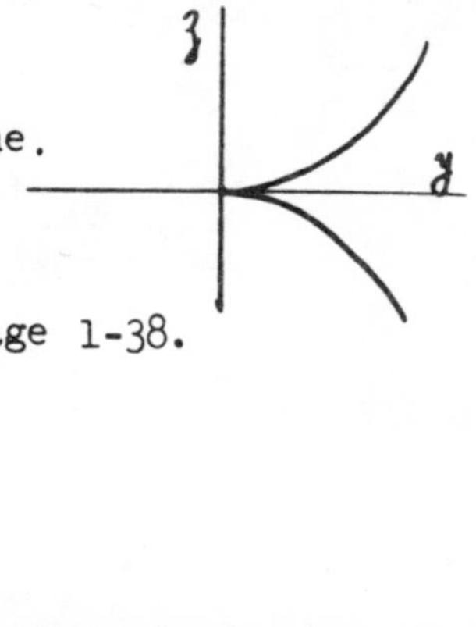

These projections are shown below for the curve on page 1-38.

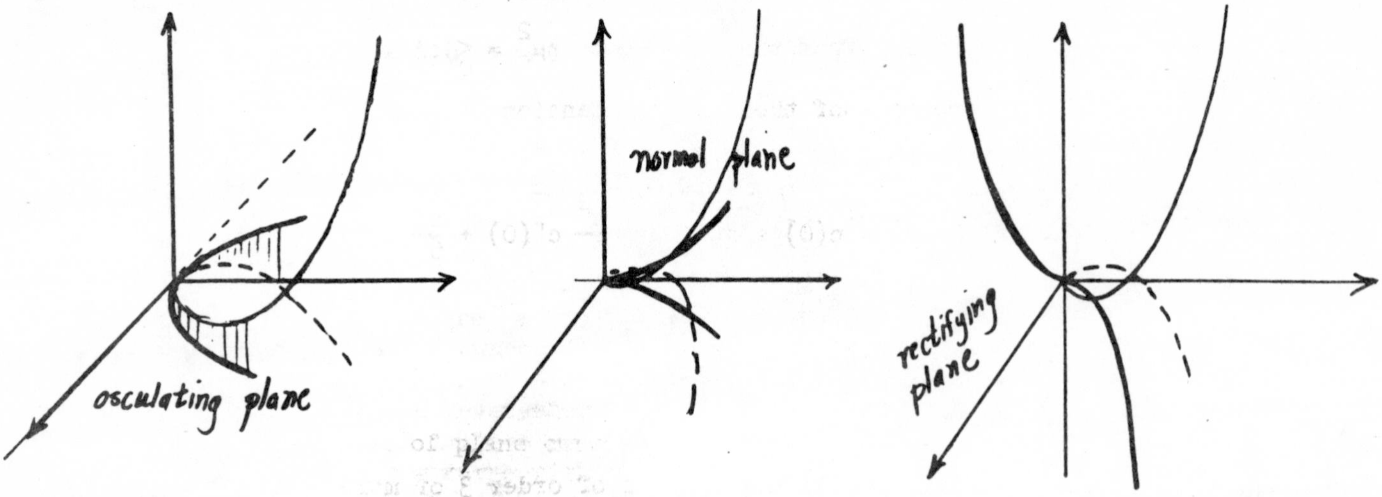

Just as in the case of plane curves, we now ask to what extent the curvature and torsion determine a curve. We note first that a curve c (parameterized by arclength) with $\kappa = 0$ everywhere, is a straight line — the proof is the same as before. Moreover, a curve with $\tau = 0$ everywhere is a plane curve. To prove this, we note that $\tau = 0$ means $\mathbf{b}' = 0$ so that $\mathbf{b}(s) = \mathbf{b}_0$, a constant vector. This implies that $\langle \mathbf{t}, \mathbf{b}_0 \rangle = 0$. But this means that

$$\frac{d}{ds} \langle c(s), \mathbf{b}_0 \rangle = 0 \quad ,$$

so $\langle c(s), \mathbf{b}_0 \rangle = a$ where a is a constant. Thus c lies in the plane $\{p \in \mathbb{R}^3 : \langle p, \mathbf{b}_0 \rangle = a\}$.

Unlike the case of plane curves, we should not expect a curve with constant curvature to be a circle, unless the torsion is 0. To get some idea of the variety of possibilities, we will examine only one special class of curves, the <u>helices</u>, given by

$$c(u) = (a \cos u, a \sin u, bu) \quad .$$

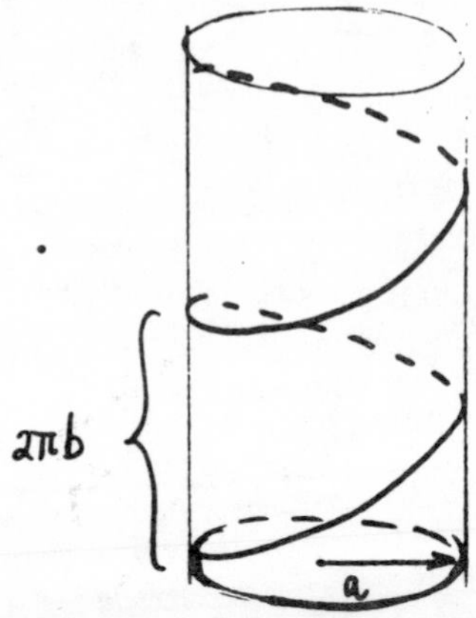

We have

$$c'(u) = (-a \sin u, a \cos u, bu) ,$$

so $|c'(u)| = \sqrt{a^2 + b^2} = D$. Then if γ is the reparameterization by arclength we have

$$\gamma(s) = c(s/D) .$$

So

$$\gamma'(s) = \left(-\frac{a}{D} \sin \frac{s}{D}, \frac{a}{D} \cos \frac{s}{D}, \frac{b}{D}\right)$$

$$\gamma''(s) = \left(-\frac{a}{D^2} \cos \frac{s}{D}, -\frac{a}{D^2} \sin \frac{s}{D}, 0\right) .$$

Thus

$$\kappa(s) = |\gamma''(s)| = \frac{|a|}{D^2}$$

$$\tau = \frac{\langle \gamma' \times \gamma'', \gamma''' \rangle}{\kappa^2} = \frac{b}{D^2} .$$

Notice that neither κ nor τ depend on the sign of a. Changing a to -a merely rotates the helix through an angle of π around its axis, which is the same as moving it a certain distance in the direction of this axis. However, changing from b to -b changes the helix from "right handed" to "left-handed" or vice-versa, and accordingly changes the sign of τ.

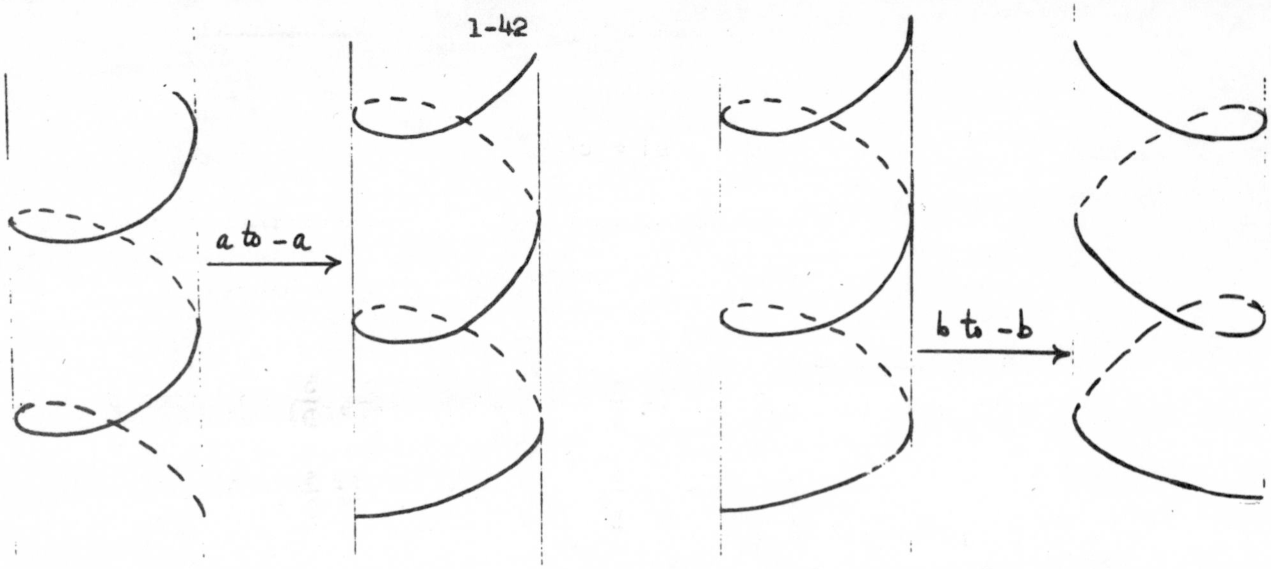

By choosing suitable a and b, we can make $|a|/D^2$ and b/D^2 equal to any desired pair (κ,τ) with $\kappa > 0$. So helices give examples of curves with any desired constant curvature (> 0) and constant torsion. Are they the only such curves? Rather than imitating the calculations for the simpler question answered previously, we will immediately ask the more general question, whether κ and τ determine c up to a Euclidean motion.

We begin with a recapitulation of the definitions:

$$\mathbf{t}' = \kappa \mathbf{n}$$
$$\mathbf{b}' = -\tau \mathbf{n} \quad .$$

Notice that we have expressed the derivatives of $\mathbf{t}$ and $\mathbf{b}$ in terms of the original vectors $\mathbf{t},\mathbf{n},\mathbf{b}$. We can do the same for $\mathbf{n}$. First, since $\langle \mathbf{n},\mathbf{n}\rangle = 1$, we obtain $\langle \mathbf{n}',\mathbf{n}\rangle = 0$, so $\mathbf{n}'$ is some linear combination of $\mathbf{t}$ and $\mathbf{b}$. Now, from $\langle \mathbf{n},\mathbf{t}\rangle = 0$ we obtain

$$\langle \mathbf{n}',\mathbf{t}\rangle = -\langle \mathbf{n},\mathbf{t}'\rangle = -\langle \mathbf{n},\kappa\mathbf{n}\rangle = -\kappa \ ,$$

(we already obtained this equation for the case of plane curves); and from $\langle \mathbf{n},\mathbf{b}\rangle = 0$ we obtain

$$\langle \mathbf{n}',\mathbf{b}\rangle = -\langle \mathbf{n},\mathbf{b}'\rangle = -\langle \mathbf{n},-\tau\mathbf{n}\rangle = \tau \quad .$$

Thus we have, altogether,

$$\begin{aligned} \mathbf{t}' &= \quad\ \kappa\mathbf{n} \\ \mathbf{n}' &= -\kappa\mathbf{t} \qquad\quad + \tau\mathbf{b} \\ \mathbf{b}' &= \quad -\tau\mathbf{n} \end{aligned}$$

These are called the Serret-Frenet formulas. They were obtained independently by Serret in 1851, and by Frenet, in his thesis of 1847, an abstract of which appeared in 1852. Before Serret and Frenet, many geometric properties of curves were investigated with great laboriousness, but afterwards many of these investigations became routine, because, as our next theorem shows, everything about space curves is contained in these formulas.

11. THEOREM. Let $\kappa,\tau\colon [a,b] \to \mathbb{R}$ be continuous, with $\kappa > 0$ on $[a,b]$. Then there is a curve $c\colon [a,b] \to \mathbb{R}^3$, parameterized by arclength, whose curvature and torsion functions are κ and τ. Any two such curves differ by a proper Euclidean motion (a translation followed by a rotation [an element of $SO(3)$]).

Proof. Let us adopt the more systematic notation $\mathbf{v}_1,\mathbf{v}_2,\mathbf{v}_3$ for $\mathbf{t},\mathbf{n},\mathbf{b}$, and define a matrix

$$a_{ij}(s) = \begin{pmatrix} 0 & \kappa(s) & 0 \\ -\kappa(s) & 0 & \tau(s) \\ 0 & -\tau(s) & 0 \end{pmatrix},$$

so that the Serret-Frenet equations become

$$(*) \qquad \mathbf{v}_i{}' = \sum_{j=1}^{3} a_{ij}\mathbf{v}_j \quad .$$

Now Theorem I.5-17 implies that on $[a,b]$ there is a function $s \mapsto (\mathbf{v}_1(s),\mathbf{v}_2(s),\mathbf{v}_3(s))$ satisfying (*). We can choose $\mathbf{v}_i(a)$ arbitrarily; choose them to be orthonormal and positively oriented. We then claim that $\mathbf{v}_i(s)$ are orthonormal for all $s \in [a,b]$. This is the only significant point in the proof; the rest of the proof is exactly like the proof of Theorem 2.

To prove that the $\mathbf{v}_i$ are always orthonormal, we note that

$$< \mathbf{v}_i,\mathbf{v}_j >' = \sum_{k=1}^{3} a_{ik} < \mathbf{v}_j,\mathbf{v}_k > + a_{jk} < \mathbf{v}_i,\mathbf{v}_k > \quad .$$

This shows that the functions $\beta_{ij} = < \mathbf{v}_i,\mathbf{v}_j >$ satisfy the differential equation

$$(**) \qquad \beta_{ij}' = \sum_{k=1}^{3} a_{ik}\beta_{jk} + a_{jk}\beta_{ik} \quad ,$$

together with the initial conditions $\beta_{ij}(a) = \delta_{ij}$. Since the solutions of (**) are determined by their initial conditions, we can prove $\mathbf{v}_i$ everywhere orthonormal by showing that the functions $\beta_{ij}(s) = \delta_{ij}$ do satisfy (**). In other words, we want to show that

$$0 = \delta_{ij}' = \sum_{k=1}^{3} a_{ik}\delta_{jk} + a_{jk}\delta_{ik}$$

$$= a_{ij} + a_{ji} \quad .$$

But this is true — the matrix (a_{ij}) is skew-symmetric. ∎

The skew-symmetry of (a_{ij}) gives an easy way to remember the Serret-Frenet formulas: the formulas for $\mathbf{t}'$ and $\mathbf{b}'$ are by definition, and the formula for $\mathbf{n}'$ is the one that makes the matrix skew-symmetric. If we think of $\mathbf{t},\mathbf{n},\mathbf{b}$ as column vectors, so that $(\mathbf{t},\mathbf{n},\mathbf{b})$ denotes a 3×3 matrix, then the Serret-Frenet equations read

$$(\mathbf{t},\mathbf{n},\mathbf{b})' = \begin{pmatrix} \mathbf{t}_1' & \mathbf{n}_1' & \mathbf{b}_1' \\ \mathbf{t}_2' & \mathbf{n}_2' & \mathbf{b}_2' \\ \mathbf{t}_3' & \mathbf{n}_3' & \mathbf{b}_3' \end{pmatrix} = \begin{pmatrix} \mathbf{t}_1 & \mathbf{n}_1 & \mathbf{b}_1 \\ \mathbf{t}_2 & \mathbf{n}_2 & \mathbf{b}_2 \\ \mathbf{t}_3 & \mathbf{n}_3 & \mathbf{b}_3 \end{pmatrix} \cdot \begin{pmatrix} 0 & \kappa & 0 \\ -\kappa & 0 & \tau \\ 0 & -\tau & 0 \end{pmatrix}.$$

Since $\mathbf{t},\mathbf{n},\mathbf{b}$ are orthonormal, the curve $\alpha(s) = (\mathbf{t}(s),\mathbf{n}(s),\mathbf{b}(s))$ is a curve in the orthogonal group $O(3)$. Recall that the tangent space $O(n)_I = \mathfrak{O}(n)$ of $O(n)$ at I is the set of skew-symmetric matrices. For $A \in O(n)$, the tangent space $O(n)_A$ is just $L_{A*}(\mathfrak{O}(n))$, which equals $L_A \cdot \mathfrak{O}(n)$, since L_A is a linear function. Thus $O(n)_A$ consists of all matrices $A \cdot M$ for $M \in \mathfrak{O}(n)$. Clearly, $A \cdot M = L_{A*}(M)$ is just $\tilde{M}(A)$, where $\tilde{M}$ denotes the left invariant vector field with value M at I. Now the curve $\alpha\colon [a,b] \longrightarrow O(3)$ must have its tangent vector $\alpha'(s)$ in $O(3)_{\alpha(s)}$, so we must have

$$\alpha'(s) = \alpha(s) \cdot (\text{skew-symmetric matrix}) .$$

As we have just seen, this skew-symmetric matrix is just

$$\begin{pmatrix} 0 & \kappa(s) & 0 \\ -\kappa(s) & 0 & \tau(s) \\ 0 & -\tau(s) & 0 \end{pmatrix} .$$

Moreover, this argument shows that skew-symmetry of this matrix is a necessary consequence of the fact that $\mathbf{t},\mathbf{n},\mathbf{b}$ are orthonormal. By the same token, we can now present a more illuminating proof that the equations

$$(*) \qquad \mathbf{v}_i' = \sum_{j=1}^{3} a_{ij}\mathbf{v}_j$$

in the proof of Theorem 11 have an everywhere orthonormal solution. The equation (*) for $\alpha(s) = (\mathbf{v}_1(s),\mathbf{v}_2(s),\mathbf{v}_3(s))$ says that

$$\alpha'(s) = a(s)\mathcal{H}(\alpha(s)) \quad .$$

This may simply be regarded as a differential equation <u>on</u> $O(3)$, of the type considered in the Addendum to Chapter 5, so its solution is a curve in $O(3)$.

The Lie group $O(3)$ is playing yet another, hitherto unmentioned, role in Theorem 11. The fact that κ and τ determine c up to a Euclidean motion is equivalent to the fact that κ and τ determine $\alpha = (\mathbf{t},\mathbf{n},\mathbf{b})$ up to an element of $O(3)$, since c is determined up to a translation by $\mathbf{t}$. In other words, if $\bar{c}$ is another curve with corresponding $\bar{\alpha} = (\bar{\mathbf{t}},\bar{\mathbf{n}},\bar{\mathbf{b}})$, and $\bar{\kappa} = \kappa$, $\bar{\tau} = \tau$, then for some $A \in O(3)$ we have

$$(\bar{\mathbf{t}},\bar{\mathbf{n}},\bar{\mathbf{b}}) = L_A \circ (\mathbf{t},\mathbf{n},\mathbf{b}) \ , \quad \text{i.e. } \bar{\alpha} = L_A \circ \alpha \quad .$$

Now we already have a theorem telling us when the relation $\bar{\alpha} = L_A \circ \alpha$ holds between two maps $\alpha,\bar{\alpha}\colon [a,b] \to O(3)$. According to Theorem I.10-18, this is the case if and only if $\alpha^*(\omega) = \bar{\alpha}^*(\omega)$ for every left invariant 1-form on $O(3)$. So κ and τ must have something to do with these left invariant 1-forms on $O(3)$. In order to see what is going on here, we begin with a review of some facts about Lie groups.

In Chapter I.10 we defined the natural $\mathfrak{g}$-valued 1-form ω on a Lie group G by $\omega(a)(\tilde{X}(a)) = X$. Thus ω is the unique left-invariant $\mathfrak{g}$-valued 1-form such that $\omega(e)\colon G_e \to \mathfrak{g} = G_e$ is the identity. If $X_1,\ldots,X_n$ is a basis of $\mathfrak{g}$, then we can write

$$\omega = \sum_{i=1}^{n} \omega^i \cdot X_i$$

for certain ordinary left-invariant 1-forms ω^i. This equation means that for any tangent vector $Y_a \in G_a$ we have

$$\omega(a)(Y_a) = \sum_{i=1}^{n} \omega^i(Y_a)\cdot X_i \in \mathfrak{g}$$

the dot denoting multiplication of $X_i \in \mathfrak{g}$ by the real number $\omega^i(a)(Y_a)$. Clearly the ω^i are a basis for the left invariant 1-forms; in fact, the $\omega^i(e)$ are the dual basis to $X_1,\ldots,X_n$. So the natural $\mathfrak{g}$-valued 1-form ω has all the left invariant 1-forms built into it.

Now for Lie groups G which are subgroups of some $GL(n,\mathbb{R})$ we have an explicit way of finding ω, and hence a basis of left invariant 1-forms. Let P (for "point") denote the inclusion map of $G \subset GL(n,\mathbb{R}) \subset \mathbb{R}^{n^2}$ into $\mathbb{R}^{n^2}$. Then $P\colon G \longrightarrow \mathbb{R}^{n^2}$ is an $\mathbb{R}^{n^2}$-valued function on G. Hence dP is an $\mathbb{R}^{n^2}$-valued 1-form on G. We can also think of dP as a matrix of ordinary 1-forms on G. This matrix is just

$$dP = (dx^{ij}) \quad ,$$

except that dx^{ij} here denotes the differential of $x^{ij}|G$. Notice that dP takes a tangent vector in G_I, i.e., an $n \times n$ matrix M, into itself, so dP corresponds to the identity map of G_I into itself. For any $A \in G$, the map $P\circ L_A\colon G \longrightarrow \mathbb{R}^{n^2}$ is

$$P\circ L_A(B) = A\cdot B \quad ,$$

from which it is easy to see that

$$\text{(1)} \qquad d(P\circ L_A) = A\cdot dP \quad .$$

On G we also have the C^∞ function $B \mapsto B^{-1}$, which we will denote by P^{-1}. For $A \in G$ the map $P^{-1} \circ L_A \colon G \to \mathbb{R}^{n^2}$ is

$$(2) \qquad P^{-1} \circ L_A(B) = (A \cdot B)^{-1} = B^{-1} \cdot A^{-1} = (P^{-1} \cdot A^{-1})(B) \quad .$$

Finally, $P^{-1} \cdot dP$ is a matrix of 1-forms (or $\mathbb{R}^{n^2}$-valued 1-form). From (1) and (2) we have

$$\begin{aligned} L_A{}^*(P^{-1} \cdot dP) &= (P^{-1} \circ L_A) \cdot L_A{}^*(dP) \\ &= (P^{-1} \circ L_A) \cdot d(L_A{}^* P) \\ &= (P^{-1} \circ L_A) \cdot d(P \circ L_A) \\ &= (P^{-1} \cdot A^{-1}) \cdot A \cdot dP \\ &= P^{-1} \cdot dP \quad , \end{aligned}$$

so $P^{-1} \cdot dP$ <u>is left invariant</u>. Moreover, for any $M \in G_I$ we have

$$P^{-1} \cdot dP(I)(M) = I^{-1} \cdot dP(I)(M) = M \quad .$$

Hence $P^{-1} \cdot dP$ <u>is the natural $\mathfrak{g}$-valued 1-form ω on</u> G.

These considerations allow us to determine ω, but it must be remembered that, unless $G = GL(n, \mathbb{R})$, the forms dx^{ij} are not linearly independent on G. Problem I.10-26 determines ω for $G = GL(n, \mathbb{R})$, and I.10-25 determines ω for the group $G \subset GL(2, \mathbb{R})$ consisting of all matrices

$$\begin{pmatrix} a & b \\ 0 & 1 \end{pmatrix} \qquad a \neq 0 \quad ;$$

for this group, x^{11} and x^{12} can be taken as a coordinate system, and ω can be expressed in terms of x^{11} and x^{12}. The more general situation is illustrated by Problem I.10-27, which we shall repeat here. The <u>special linear group</u> $SL(n,\mathbb{R}) \subset GL(n,\mathbb{R})$ is the set of all matrices of determinant 1. Its Lie algebra, $\mathfrak{sl}(n,\mathbb{R})$, consists of all matrices with trace 0. We can prove this from the fact that

$$\det \exp M = e^{\text{trace } M} ,$$

in the same way that we found $\mathfrak{o}(n)$; the formula just given is proved in Problem I.10-15. Now we see that, if $x^{11}, x^{12}, x^{21}, x^{22}$ are denoted by x, y, u, v, then for $SL(2,\mathbb{R})$ we have

$$P^{-1} \cdot dP = \begin{pmatrix} v & -y \\ -u & x \end{pmatrix} \cdot \begin{pmatrix} dx & dy \\ du & dv \end{pmatrix} = \begin{pmatrix} v\,dx - y\,du\,, & v\,dy - y\,dv \\ -u\,dx + x\,du\,, & -u\,dy + x\,dv \end{pmatrix} .$$

Remember that x, y, u, v are <u>not</u> linearly independent, for the dimension of $\mathfrak{sl}(2, \mathbb{R})$, and hence of $SL(2,\mathbb{R})$, is clearly 3. In fact, we know that $xv - yu = 1$ on $SL(2,\mathbb{R})$, which shows that

$$0 = d(xv - yu) = x\,dv + v\,dx - y\,du - u\,dy$$

on $SL(2,\mathbb{R})$, i.e., the right side gives 0 when applied to a matrix $M \in SL(2,\mathbb{R})_A$ for any $A \in SL(2,\mathbb{R})$. Looking at the formula for $P^{-1} \cdot dP$, this shows that $P^{-1} \cdot dP$ takes such a matrix M to a matrix of trace 0, as it must.

We will now use this circle of ideas to rederive our results about curves, in a more systematic, if less geometric, way. In order to clarify the general nature of the results, we begin once again with curves in the plane, but we now seek the answer to a different classification problem. We already know that if

A is a proper Euclidean motion, and $c\colon [a,b] \longrightarrow \mathbb{R}^2$ is any curve, then

(a) $A \circ c$ is parameterized by arclength whenever c is; we express this fact by saying that arclength is a "natural parameter for curves under the group of Euclidean motions."

(b) For curves c parameterized by arclength, the curvature function is invariant under proper Euclidean motions, i.e., the function κ for $A \circ c$ equals the function κ for c, whenever A is a Euclidean motion. Moreover, the curvature is a complete set of invariants for curves parameterized by arclength: if $\bar{\kappa}$ for $\bar{c}$ equals κ for c, then $\bar{c} = A \circ c$ for some Euclidean motion A.

We now ask for similar results when A is allowed to be any "special affine motion" [a translation followed by any member of $SL(2,\mathbb{R})$] .

Consider a curve $c\colon [a,b] \longrightarrow \mathbb{R}^2$ for which c' and c'' are always linearly independent. Then $\det(c',c'') \neq 0$ at all points. We assume that $\det(c',c'') > 0$, to avoid writing absolute value signs in various formulas. We then have a curve $\alpha_c\colon [a,b] \longrightarrow SL(2,\mathbb{R})$ defined by

$$\alpha_c(t) = \frac{(c'(t),c''(t))}{\sqrt{\det(c'(t),c''(t))}}$$

$$= \frac{1}{\sqrt{\det(c'(t),c''(t))}} \cdot \begin{pmatrix} c_1'(t) & c_1''(t) \\ c_2'(t) & c_2''(t) \end{pmatrix}.$$

If $\bar{c} = A \circ c$ for some special affine motion $A = B \circ \tau$, where $B \in SL(2,\mathbb{R})$ and τ is a translation, then clearly

$$\bar{c}' = B \cdot c'$$

$$\bar{c}'' = B \cdot c''$$

which implies that

$$(\bar{c}',\bar{c}'') = B\cdot(c',c'')$$

$$\det(\bar{c}',\bar{c}'') = \det(c',c'') \quad ,$$

and hence that

$$\alpha_{\bar{c}}(t) = B\cdot\alpha_c(t) \quad , \quad \alpha_{\bar{c}} = L_B\circ\alpha_c \quad .$$

Conversely, if $c,\bar{c}$ are two curves for which α_c and $\alpha_{\bar{c}}$ can be defined, and $\alpha_{\bar{c}} = L_B\circ\alpha_c$ for some $B \in SL(2,\mathbb{R})$, then, in particular, $\bar{c}' = B\cdot c'$, which implies that $\bar{c} = A\circ c$ for some special affine motion A. Now we already know that there is some $B \in SL(2,\mathbb{R})$ with $\alpha_{\bar{c}} = L_B\circ\alpha_c$ if and only if $\alpha_{\bar{c}}^*(\omega^i) = \alpha_c^*(\omega^i)$ for every left invariant 1-form ω^i on $SL(2,\mathbb{R})$, by Theorem I.10-18. This is equivalent to sayin that $\alpha_{\bar{c}}^*(\omega) = \alpha_c^*(\omega)$ for the natural $\mathfrak{sl}(2,\mathbb{R})$-valued form $\omega = P^{-1}\cdot dP$ on $SL(2,\mathbb{R})$. We therefore begin by calculating $\alpha_c^*(P^{-1}\cdot dP)$, which will be a matrix of 1-forms on $[a,b]$

$$\begin{pmatrix} \_\; dt & \_\; dt \\ \_\; dt & \_\; dt \end{pmatrix} \quad ;$$

for simplicity we will write all the dt's after the matrix.

To calculate $\alpha_c^*(P^{-1}\cdot dP)$ we can either use our formula for $P^{-1}\cdot dP$ and write

$$\alpha_c^*(P^{-1}\cdot dP) = \alpha_c^*\begin{pmatrix} v\ dx - y\ du\ , & v\ dy - y\ dv \\ -u\ dx + x\ du & -u\ dy + x\ dv \end{pmatrix}$$

$$= \begin{pmatrix} v\circ\alpha_c d(x\circ\alpha_c) - \ldots & \cdots \\ \cdots & \cdots \end{pmatrix}$$

or, what amounts to the same thing, calculate

$$\alpha_c^*(P^{-1}\cdot dP) = \alpha_c^{-1}\cdot d\alpha_c \quad ,$$

the entries of which are dt times

$$[\alpha_c(t)]^{-1}\cdot\alpha_c'(t) \quad .$$

To carry through the latter calculation, we first compute

$$[\alpha_c(t)]^{-1} = \sqrt{\det(\ ,\ )}\cdot\frac{1}{\det(\ ,\ )}\begin{pmatrix} c_2''(t), & -c_1''(t) \\ -c_2'(t) & c_1'(t) \end{pmatrix}$$

$$= \frac{1}{\sqrt{\det(\ ,\ )}}\begin{pmatrix} c_2''(t)\ , & -\ c_1''(t) \\ -c_2'(t) & c_1'(t) \end{pmatrix}$$

and

$$[\det(c',c'')]' = \det(c'',c'') + \det(c',c''') = \det(c',c''') \quad ,$$

so that

$$\alpha_c'(t) = \frac{1}{\sqrt{\det(\ ,\ )}}\begin{pmatrix} c_1''(t)\ , & c_1'''(t) \\ c_2''(t)\ , & c_2'''(t) \end{pmatrix}$$

$$- \frac{\det(c',c''')}{2[\det(\ ,\ )]^{3/2}}\begin{pmatrix} c_1'(t) & c_1''(t) \\ c_2'(t) & c_2''(t) \end{pmatrix} \quad .$$

After some calculation we arrive at

$$\alpha_c{}^*(P^{-1}\cdot dP) = [\alpha_c(t)]^{-1}\cdot\alpha_c{}'(t)\ dt$$

$$= \begin{pmatrix} -\frac{1}{2}\frac{\det(c'c''')}{\det(c',c'')} & -\frac{\det(c'',c''')}{\det(c',c''')} \\ 1 & \frac{1}{2}\frac{\det(c',c''')}{\det(c',c'')} \end{pmatrix} dt .$$

It should be clear why the trace of this matrix has to be 0.

We now have an answer, of sorts, to one of our questions: Two curves $c, \bar{c}\colon [a,b] \to \mathbb{R}^2$ differ by a special affine motion if and only if

$$\frac{\det(c',c''')}{\det(c',c'')} = \frac{\det(\bar{c},\bar{c}''')}{\det(\bar{c}',\bar{c}'')}$$

$$\frac{\det(c'',c''')}{\det(c',c'')} = \frac{\det(\bar{c}'',\bar{c}''')}{\det(\bar{c}',\bar{c}'')} .$$

This answer doesn't look very nice because we have not picked a nice parameterization. We now ask whether there is a function $\sigma\colon [a,b] \to [0,\ell]$ such that the reparameterization $\gamma = c\circ\sigma^{-1}$ satisfies

$$\det(\gamma',\gamma''') = 0 ,$$

so that the diagonal elements of $\alpha_\gamma{}^*(P^{-1}\cdot dP)$ vanish. Since $\det(\gamma',\gamma''') = [\det(\gamma',\gamma'')]'$, it suffices to have $\det(\gamma',\gamma'') = 1$. Now

$$c = \gamma\circ\sigma$$

$$c' = \sigma'\cdot(\gamma'\circ\sigma)$$

$$c'' = (\sigma')^2\cdot(\gamma''\circ\sigma) + \sigma''\cdot(\gamma'\circ\sigma) ,$$

so we want

$$\det(c',c") = \det(\sigma'\cdot(\gamma'\circ\sigma),(\sigma')^2\cdot(\gamma"\circ\sigma) + \sigma"\cdot(\gamma'\circ\sigma))$$

$$= (\sigma')^3\det(\gamma'\circ\sigma,\gamma"\circ\sigma)$$

$$= (\sigma')^3 \quad ;$$

we can thus define σ, the "special affine arclength" by

$$\sigma(t) = \int_a^t \sqrt[3]{\det(c'(u),c"(u))}\cdot du \quad .$$

It is clear that σ is a "natural parameter for curves under the group of special affine motions": if c is parameterized by σ, then so is $A\circ c$ for any special affine motion A. Moreover, we now see that curves parameterized by σ are determined, up to a special affine motion, by one function, the "special affine curvature"

$$\varkappa = \det(c",c"') \quad .$$

We notice, too late to do us any good, that we ought to have introduced the parameter σ at the very beginning, when we first found the matrix $\alpha_c(t)$.

It is also possible to give a geometric interpretation of the curvature, which we only briefly indicate. We first note that a curve c, parameterized by σ, with constant curvature $\varkappa$ satisfies

$$[\alpha_c(\sigma)]^{-1}\cdot\alpha_c'(\sigma) = \begin{pmatrix} 0 & -\varkappa \\ 1 & 0 \end{pmatrix}$$

$$\alpha_c'(\sigma) = \alpha_c(\sigma)\cdot\begin{pmatrix} 0 & -\varkappa \\ 1 & 0 \end{pmatrix}.$$

One solution of this matrix differential equation is

$$\alpha_c(\sigma) = \exp \sigma \cdot \begin{pmatrix} 0 & -\kappa \\ 1 & 0 \end{pmatrix}$$

$$= \begin{pmatrix} 1 & 0 \\ 0 & 1 \end{pmatrix} + \sigma \begin{pmatrix} 0 & -\kappa \\ 1 & 0 \end{pmatrix} + \frac{\sigma^2}{2!} \begin{pmatrix} -\kappa & 0 \\ 0 & -\kappa \end{pmatrix} + \frac{\sigma^3}{3!} \begin{pmatrix} 0 & -\kappa^2 \\ \kappa & 0 \end{pmatrix}$$

$$+ \frac{\sigma^4}{4!} \begin{pmatrix} \kappa^2 & 0 \\ 0 & \kappa^2 \end{pmatrix} + \ldots$$

$$= \begin{pmatrix} \cos \kappa^{1/2}\sigma & -\kappa^{-1/2}\sin \kappa^{1/2}\sigma \\ \kappa^{1/2}\sin \kappa^{1/2}\sigma & \cos \kappa^{1/2}\sigma \end{pmatrix}$$

for $\kappa > 0$, with a similar result involving hyperbolic trigonometric functions for $\kappa < 0$, and

$$\alpha_c(\sigma) = \begin{pmatrix} 1 & 0 \\ \sigma & 1 \end{pmatrix} \quad , \quad \text{for} \quad \kappa = 0 \ .$$

The first column of these solutions give

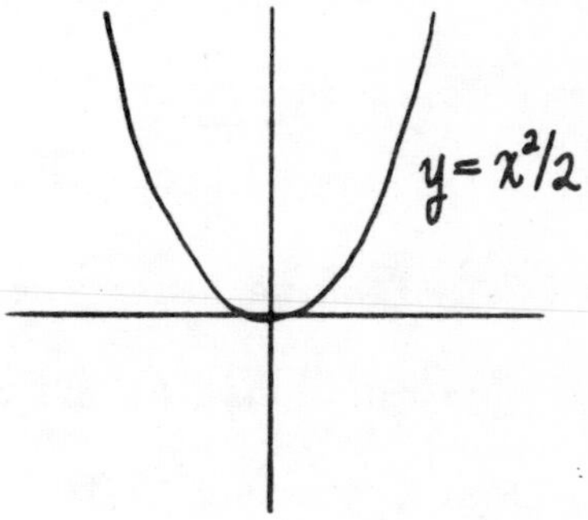

$\kappa = 0$: $\quad c'(\sigma) = (1,\sigma)$

$\quad c(\sigma) = \text{constant} + (\sigma,\sigma^2/2)$,

a parabola

$\kappa > 0$: $\quad c'(\sigma) = (\cos \kappa^{1/2}\sigma, \kappa^{1/2}\sin \kappa^{1/2}\sigma)$

$\quad c(\sigma) = \text{constant} + (\kappa^{-1/2}\sin \kappa^{1/2}\sigma, -\kappa^{-1}\cos \kappa^{1/2}\sigma)$,

an ellipse

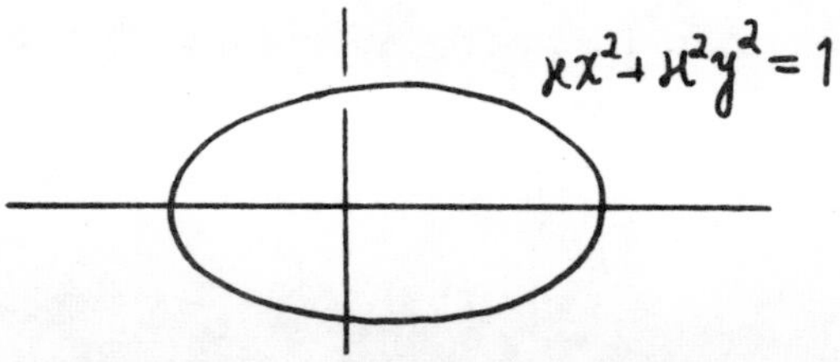

$\varkappa < 0$: $\quad c(\sigma) = \text{constant} + (|\varkappa|^{1/2}\cosh|\varkappa|^{1/2}\sigma, |\varkappa|^{-1}\sinh|\varkappa|^{1/2}\sigma)$,

a hyperbola.

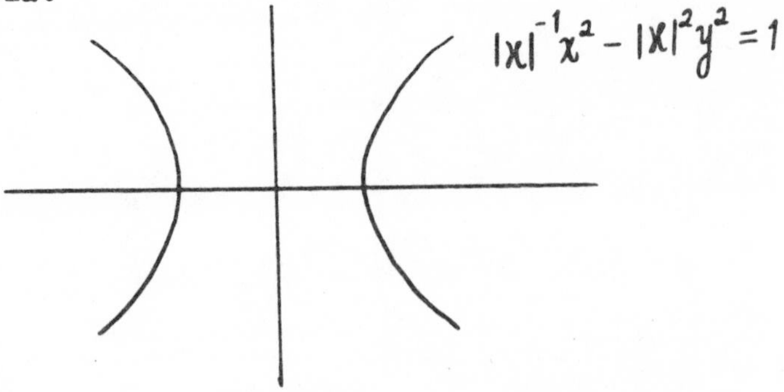

All other solutions $\alpha_c\colon \mathbb{R} \longrightarrow SL(2,\mathbb{R})$ are special linear transformations times these, and hence are still conic sections. Now it turns out that as $\sigma_1, \sigma_2, \sigma_3 \longrightarrow \sigma$, the parabola through $c(\sigma_1), c(\sigma_2), c(\sigma_3)$ approaches a given parabola, the osculating parabola, whose axis lies in the direction of $c''(s)$. And as $\sigma_1, \sigma_2, \sigma_3, \sigma_4 \longrightarrow \sigma$, the conic through the four points $c(\sigma_i)$ approaches a given conic, the hyperosculating conic, whose curvature $\varkappa$ is that of c at s. However, we will not prove these facts.

We now vary our question a little and ask for a classification of plane curves under the group of proper affine motions of the plane (translations followed by linear transformations of positive determinant). Any proper affine motion is the composition of a special affine motion and a "dilation", a map $\overline{M}$ of the form $x \longmapsto Mx$, for some number $M > 0$. For this group of motions, there is no natural parameter (if c is parameterized by affine arclength, then $\overline{M} \circ c$ is not, if $M \neq 1$). Instead we have merely a natural class of parameterizations, the parameterizations proportional to affine arclength. The simplest way to classify curves under the proper affine group is to give the curvature function $\varkappa$ which the curve has when it is parameterized by affine arclength, it being understood that this function changes by a constant multiple when a different parameterization in the natural class of parameterizations is used. If this is aesthetically unsatisfactory we can instead use the following approach. If c is parameterized by affine arclength, then $\bar{c} = \overline{M} \circ c$ satisfies

$$\bar{c}' = M\cdot c'$$
$$\bar{c}'' = M\cdot c'' \quad ,$$

so

$$\bar{\sigma}(t) = \int_a^t \sqrt[3]{\det(\bar{c}',\bar{c}'')} = M^{2/3}\sigma(t) \quad .$$

This means that to reparameterize $\bar{c}$ by affine arclength, we must introduce the curve

$$\bar{\gamma}(t) = \bar{c}(M^{-2/3}t) = Mc(M^{-2/3}t),$$

with

$$\bar{\gamma}'(t) = M^{1/3}c'(M^{-2/3}t)$$
$$\bar{\gamma}'(t) = M^{-1/3}c''(M^{-2/3}t)$$
$$\bar{\gamma}'''(t) = M^{-1}c'''(M^{-2/3}t) \quad .$$

Hence the curvature $\bar{\kappa}$ of $\bar{c}$ at t equals the curvature of $\bar{\gamma}$ at $M^{2/3}t$, so

$$\begin{aligned}\bar{\kappa}(t) &= \det(\bar{\gamma}''(M^{2/3}t),\bar{\gamma}'''(M^{2/3}t)) \\ &= M^{-4/3}\det(c''(t),c'''(t)) \\ &= M^{-4/3}\kappa(t) \quad .\end{aligned}$$

For a curve parameterized proportionally to affine arclength we introduce the "proportionality constant"

$$C = \frac{d\sigma}{dt} = \sqrt[3]{\det(c',c'')} \quad .$$

Then our calculations show that $c^2\chi$ is invariant under the group of proper affine motions. Clearly, curves parameterized proportionally to affine arclength are determined up to proper affine motion by $c^2\chi$.

It is of interest to see how we might be led to these same conclusions if we had investigated the group of proper affine motions directly, using the same methods that were used for the special affine motions. In this case, if $c\colon [a,b] \to \mathbb{R}^2$ is a curve with $\det(c',c'') > 0$, we have an associated curve $\alpha_c\colon [a,b] \to \{A \in GL(2,\mathbb{R})\colon \det A > 0\}$ defined simply by

$$\alpha_c(t) = (c'(t), c''(t)) \quad .$$

This leads to much easier calculations:

$$\begin{aligned}
\alpha_c{}^*(P^{-1}\cdot dP) &= [\alpha_c(t)]^{-1}\cdot\alpha_c{}'(t)\, dt \\
&= \frac{1}{\det(c'(t),c''(t))}\begin{pmatrix} c_2''(t) & -c_1''(t) \\ -c_2'(t) & c_1'(t)\end{pmatrix}\cdot\begin{pmatrix} c_1''(t) , & c_1'''(t) \\ c_2''(t) , & c_2'''(t)\end{pmatrix} dt \\
&= \begin{pmatrix} 0 & -\dfrac{\det(c'',c''')}{\det(c',c'')} \\ 1 & \dfrac{[\det(c',c'')]'}{\det(c',c'')}\end{pmatrix} dt \quad .
\end{aligned}$$

The introduction of parameterizations proportional to affine arclength would then be made in order to reduce the bottom term to 0.

We can now return to curves in space, and apply these same ideas to classify curves under the group of proper Euclidean motions. With each curve $c\colon [a,b] \to \mathbb{R}^3$ we want to associate a curve $\alpha_c\colon [a,b] \to SO(3)$. Assuming that $\det(c',c'',c''') > 0$, the obvious choice is to let $\alpha_c(t)$ be the result of applying the Gram-Schmidt orthogonolization process to these three vectors:

$$\alpha_c(t) = \left(\frac{c'(t)}{|c'(t)|} \; , \; \frac{c''(t) - \langle c''(t), c'(t) \rangle / |c'(t)|^2}{|c''(t) - \langle c''(t), c'(t) \rangle / |c'(t)|^2|} \; , \text{ etc.} \right)$$

Rather than actually computing $\alpha_c{}^*(P^{-1} \cdot dP)$, we immediately introduce the parameterization by (ordinary) arclength (which otherwise would be introduced after the calculation, in order to make certain entries of the matrix vanish). Then

$$\alpha_c(s) = \left(c'(s), \; \frac{c''(s)}{|c''(s)|} \; , \; c'(s) \times \frac{c''(s)}{|c'(s)|} \right) .$$

We now have

$$\alpha_c{}'(s) = \alpha_c(s) \cdot \begin{pmatrix} 0 & k(s) & 0 \\ -k(s) & 0 & t(s) \\ 0 & -t(s) & 0 \end{pmatrix} \quad ;$$

the zero in position (1,3) comes about because we choose $c', c''/|c''|$ as the first two columns of α_c, while all other features of the matrix are accounted for by the fact that $\alpha_c{}^{-1} \cdot \alpha'{}_c = \alpha_c{}^*(P^{-1} \cdot dP)$ is skew-symmetric, since $\alpha_c \colon [a,b] \longrightarrow SO(3)$. Clearly, curves parameterized by arclength are determined up to proper Euclidean motions by k and t, and these functions are obviously just our old κ and τ. If we had bothered to compute $\alpha_c{}^{-1} \cdot \alpha_c{}'$ for the original curve, not parameterized by arclength, then the entries in positions (1,2) and (2,3) would have given us the formulas for κ and τ, derived earlier, for curves with an arbitrary parameterization.

We can classify curves $c \colon [a,b] \longrightarrow \mathbb{R}^n$ under the group of proper Euclidean motions in a similar way. We assume first of all that $\det(c', \ldots, c^{(n)}) > 0$. To obtain a curve $\alpha_c \colon [a,b] \longrightarrow SO(n)$, we first introduce the parameterization by arclength. The first two columns V_1, V_2 of α_c will be $c', c''/|c''|$. So the first row of $\alpha_c{}^{-1} \cdot \alpha_c{}'$, which expresses $V_1{}'$ in terms of the V_j, will be

$$0, k_1, 0 \ldots 0$$

for $k_1 = |c''|$. Using skew-symmetry, $\alpha_c^{-1} \cdot \alpha_c'$ looks like

$$\alpha_c^{-1} \cdot \alpha_c' = \begin{pmatrix} 0 & k_1 & 0 \ldots 0 \\ -k_1 & 0 & \times \\ 0 & & 0 \\ \vdots & \times & & \ddots \\ 0 & & & & 0 \end{pmatrix} .$$

Since $\mathbf{V}_3$, the third column of α_c, is obtained by applying the Gram-Schmidt orthogonalization process to $\mathbf{V}_1, \mathbf{V}_2, c'''$, it is clear that $\mathbf{V}_2'$ will be a linear combination of $\mathbf{V}_1$ and $\mathbf{V}_3$. So the second row of $\alpha_c^{-1} \cdot \alpha_c'$ will be

$$-k_1 \; 0 \; k_2 \; 0 \ldots 0 \quad ,$$

for some function k_2. Thus $\alpha_c^{-1} \cdot \alpha_c'$ looks like

$$\alpha_c^{-1} \cdot \alpha_c' = \begin{pmatrix} 0 & k_1 & 0 \ldots 0 \\ -k_1 & 0 & k_2 \ldots 0 \\ 0 & -k_2 & 0 & \times \\ \vdots & \times & & \ddots \\ 0 & & & & 0 \end{pmatrix} .$$

Continuing in this way, we find that

$$\alpha_c^{-1} \cdot \alpha_c' = \begin{pmatrix} 0 & k_1 & & & \\ -k_1 & 0 & k_2 & & \times \\ & -k_2 & \cdot & & \\ \times & & & \cdot & k_{n-1} \\ & & & -k_{n-1} & 0 \end{pmatrix}$$

for n-1 different "curvature functions" $k_1,\ldots,k_{n-1}$. These n-1 functions classify c up to proper Euclidean motion.

Finally, we insert a remark about one feature of this whole approach, which may have been bothering the reader all along. To classify curves in $\mathbb{R}^n$ we have had to restrict our attention to curves with $c',\ldots,c^{(n)}$ everywhere linearly independent. If we have a curve c such that $c^{(k)}$ is linearly dependent on $c',\ldots,c^{(k-1)}$ along a whole interval $[a,b]$, then it is easy to see that on this interval c lies in some (k-1)-dimensional subspace of $\mathbb{R}^n$, so that we actually have an easier classification problem on this interval. The difficulties arise when we try to piece together the information we obtain for the separate intervals, or if $c^{(k)}$ is linearly dependent on $c',\ldots,c^{(k-1)}$ at only isolated points (or at some sequence of points, etc.) For example, how can one use functions like curvature and torsion to distinguish between the following two curves?

$$c(t) = \begin{cases} 0 & t = 0 \\ (t, e^{-1/t^2}, 0) & t > 0 \\ (t, 0, e^{-1/t^2}) & t < 0 \end{cases}$$

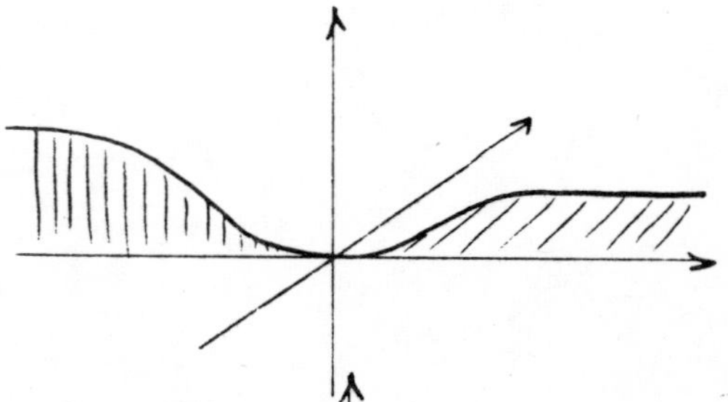

$$c(t) = \begin{cases} 0 & t = 0 \\ (t, 0, e^{-1/t^2}) & t \neq 0 \end{cases}$$

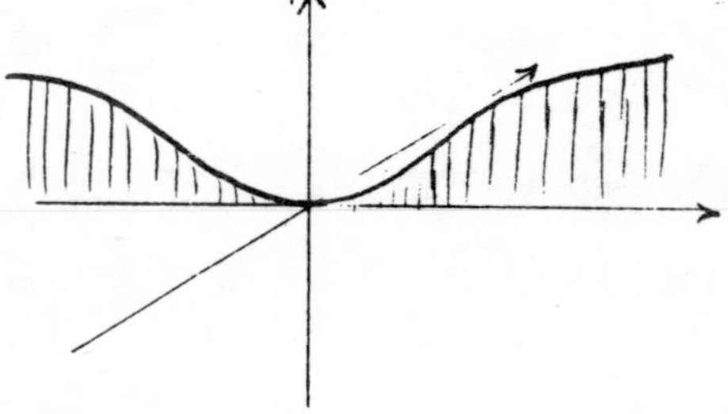

Of course, for analytic curves these difficulties cannot arise. If any small portion of one analytic curve differs by a proper Euclidean motion from a small portion of a second, then the two analytic curves themselves differ by this proper Euclidean motion. But for C^∞ curves, our restrictions are natural ones to make.

Chapter 2. What they knew about Surfaces before Gauss

Having traversed nearly the whole history of the study of curves, in one chapter, we now turn back to the beginnings of surface theory, and start the long journey toward the modern theory of higher dimensional manifolds. Even though the study of surfaces had begun long before the Serret-Frenet formulas appeared in 1847, the theory of plane curves, at least, was well understood.

The initial study of surfaces in $\mathbb{R}^3$ began in a way that seems natural enough. Since we have a theory of curves in the plane, we may hope to describe surfaces by investigating the curves in which the surface intersects various planes. Indeed it turns out that we can describe the curvature of such curves in a suprisingly nice and precise way.

The first results along this line, due to Euler, date from 1760. Through a point p on a surface $M \subset \mathbb{R}^3$ we construct the line ℓ which is perpendicular to M_p. For each unit vector $X \in M_p$ we can then consider the plane through

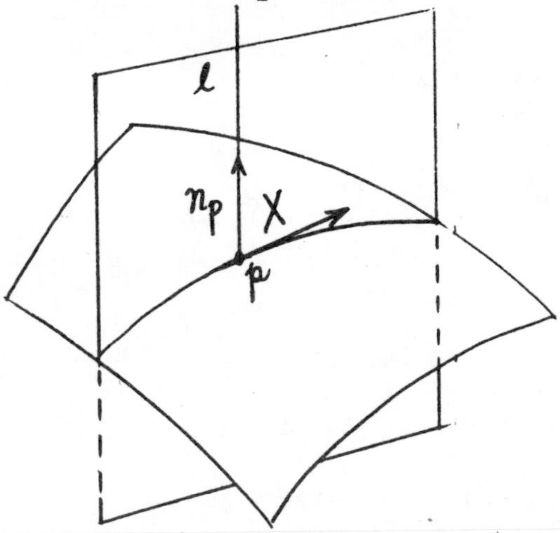

p which contains both X and ℓ. The intersection of this plane and M is the image of a curve c_X with $c_X(0) = p$; we will also suppose c parameterized by arclength, so that $c_X'(0) = X$. We orient all these planes through ℓ by choosing a vector n_p perpendicular to M_p and orienting the plane through X and ℓ so that X, n_p is positively oriented. Then c_X has a signed curvature at 0, which will be denoted by κ_X .

When we replace X by $-X$, the curve c_{-X} is just c_X traversed in the opposite direction. Since the plane through X and n_p now receives the opposite orientation, the curvature κ_{-X} equals the curvature κ_X. Thus, $X \mapsto \kappa_X$ may be thought of as a function of directions in M_p. Euler discovered a striking fact about the curvatures in these different directions:

1. THEOREM (EULER). If the κ_X are not all equal, then there is precisely one direction, represented by a unit vector X_1, say, in which κ_X has a minimum value $\kappa_1 = \kappa_{X_1}$, and one in which it has a maximum value $\kappa_2 = \kappa_{X_2}$. These two directions are perpendicular, and if X makes an angle of θ with X_1, then

$$\kappa_X = \kappa_1 \cos^2\theta + \kappa_2 \sin^2\theta \quad .$$

(Notice that we can just as well write

$$\kappa_X = \kappa_2 \cos^2\varphi + \kappa_1 \sin^2\varphi \quad ,$$

where φ is the angle which X makes with X_2, since $\varphi = \pi/2 - \theta$, and consequently $\cos^2\theta = \sin^2\varphi$ and $\sin^2\theta = \cos^2\varphi$. There is also a certain ambiguity in the statement of Euler's theorem which does not affect the final result: if we change the vector n_p so that it points in the opposite direction, then all curvatures κ_X are changed to their negatives.)

I do not know Euler's proof of his theorem, and for reasons that will appear later, I am sure that the proof to be presented here is much simpler than the original. Nevertheless, it is sufficiently classical in spirit to serve as an historical substitute for Euler's.

Proof. We begin by choosing our coordinate system so that $p = (0,0,0)$ and so that the tangent plane at p is the x-y plane, which means that in a neighborhood of p the surface M is $\{(x,y,z): z = f(x,y)\}$, where

$$f(0,0) = 0$$

$$\frac{\partial f}{\partial x}(0,0) = 0$$

$$\frac{\partial f}{\partial y}(0,0) = 0 \quad .$$

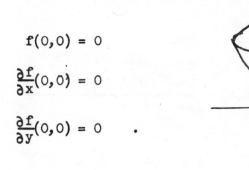

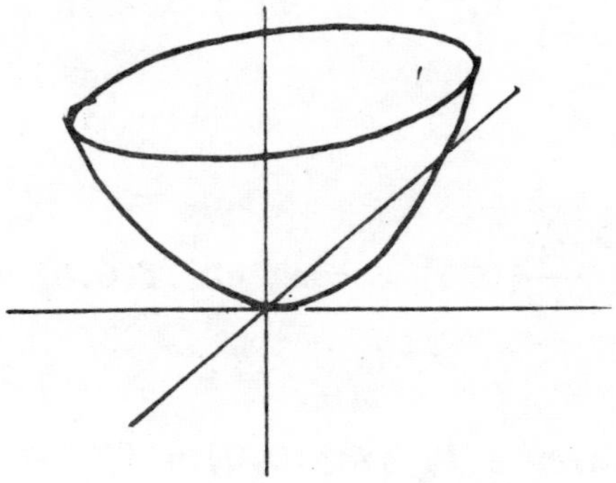

We now maintain that by rotating the x-y plane we can arrange to have

$$\frac{\partial^2 f}{\partial x \partial y}(0,0) = 0 \quad .$$

To see this, we first recall that rotation through an angle of θ radians is the linear transformation R_θ with matrix

$$\begin{pmatrix} \cos\theta & -\sin\theta \\ \sin\theta & \cos\theta \end{pmatrix} \quad .$$

If we rotate the x-y plane by an angle θ, then M becomes the graph of

$$f_\theta = f \circ R_{-\theta} \quad ,$$

$$f_\theta(x,y) = f(x\cos\theta - y\sin\theta,\ x\sin\theta + y\cos\theta) \quad .$$

So

$$\frac{\partial f_\theta}{\partial y}(x,y) = D_1f(\ ,\ )(-\sin\theta) + D_2f(\ ,\ )(\cos\theta)$$

$$\frac{\partial^2 f_\theta}{\partial x\partial y}(x,y) = D_{11}f(\ ,\ )(-\sin\theta\cos\theta) + D_{12}(\ ,\ )[-\sin^2\theta + \cos^2\theta]$$

$$+ D_{22}f(\ ,\ )(\sin\theta\cos\theta)$$

$$\frac{\partial^2 f_\theta}{\partial x\partial y}(0,0) = \cos 2\theta D_{12}f(0,0) + \frac{(D_{22}f(0,0) - D_{11}f(0,0))}{2}\sin 2\theta \quad .$$

In order to have $\partial^2 f_\theta/\partial x\partial y(0,0) = 0$, we just choose θ so that

$$\tan 2\theta = \frac{2D_{12}f(0,0)}{D_{11}f(0,0) - D_{22}f(0,0)}, \quad \text{if } D_{11}f(0,0) \neq D_{22}f(0,0)$$

$$\theta = \pi/4, \quad \text{if } D_{11}f(0,0) = D_{22}f(0,0) \quad .$$

Having made this choice of coördinates, we now look at the various planes containing ℓ i.e., containing the z-axis. First, the x-z plane intersects the surface in the curve

$$c(t) = (t, f(t,0)) \quad .$$

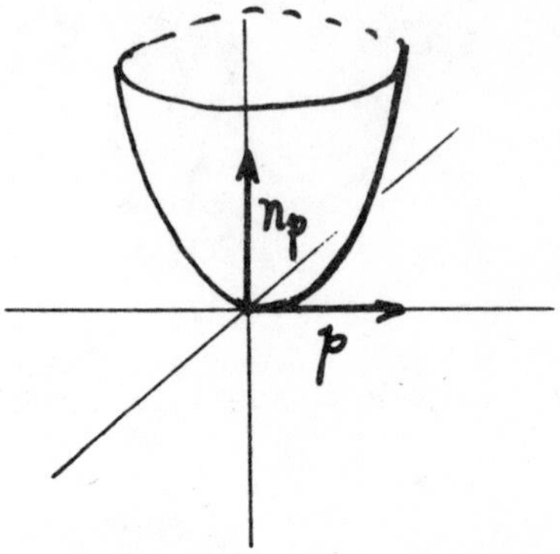

It's curvature at 0 is therefore

$$\kappa_1 = \frac{\dot c_1\ddot c_2 - \dot c_2\ddot c_1}{(\dot c_1{}^2 + \dot c_2{}^2)^{3/2}} = \frac{\partial^2 f}{\partial x^2}(0,0) \quad .$$

This result holds when we give the x-z plane its usual orientation, which is the same as the orientation making $(e_1)_p, n_p$ positively oriented when we choose n_p to be e_{3_p}. Similarly, the y-z plane intersects the surface in the curve $c(t) = (t, f(0,t))$ with curvature

$$\kappa_2 = \frac{\partial^2 f}{\partial y^2}(0,0) \quad .$$

Finally, the plane through the z-axis which makes an angle of θ with the x-axis intersects the surface in the curve

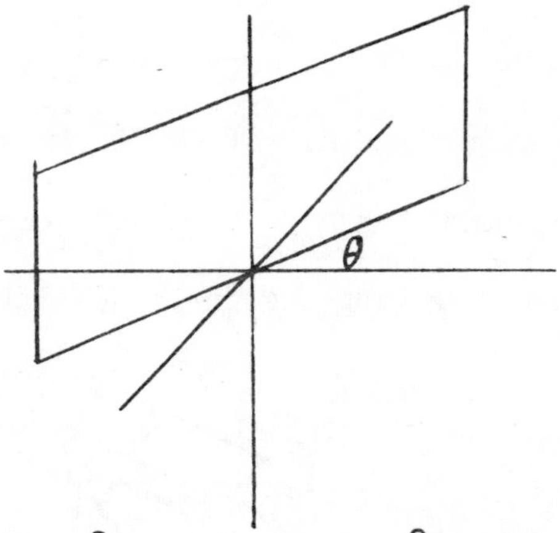

$$c(t) = (t, f(t\cos\theta, t\sin\theta)) \quad ,$$

with curvature

$$\kappa = \frac{d^2}{dt^2} f(t\cos\theta, t\sin\theta) = \cos^2\theta\, \frac{\partial^2 f}{\partial x^2}(0,0) + \sin^2\theta\, \frac{\partial^2 f}{\partial y^2}(0,0)$$

$$= \kappa_1 \cos^2\theta + \kappa_2 \sin^2\theta \quad .$$

This formula shows that κ always lies between κ_1 and κ_2, the curvatures in two perpendicular directions, and thus proves the whole theorem. ∎

Notice that if $\partial^2 f/\partial x^2, \partial^2 f/\partial y^2 > 0$, so that the surface lies above the x-y plane near p, then our choice of n_p makes $\kappa_1, \kappa_2 > 0$. In general, if a surface locally lies on one side of its tangent plane through p, then the choice of n_p as a vector pointing toward this side makes $\kappa_1, \kappa_2 > 0$. If our surface is the boundary of a convex set in $\mathbb{R}^3$, we must therefore choose the <u>inward</u> pointing normal to obtain positive curvatures. If κ_1 and κ_2 are

of different signs, there is generally no such way to distinguish a direction for n_p.

The other main result about curves on surfaces, due to Meusnier, came shortly afterwards, in 1776. Meusnier completed Euler's investigations by finding the curvatures of the curves obtained by intersecting the surface M with <u>any</u> plane through $p \in M$. If P is the plane through p which contains ℓ and a unit vector $X \in M_p$, then any other plane P_φ which contains X can be described by giving the angle φ which it makes with P. Let c be the

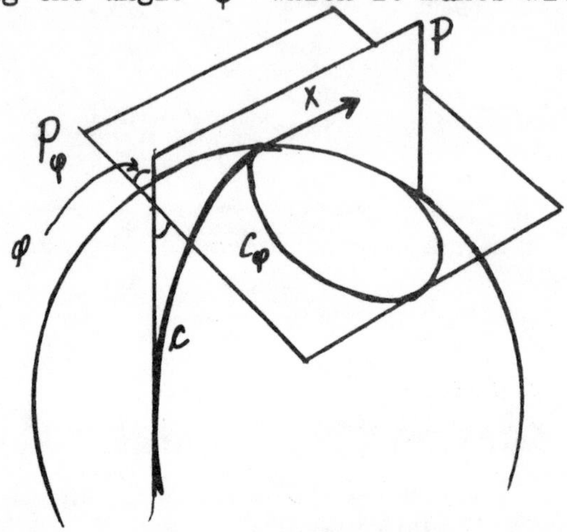

intersection of P and M, with $c(0) = p$ and c parameterized by arclength, so that $c'(0) = X$, and let c_φ be the corresponding intersection of P_φ and M. Meusnier's theorem describes the curvature κ_φ of c_φ at 0 in terms of the curvature κ_X of c at 0:

$$\kappa_\varphi \cdot \cos\varphi = \kappa_X \quad .$$

For the case of the unit sphere, pictured above, the curve c is a great circle, of radius 1, while c_φ is a circle of radius $\cos\varphi$, so that $\kappa_\varphi = 1/\cos\varphi$. Naturally, in Meusnier's theorem we must restrict φ to be less than $\pi/2$. At this angle, the plane P_φ is just the tangent plane, and does not generally intersect M in a curve at all; for φ close to $\pi/2$, the plane P_φ intersects M in a curve of very large curvature (if $\kappa_X \neq 0$).

For Meusnier's theorem we supply a proof which is decidedly non-classical in spirit, but which will be useful to have later on. We first define a function n in a neighborhood $U \subset M$ of p such that $n(q) \in \mathbb{R}^3$ is a unit vector and $n(q)_q \in \mathbb{R}^3_q$ is perpendicular to M_q for all $q \in U$. There are two choices for each $n(q)$; in order to obtain a continuous function $n\colon U \longrightarrow M$, we orient U, and then pick $n(q)$ so that $n(q), v, w$ is positively oriented for v_q, w_q positively oriented in M_q.

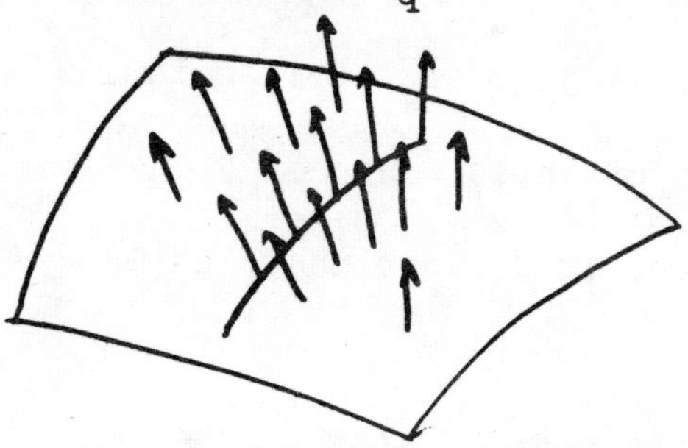

For any curve c in M with $c(0) = p$ we have

$$< c'(s), n(c(s)) > = 0 \qquad \text{for all } s.$$

Differentiating this equation, we have

$$< c''(0), n(p) > = - \left\langle c'(0) , \left. \frac{dn(c(s))}{ds} \right|_{s=0} \right\rangle .$$

Now the vector $dn(c(s))/ds|_{s=0}$, with components $dn^i(c(s))/ds|_{s=0}$, depends <u>only on</u> $c'(0) = X$; in fact it equals $(X(n^1), X(n^2), X(n^3))$. To see this, just remember that to operate on a function $f\colon M \longrightarrow \mathbb{R}$ with a tangent vector $X_p \in M_p$, we can take any curve c with $c'(0) = X_p$ and then $X_p(f) = df(c((t))/dt|_{t=0}$. We can thus write

$$(*) \quad < c''(0), n(p) > = a(X) \qquad X = c'(0).$$

Meusnier's theorem follows directly from this equation:

2. THEOREM (MEUSNIER). Let P be the plane through $n(p)$ and $X \in M_p$, and let κ_φ be the curvature of the curve on M cut out by a plane P_φ containing X and making an angle φ with the plane P. Then

$$\kappa_\varphi \cdot \cos\varphi = \kappa_X .$$

Proof. First apply (*) to the curve c cut out by P. The second derivative $c''(0)$ of this curve is in P, and is also perpendicular to $c'(0) = X$. Consequently, it is a multiple of $n(p)$. Since P has been oriented so that X, $n(p)$ is positively oriented, it follows that $c''(0) = \kappa_X \cdot n(p)$, so

$$a(X) = \langle c''(0), n(p) \rangle = \kappa_X .$$

On the other hand, since $c_\varphi'(0) = c'(0) = X$, equation (*) also gives

$$(1) \qquad \kappa_X = a(X) = \langle c_\varphi''(0), n(p) \rangle .$$

We can write $c_\varphi''(0) = \kappa_\varphi \cdot \nu$ where ν is a unit vector which lies in P_φ and is perpendicular to X. Then ν makes an angle of φ with $n(p)$, which means that

$$(2) \qquad \langle \nu, n(p) \rangle = \cos\varphi .$$

Combining equations (1) and (2), we obtain

$$\begin{aligned} \kappa_X &= \langle c_\varphi''(0), n(p) \rangle \\ &= \langle \kappa_\varphi \cdot \nu, n(p) \rangle \\ &= \kappa_\varphi \cdot \cos\varphi \quad . \blacksquare \end{aligned}$$

Despite the appealing simplicity of these results, there is something dissatisfying about this whole approach of dissecting a surface into curves; we never seem to really get our hands on the surface itself. To do this, we must move forward 50 years in time.

Chapter 3A. How to Read Gauss

The single most important work in the history of differential geometry is Gauss' paper of 1827, Disquisitiones generales circa superficies curvas (in Latin). An English translation is available under the title General Investigations of Curved Surfaces, published by the Raven Press. Although the second part of this Chapter is an exposition of Gauss' results, in modern notation, a preliminary reading of Gauss' great work is heartily recommended.

The paper contains 29 sections, of which we want to read only the first 20 (pages 3 to 30). The footnotes for these sections can be ignored; most of them add only insignificant details about errors in various editions, etc. Although the courageous reader may attempt a reading of Gauss unaided, the following remarks will probably make this first confrontation with classical differential geometry much less painful. Since some of the difficulties will be clarified in the second part of this Chapter, as a general rule it is a good idea to read on, even if a particular section makes very little sense.

§1. The picture for this section is shown below. Notice that (1), (2), (3) are used as the names of certain points [(1) = (1,0,0), etc.], a circumstance that is easy to forget later on.

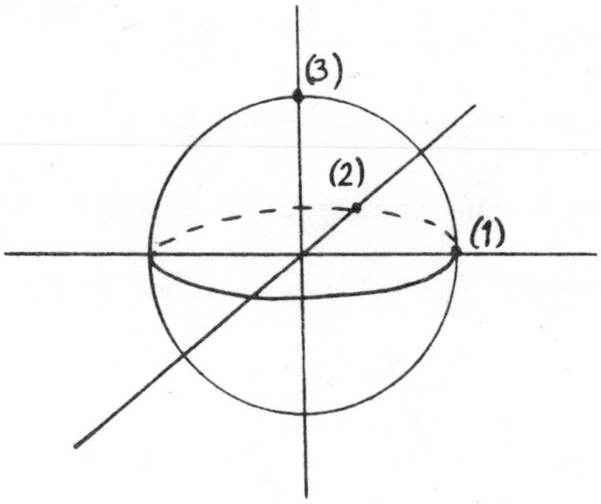

§2. This section may be skipped entirely. It gives a complicated proof, using spherical trigonometry, that the volume of the pyramid shown below is

$$\frac{1}{6}\left|\det\begin{pmatrix} x & y & z \\ x' & y' & z' \\ x'' & y'' & z'' \end{pmatrix}\right|,$$

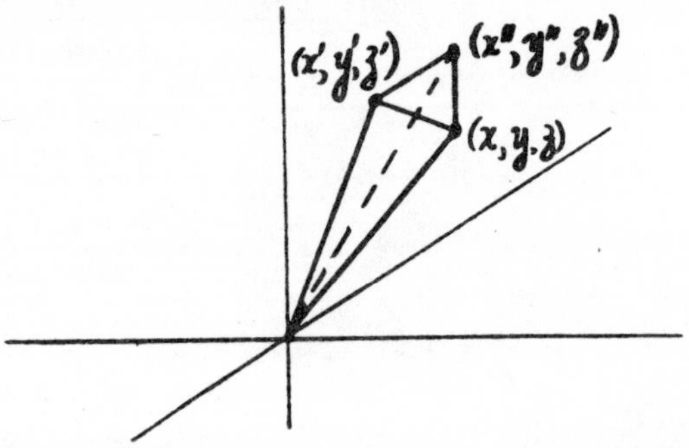

and also includes remarks about the significance of the sign of the determinant. This result is equivalent to the well-known fact that $|\det A|$ is the volume of the parallelipiped spanned by the rows of A.

§3. This section merely defines (or tries to define) a differentiable surface, and its tangent plane at a point.

§4. At a point $A = (x,y,z)$ in the surface we have a unit normal vector v_A, and $v \in S^2 \subset \mathbb{R}^3$ is what Gauss calls L.

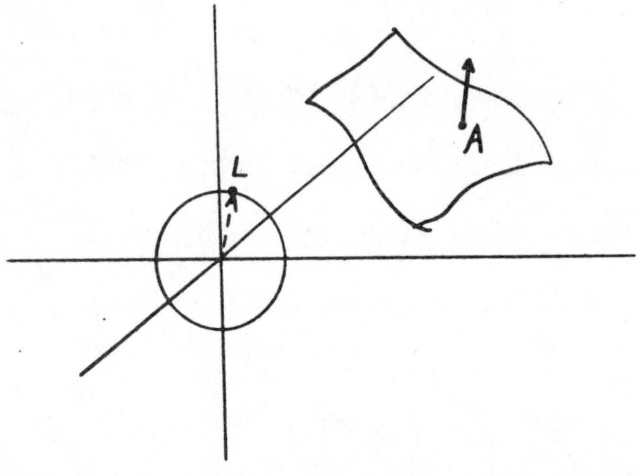

The expression cos (1)L means the cosine of the angle between the rays from (0,0,0) through L and through (1) = (1,0,0). So X,Y,Z are just the

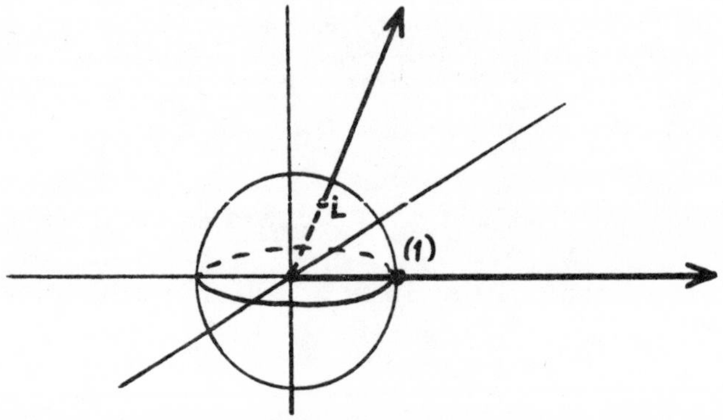

components of L. Thus X,Y,Z can be considered as functions on the surface [X(A) = first component of v, for v_A a unit normal at A, etc.]

Gauss now nonchalantly introduces infinitely small quantities. The goal is the equation

$$X\,dx + Y\,dy + Z\,dz = 0 \quad .$$

If x,y,z are considered as functions on the surface (that is, as the restriction to the surface of the standard coordinate functions on $\mathbb{R}^3$), then this equation is literally true, interpreting dx, dy, dz as modern differentials. It should be easy to see this (remember how X, Y, Z are defined). Also try to follow Gauss' argument.

The rest of section 4 gives formulas for X, Y, Z in terms of different descriptions of the surface; in each case the formulas are paired with their negatives, since there are two different choices for the unit normal vector:

(a) If the surface is $\{p \in \mathbb{R}^3 : W(p) = 0\}$, for $W: \mathbb{R}^3 \to \mathbb{R}$, then

$$X = \frac{P}{\sqrt{P^2 + Q^2 + R^2}} \quad , \qquad \text{where } P = D_1W,\ Q = D_2W,\ R = D_3W \quad .$$

etc.

(b) If the surface is the image of $f: \mathbb{R}^2 \to \mathbb{R}^3$, then

$$X = \frac{bc' - cb'}{\Delta} \quad , \qquad \text{where } b = D_1f^2,\ b' = D_2f^2 \ , \text{ etc.}$$

Notice that on the bottom of page 7 Gauss writes dx for $d(x \circ f) = df^1$, etc.

(c) If the surface is $\{(x,y,z): z = f(x,y)\}$ for $f: \mathbb{R}^2 \rightarrow \mathbb{R}^3$, then

$$X = \frac{t}{\sqrt{1 + t^2 + u^2}}, \quad \text{where } t = D_1 f, \quad u = D_2 f \ .$$

etc.

It should not be hard to work out these results, using our terminology. Again, it is instructive to follow Gauss' derivations as well.

§5. This section talks about orienting the surface, so that one can choose between the two unit normals.

§6. In this section Gauss considers the map n, from the surface to S^2, which takes A to the unit vector v which is normal to the surface at that point. The map n can be used to take any subset R of the surface to a subset $n(R)$ of S^2. The area of $n(R)$ is referred to by Gauss as the total curvature of R. The curvature at a point A in the surface is defined as

$$\frac{\text{total curvature of } R}{\text{area of } R},$$

when R is the "surface element" at A, which is supposed to have infinitely small area. As a first approximation to what Gauss is trying to say, we might define the curvature as

$$\lim \frac{\text{total curvature of } R}{\text{area of } R}$$

where the limit is taken as R approaches the point A. It is not *a priori* so clear whether this limit exists, or if it depends on the way in which R

"approaches" A.

Gauss also give considerable discussion to the sign of the curvature.

§7. In this section Gauss finds a formula for the curvature k at A. His answer, at the very bottom of page 12, is given for a surface which is the graph of $f\colon \mathbb{R}^2 \to \mathbb{R}^3$. In this case, the functions X, Y can be thought of as functions on $\mathbb{R}^2$ (that is, we consider $X \circ f$ and $Y \circ f$) and Gauss' answer is

$$k = D_1(X \circ f)D_2(Y \circ f) - D_2(X \circ f)D_1(Y \circ f)$$

$$= \frac{\partial X}{\partial x}\frac{\partial Y}{\partial y} - \frac{\partial X}{\partial y}\frac{\partial Y}{\partial x} \quad .$$

He obtains this answer by considering an infinitesimal triangle "$d\sigma$" with one vertex at $(x,y,f(x,y))$ and the other two vertices at $(x + dx,\ y + dy,\ f(x + dx,\ y + dy))$ and $(x + \delta x,\ y + \delta y,\ f(x + \delta x,\ y + \delta y))$. It is a challenge both to follow Gauss' reasoning, and to put it in modern terms. Either way, one needs Gauss' preliminary observation that

$$\frac{\text{area } n(d\sigma)}{\text{area } d\sigma} = \frac{\text{area of projection on x-y plane of } n(d\sigma)}{\text{area of projection on x-y plane of } d\sigma} \quad .$$

This mysterious equation really says that the tangent plane of M at A is parallel to the tangent plane of S^2 at $n(A)$. If this hint does not help, simply accept the formula for k, which will be derived later, using modern terminology.

The remainder of section 7 evaluates k in terms of partial derivatives of f (which Gauss denotes by z).

§8. This section, except for the last theorem, was already done in Chapter 2.

§9, 10, 11. These sections are essentially calculations, involving no new ideas. Every once in a while Gauss calculates a differential instead of some partial derivatives, but this should cause no difficulties.

§12. If $M, N \subset \mathbb{R}^3$ are surfaces, then a <u>development</u> of M on N is simply a map $f\colon M \to N$ which is an isometry (with respect to the induced Riemannian metrics).

§14. Throughout this section Gauss uses x, y, z to denote $x \circ c$, $y \circ c$, $z \circ c$, for the curve c under consideration.

On page 22, the second integral (appearing in line 7) is what we would write as

$$\left.\frac{dL(\bar{\alpha}(u))}{du}\right|_{u=0} = \left.\frac{d}{du}\right|_{u=0} \int_a^b \sqrt{\left(\frac{\partial \alpha^1(u,t)}{\partial t}\right)^2 + \cdots}\; dt$$

$$= \int_a^b \frac{\dfrac{\partial \alpha^1(0,t)}{\partial u}\,\dfrac{\partial^2 \alpha^1(0,t)}{\partial u \partial t} + \cdots}{\sqrt{\left(\dfrac{\partial \alpha^1(0,t)}{\partial u}\right)^2 + \cdots}}\, dt$$

$$= \int_a^b \frac{\dfrac{dc^1}{dt}\,\dfrac{\partial^2 \alpha^1(0,t)}{\partial u \partial t} + \cdots}{\sqrt{}}\, dt.$$

Thus

$$\frac{dc^1}{dt} \text{ is } \frac{dx}{[dt]} \quad\text{and}\quad \frac{\partial^2 \alpha^1(0,t)}{\partial u \partial t} = \frac{\partial^2 \alpha^1(0,t)}{\partial t \partial u} \text{ is } \frac{d\delta x}{[\partial t \partial u]}.$$

Lines 9 and 10 show what this becomes after integration by parts. The term on line 10 is

$$-\int_a^b \frac{\partial\alpha^1}{\partial u}(0,t)\ \frac{d}{dt}\left(\frac{\frac{dc^1}{dt}}{\sqrt{\qquad\qquad}}\right) + \ \ldots \quad dt;$$

here

$$\frac{\partial\alpha^1}{\partial u}(0,t) \quad \text{is} \quad \frac{\delta x}{[\partial u]}.$$

Notice that Gauss has given $dL(\bar{\alpha}(u))/du\big|_{u=0}$ for an arbitrary variation in 3-space, not just a variation through curves in the surface. His $x = x \circ c$ is a coordinate function of c in three space, not a coordinate function with respect to some coordinate system on the surface. If the surface is $\{p: W(p) = 0\}$ for $W: \mathbb{R}^3 \to \mathbb{R}$, so that on the surface we have

$$0 = dW = P\,dx + Q\,dy + R\,dz \qquad\qquad P = D_1W,\ Q = D_2W,\ R = D_3W,$$

then for variations α through curves on the surface we will have

$$dW(\delta x, \delta y, \delta z) = dW\left(\frac{\partial\alpha^1}{\partial u}(0,t), \frac{\partial\alpha^2}{\partial u}(0,t), \frac{\partial\alpha^3}{\partial u}(0,t)\right) = 0 \quad ,$$

and any set of $\partial\alpha^i/\partial u(0,t)$ with this property comes from some variation on the surface. Using this, Gauss deduces a necessary and sufficient condition for a curve γ, parameterized by arclength, to be a geodesic on the surface. Unlike our equations for geodesics, this condition, which appears at the very bottom of page 22, is expressed in terms of quantities which make sense only in $\mathbb{R}^3$:

$$\frac{\gamma^{1\prime\prime}(t)}{X(\gamma(t))} = \frac{\gamma^{2\prime\prime}(t)}{Y(\gamma(t))} = \frac{\gamma^{3\prime\prime}(t)}{Z(\gamma(t))} \quad ,$$

i.e., $\gamma''(t)$ is a multiple of the normal vector at $\gamma(t)$. It takes a little detective work to see that Gauss is really considering a curve parameterized by arclength. Try to prove Gauss' result by modifying our proof of Euler's equations.

§15. The proof in this section is essentially our (first) proof of Gauss' Lemma (I.9-12). There are two main differences. First, Gauss uses the condition of section 14 rather than our equations. Second, for a surface it is unnecessary to choose a curve $v\colon \mathbb{R} \to M_q$ and manufacture the variation α which occurs in the proof of Lemma I.9-12. Instead, we just use

$$\alpha(r,\varphi) = \text{point with "polar coordinates" } (r,\varphi) \quad .$$

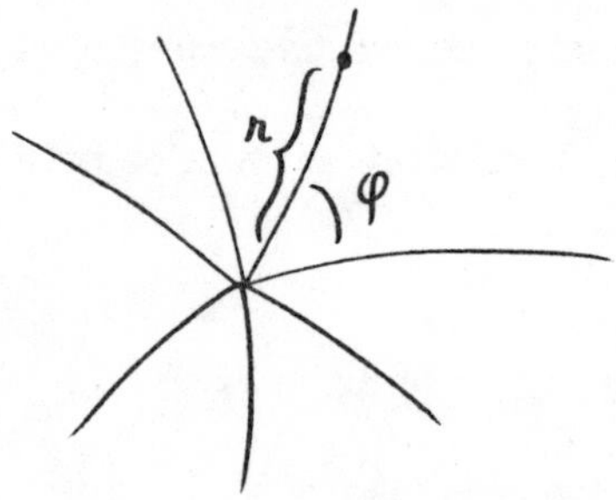

Gauss also gives a "geometric" proof of the lemma, using infinitesimal triangles. I think the easiest way to make this rigorous would be to use our second proof of Gauss' Lemma.

§16. This section states a generalization of Gauss' Lemma, which is also given in Problem I.9-28.

§17. In terms of a coordinate system (p,q) on a surface, the Riemannian metric which it acquires as a subset of $\mathbb{R}^3$ has the expression

$$\langle\ ,\ \rangle = E\, dp \otimes dp + F\, dp \otimes dq + F\, dq \otimes dp + G\, dq \otimes dq \quad ,$$

so that

$$\| \; \| = \sqrt{E\ dp\cdot dp + 2F\ dp\cdot dq + G\ dq\cdot dq} \quad .$$

Gauss uses ω to denote the angle between $\partial/\partial p$ and $\partial/\partial q$ (thus, ω is a function on the surface). Gauss' formula for $\cos\omega$ should be clear. Gauss also mentions that

$$dV = \sqrt{EG - F^2}\ dp \wedge dq \quad ,$$

a special case of the formula on page I.9-17.

To interpret the last two formulas in this section, we must divide ds, dp, and dq by dt in all places; it is to be understood that $dp/dt = (p\circ c)'(t)$, etc., where c is the curve we are considering. It is simplest to assume that c is parameterized by arclength, so that the terms ds/dt are 1. If

$$\theta(s) = \text{angle between } c'(s) \text{ and } \left.\frac{\partial}{\partial p}\right|_{c(s)} \quad ,$$

then

$$\cos\theta = \frac{\langle c', \frac{\partial}{\partial p}\rangle}{\|\frac{\partial}{\partial p}\|} = \frac{E\,\frac{d(p(c(s))}{ds} + F\,\frac{dq(c(s))}{ds}}{\sqrt{E}} \quad ,$$

since

$$c' = \frac{dp(c(s))}{ds}\,\frac{\partial}{\partial p} + \frac{dq(c(s))}{ds}\,\frac{\partial}{\partial q} \quad .$$

Moreover, the area of the parallelogram spanned by c' and $\partial/\partial p$ is

$$\sin\theta\cdot\left\|\frac{\partial}{\partial p}\right\| , \quad \text{and also} \quad dV\left(\frac{\partial}{\partial p}, c'\right) ,$$

from which we obtain

$$\sin\theta = \frac{\sqrt{EG - F^2}\,\dfrac{dq(c(s))}{ds}}{\sqrt{E}} .$$

§18. In this section Gauss deduces the conditions for a curve γ (with component functions $\gamma^1 = p\circ\gamma$, $\gamma^2 = q\circ\gamma$) to be a critical point for the length function. Unlike the condition in section 14, the result is expressed totally in terms of the Riemannian metric $\langle\ ,\ \rangle$ on the surface, and is essentially the condition for a geodesic which we obtained in Chapter I.9. However, the derivation is different, because the geodesic is assumed to satisfy $q(\gamma(t)) = \gamma^2(t) = t$ ["we regard p as a function of q"]. It is not necessary to actually follow the derivation; the important point is the equation on line 4 of page 27. For a curve parameterized by arclength, this equation says that

$$\frac{\partial E}{\partial p}(c(s))\left(\frac{dc^1}{ds}\right)^2 + 2\frac{\partial F}{\partial p}(c(s))\,\frac{dc^1}{ds}\,\frac{dc^2}{ds} + \frac{\partial G}{\partial p}(c(s))\left(\frac{dc^2}{ds}\right)^2 = 2\frac{d\left(E(c(s))\dfrac{dc^1}{ds} + F(c(s))\dfrac{dc^2}{ds}\right)}{ds}$$

It is a very useful exercise to write out the equations on page I.9-38 for the case of a 2-dimensional manifold, with $g_{11} = E$, $g_{12} = F$, $g_{22} = G$, and show that the first of these equations (the equation for $\ell = 1$) yields the above equation (it will be necessary to perform the differentiation on the right side).

Although Gauss performs various further manipulations, it is only necessary to follow the next step,

$$2\,\frac{d\left(E\dfrac{dc^1}{ds} + F\dfrac{dc^2}{ds}\right)}{ds} = 2\,\frac{d\left(\sqrt{E}\cos\theta\right)}{ds} ,$$

where θ is defined in the previous section.

§19. In this section Gauss rewrites formulas from preceding sections for the case of a coordinate system (p,q) which is "orthogonal" ($\langle \partial/\partial p, \partial/\partial q \rangle = F = 0$). The important case for us is the last he considers, in which the coordinates are the "polar coordinates" (r,φ) defined in terms of the geodesics emanating from a point A of the surface. Here Gauss obtains the formula

$$k = -\frac{1}{\sqrt{G}}\,\frac{\partial^2 \sqrt{G}}{\partial r^2}\ ,$$

and

$$\frac{d\theta}{ds} = -\frac{\partial\sqrt{G}}{\partial r}\,\frac{d\varphi(c(s))}{ds}\ ,$$

where θ is the angle the geodesic c makes with the lines φ = constant. Notice that (r,φ) is not a coordinate system on a whole neighborhood of A; we must delete one geodesic ray, including the point A itself. Consequently, $\sqrt{G}$ and $\partial\sqrt{G}/\partial r$ are not even defined at A. Gauss' final assertions in this section should be interpreted as saying that

$$\lim_{B\to A} \sqrt{G}(B) = 0$$

$$\lim_{B\to A} \frac{\partial\sqrt{G}}{\partial r}(B) = 1\ .$$

§20. This is the final section that we want to read. If you have come this far, there should be no problems with this section. Here is the picture.

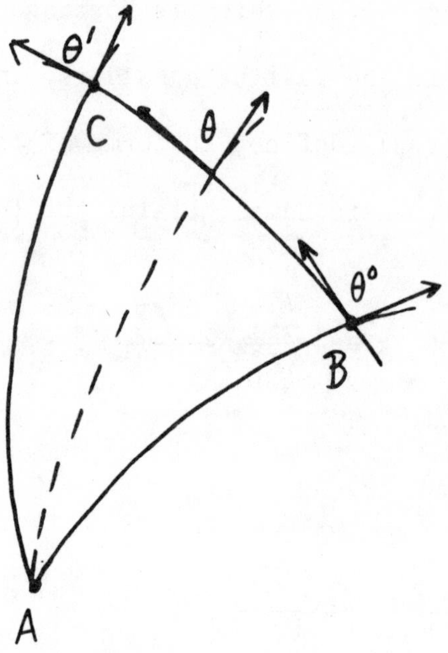

Chapter 3B. Gauss' Theory of Surfaces

This part of the Chapter presents Gauss' results in modern dressing. It can be read completely independently of the first part, but there are frequent comparisons with Gauss' original paper.

We consider a 2-dimensional submanifold M of $\mathbb{R}^3$, with $i: M \to \mathbb{R}^3$ the inclusion map. We also assume that M has been oriented. Since all our results will be local ones, we merely need an orientation in a neighborhood of each point, so this assumption does not place any real restriction on M. (However, we must still investigate to what extent our results depend on the choice of the orientation.)

At each point $p \in M$ there is a unique unit vector $n(p) \in \mathbb{R}^3$ such that

(1) $n(p)_p \in \mathbb{R}^3{}_p$ is perpendicular to M_p

(2) $n(p), v, w$ is positively oriented in $\mathbb{R}^3$ whenever $v_p, w_p \in M_p$ is positively oriented.

We thus have the normal map $n: M \to \mathbb{R}^3$, which actually goes to the unit sphere, $n: M \to S^2 \subset \mathbb{R}^3$. Notice that in his paper Gauss uses X, Y, Z

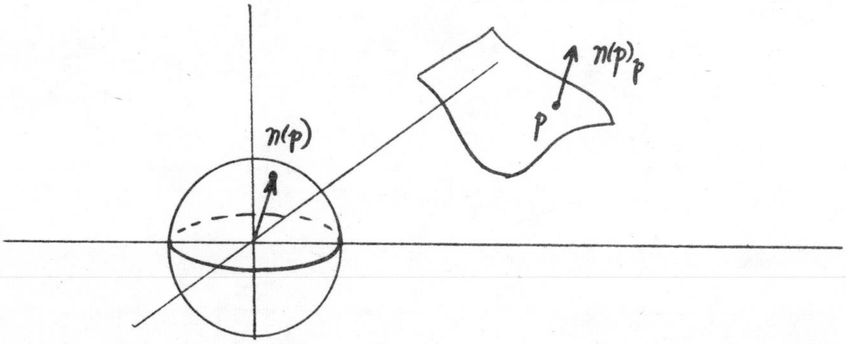

for $n^1(p)$, $n^2(p)$, $n^3(p)$. The idea of using this map may have been suggested to Gauss by astronomical practices, as he indicates in the abstract of the paper. At any rate, the map n turns out to play such a crucial role that it is often called the Gauss map.

An explicit formula for $n: M \to S^2$ can be obtained from various explicit descriptions of M. For example, if $M = \{p \in \mathbb{R}^3 : W(p) = 0\}$ for some function $W: \mathbb{R}^3 \to \mathbb{R}$, then $dW = 0$ on M, i.e., $dW(v_q) = 0$ for all $v_q \in M_q$. This means that

$$\left.\begin{aligned} &D_1W(q)\cdot v^1 + D_2W(q)\cdot v^2 + D_3W(q)\cdot v^3 = 0 \\ &\frac{\partial W}{\partial x}(q)\cdot v^1 + \frac{\partial W}{\partial y}(q)\cdot v^2 + \frac{\partial W}{\partial z}(q)\cdot v^3 = 0 \end{aligned}\right\} \quad \text{for all } v_q \in M_q .$$

This equation can be written

$$< \left(\frac{\partial W}{\partial x}(q), \frac{\partial W}{\partial y}(q), \frac{\partial W}{\partial z}(q)\right), v > = 0 \qquad \text{for all } v_q \in M_q .$$

Consequently

$$n(q) = \text{normalized} \left(\frac{\partial W}{\partial x}(q), \frac{\partial W}{\partial y}(q), \frac{\partial W}{\partial z}(q)\right) ,$$

which is precisely the formula Gauss gives.

We can also find n in terms of a coordinate system χ. To avoid confusion, we will denote the standard coordinate system in $\mathbb{R}^2$ by (s,t)

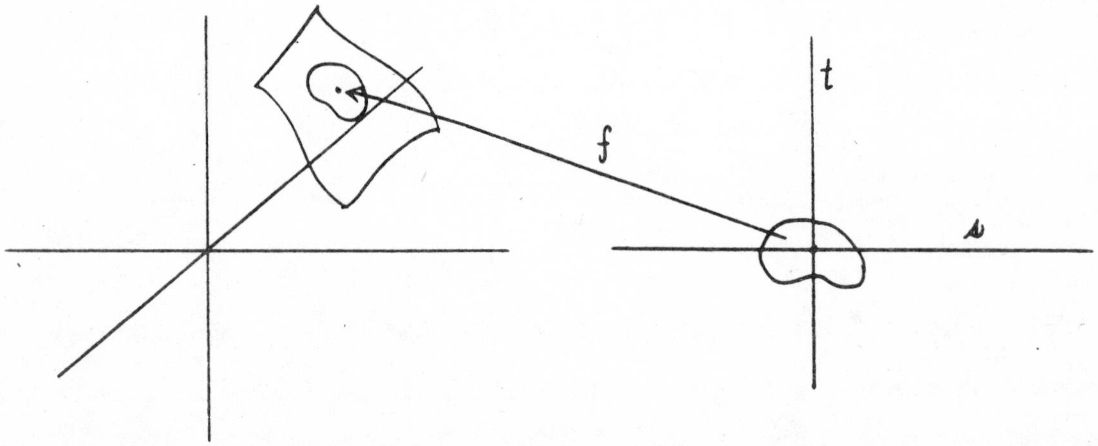

[Gauss uses (p,q)] and we will denote the inverse function $\chi^{-1}: \mathbb{R}^2 \to M \subset \mathbb{R}^3$ by f. It is naturally necessary to consider the component functions of f, considered as a map into $\mathbb{R}^3$, in order to obtain a formula for n; we cannot obtain a formula for n totally in terms of χ, since this coordinate system tells us nothing about the way M is situated in $\mathbb{R}^3$. Note that if $q = f(s,t)$, then

$$\left.\frac{\partial}{\partial x^1}\right|_q = \left(\frac{\partial f}{\partial s}(s,t)\right)_q = \left(\frac{\partial f^1}{\partial s}(s,t), \frac{\partial f^2}{\partial s}(s,t), \frac{\partial f^3}{\partial s}(s,t)\right)_q$$

$$\left.\frac{\partial}{\partial x^2}\right|_q = \left(\frac{\partial f}{\partial t}(s,t)\right)_q = \left(\frac{\partial f^1}{\partial t}(s,t)\ \frac{\partial f^2}{\partial t}(s,t)\ \frac{\partial f^3}{\partial t}(s,t)\right)_q .$$

Consequently,

$$n(f(s,t)) = \text{normalized cross product}$$

$$\left(\frac{\partial f^1}{ds}, \frac{\partial f^2}{\partial s}, \frac{\partial f^3}{\partial s}\right) \times \left(\frac{\partial f^1}{\partial t}, \frac{\partial f^2}{\partial t}, \frac{\partial f^3}{\partial t}\right)$$

$$= \pm \frac{\dfrac{\partial f^2}{\partial s}\dfrac{\partial f^3}{\partial t} - \dfrac{\partial f^2}{\partial t}\dfrac{\partial f^3}{\partial s}, \ \text{—}, \ \text{—}}{\Delta},$$

for

$$\Delta = \text{norm of the cross-product},$$

exactly as in Gauss.

Finally, if M is the graph of $g: \mathbb{R}^2 \longrightarrow \mathbb{R}$, so that $M = \{(x,y,g(x,y)): x,y \in \mathbb{R}^2\}$, then M is the image of $f: \mathbb{R}^2 \longrightarrow \mathbb{R}^3$ defined by

$$f(s,t) = (s,t,g(s,t)) .$$

It follows that

$$n(x,y,g(x,y)) = \text{normalized cross product}$$
$$(1,0\,\frac{\partial g}{\partial x}) \times (0,1,\,\frac{\partial g}{\partial y}) \quad .$$

We are now ready for a preliminary, non-rigorous, definition of the curvature $k(p)$ of M at p:

$$k(p) = \lim_{A \to p} \frac{\text{area } n(A)}{\text{area } A} \quad ,$$

where the limit is taken as the region A around p becomes smaller and smaller. There would be considerable difficulties involved in making this definition rigorous. In the first place, we would have to prove that the limit exists. More crucial, the "area of $n(A)$" needs some interpretation; in the figure below we want "area of $n(A)$" to be negative, because n is

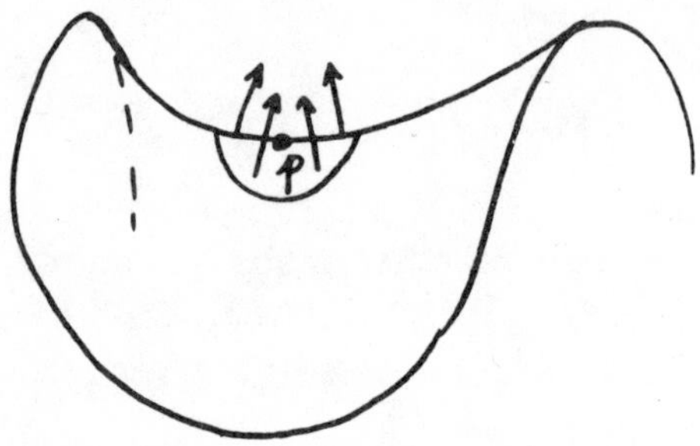

orientation reversing near p. However, even with this non-rigorous definition we can find the curvature of certain surfaces.

Consider first the surface S^2. The map $n\colon S^2 \to S^2$ is just the identity, so at each point $p \in S^2$ we have

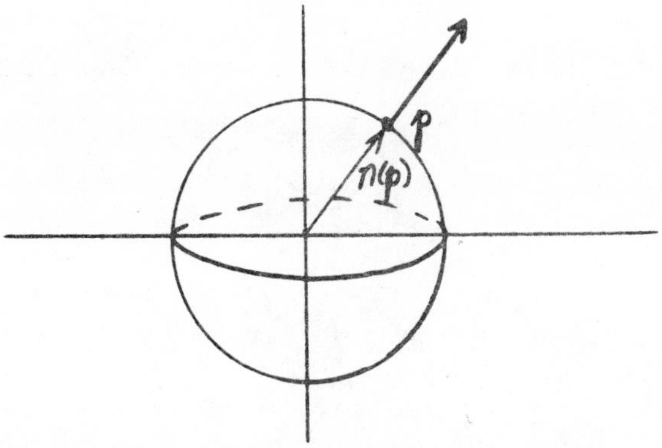

$$k(p) = \lim_{A \to p} \frac{\text{area } n(A)}{\text{area } A} = \lim_{A \to p} \frac{\text{area } A}{\text{area } A} = 1.$$

For the case of the sphere $S^2(r)$ of radius r, we have

$$k(p) = \lim_{A \to p} \frac{\text{area } n(A)}{\text{area } A}$$

$$= \frac{1/r^2 \cdot \text{area } A}{\text{area } A}$$

$$= \frac{1}{r^2} \quad ,$$

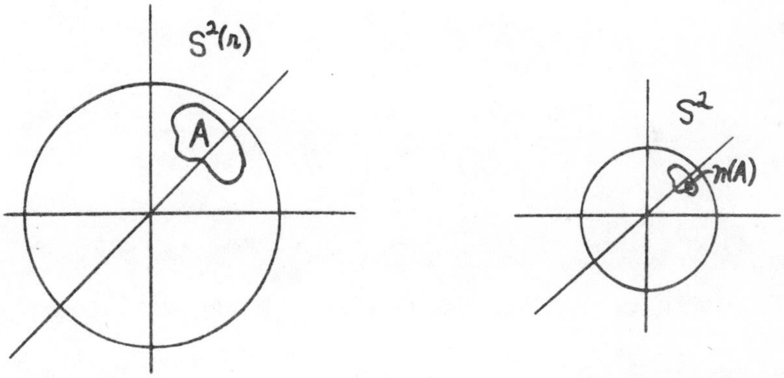

which certainly seems reasonable.

Next we consider a plane P. The function $n: P \to S^2$ is constant, so

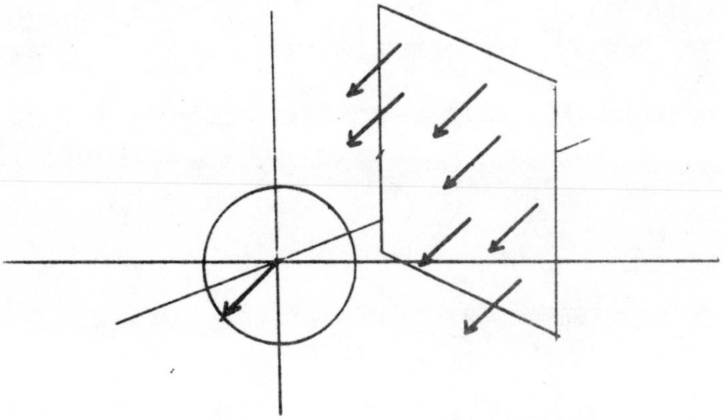

$$k(p) = \lim_{A \to p} \frac{\text{area } n(A)}{\text{area } A} = \lim_{A \to p} \frac{0}{\text{area } A} = 0 \quad ;$$

the plane does not curve.

Finally we consider a cylinder Z. In this case, the function $n: Z \to S^2$

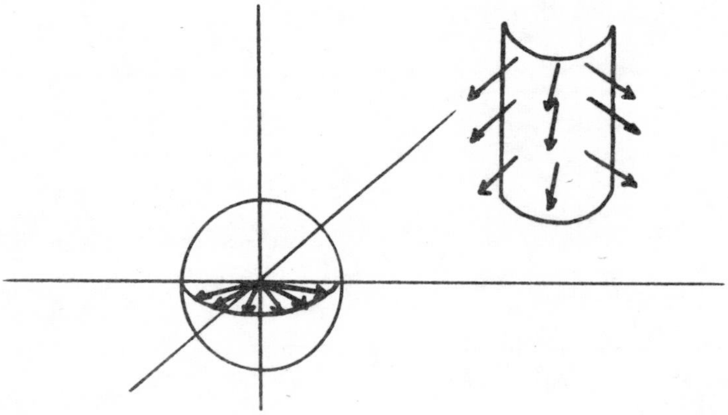

is not constant, but the image of n always lies along a certain arc in S^2, so for all $p \in Z$ we have

$$k(p) = \lim_{A \to p} \frac{\text{area } n(A)}{\text{area } A} = \lim_{A \to p} \frac{0}{\text{area } A} = 0 \quad .$$

The cylinder, too, does not curve! It begins to look as if we have the "wrong" definition of curvature; only later will Gauss explain why this is the "right" definition.

The manner in which we produce a rigorous definition of curvature is really very simple. Since M is a submanifold of $\mathbb{R}^3$, which has the usual Riemannian metric $\langle\ ,\ \rangle$, we can give M the induced Riemannian metric $i^*\langle\ ,\ \rangle$. Together with the orientation which we have given M, this metric determines a 2-form "dV" on M, namely

$$dV(q)(v_q, w_q) = \text{signed area of the parallelogram spanned by } v \text{ and } w.$$

On the sphere S^2 we also have a volume element, coming from its induced Riemannian metric and its usual orientation. As a glance at page I.8-33 will show, this is just the 2-form which we have denoted by σ'. We now define the <u>Gaussian curvature</u> $k(p)$ of M at p to be

$$k(p) = \frac{n^*(\sigma')(p)}{dV(p)} \quad ;$$

in this equation we are dividing two forms, but this makes sense, since any 2-form on the 2-dimensional manifold M is a multiple of the non-zero 2-form dV. If $v_p, w_p \in M_p$ are orthonormal, then our definition says that

$$\begin{aligned} k(p) &= n^*(\sigma')(p)(v_p, w_p) \\ &= \sigma'(n(p))(n_*v_p, n_*w_p) \qquad \text{for orthonormal } v_p, w_p . \end{aligned}$$

This definition of $k(p)$ involves n, and hence the orientation μ which we picked for a neighborhood of p. Choosing the opposite orientation $-\mu$ changes n to $-n = A \circ n$, where $A: S^2 \to S^2$ is the antipodal map, and consequently changes

$$n^*(\sigma') \quad \text{to} \quad n^*(A^*(\sigma')) = -n^*(\sigma') \quad .$$

On the other hand, dV is also changed to $-dV$, so $k(p)$ does not depend on the choice of orientation.

Notice that if n is one-one in a neighborhood of p, then for every region A contained in that neighborhood we have

$$\begin{aligned} \text{area } n(A) &= \int_{n(A)} \sigma' \\ &= \pm \int_A n^*(\sigma') \qquad \text{depending on whether } n \text{ is orientation preserving or reversing.} \end{aligned}$$

Consequently,

$$\frac{\text{area } n(A)}{\text{area } A} = \frac{\pm\int_A n^*(\sigma')}{\int_A dV} .$$

We thus recover our original "definition", provided that

$$\lim_{A \to p} \frac{\pm\int_A n^*(\sigma')}{\int_A dV} = \pm\frac{n^*(\sigma')(p)}{dV(p)} = \pm k(p) \quad ;$$

although this result seems reasonable, by continuity of $n^*(\sigma')$ and dV, we will not worry about the exact manner in which A must approach p in order for the limit to work out. If n is not one-one in a neighborhood of p, then we must have $n_*(p) = 0$, so the rigorous definition gives $k(p) = 0$, which is more or less what one would expect from the intuitive definition.

Although we will eventually obtain a neater expression for k, we begin by deriving Gauss' first formula for k, using essentially the same reasoning as Gauss uses. We first observe that the tangent plane M_p is <u>parallel</u> to the tangent plane $S^2_{n(p)}$ of S^2 at n(p). The reason is very simple: the tangent

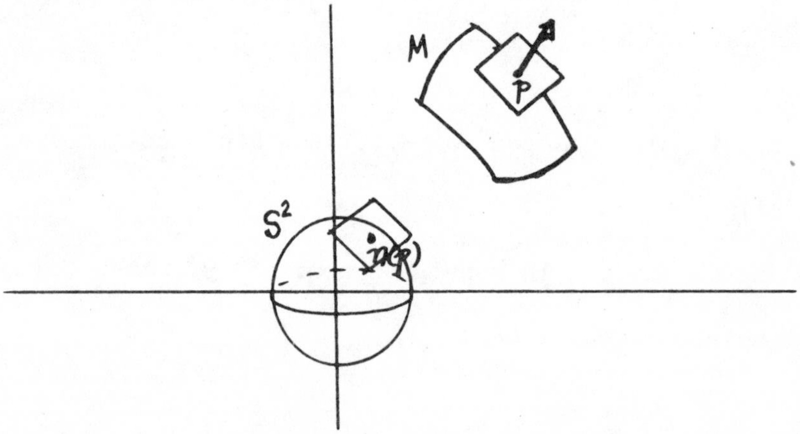

plane $S^2_{n(p)}$ is perpendicular to n(p), and so is M_p, by the very definition of n(p). In the previous two sentences we have used the identification of $\mathbb{R}^3_p$ with $\mathbb{R}^3$, the identification of M_p with $i_*M_p \subset \mathbb{R}^3_p$, etc. Without further warning, we shall continue to do so, to avoid cluttering up the page with extra symbolism.

If $v_p, w_p \in M_p$ are linearly independent, then

$$k(p) = \frac{\text{area of parallelogram } P \text{ spanned by } n_* v_p, n_* w_p}{\text{area of parallelogram } Q \text{ spanned by } v_p, w_p} \quad .$$

Since M_p is parallel to $S^2{}_{n(p)}$, this implies that

$$k(p) = \frac{\text{area of projection of } P \text{ on x-y plane}}{\text{area of projection of } Q \text{ on x-y plane}}$$

(provided the denominator is not zero). In particular, suppose M is the graph of $g: \mathbb{R}^2 \longrightarrow \mathbb{R}$, and consequently the image of $f: \mathbb{R}^2 \longrightarrow \mathbb{R}^3$ defined by

$$f(s,t) = (s,t,g(s,t)) \quad .$$

We choose

$$v = \frac{\partial f}{\partial s} = (1,0, \frac{\partial g}{\partial s})$$

$$w = \frac{\partial f}{\partial t} = (0,1, \frac{\partial g}{\partial t})$$

and consider $v_p, w_p \in M_p$, for $p = f(s,t)$. Then

area of projection of Q on x-y plane

= area of parallelogram spanned by $(1,0)$ and $(0,1)$

= 1 .

On the other hand,

$$n_*(v_p) = n_*\left(\frac{\partial f}{\partial s}\bigg|_p\right) = n_* f_*\left(\frac{\partial}{\partial s}\right)$$

$$= (n \circ f)_*\left(\frac{\partial}{\partial s}\right)$$

$$= \left(\frac{\partial(n\circ f)}{\partial s}\right)_{n(p)} = \left(\frac{\partial(n^1\circ f)}{\partial s}, \frac{\partial(n^2\circ f)}{\partial s}, \frac{\partial(n^3\circ f)}{\partial s}\right)_{n(p)} .$$

(Here all partial derivatives are to be evaluated at the point (s,t).) Similarly,

$$n_*(w_p) = \left(\frac{\partial(n^1\circ f)}{\partial t}, \frac{\partial(n^2\circ f)}{\partial t}, \frac{\partial(n^3\circ f)}{\partial t}\right)_{n(p)} .$$

Consequently,

$$k(f(s,t)) = \text{area of projection of } P \text{ on } x\text{-}y \text{ plane}$$

$$= \text{area of parallelogram spanned by}$$

$$\left(\frac{\partial(n^1\circ f)}{\partial s}, \frac{\partial(n^2\circ f)}{\partial s}\right) \text{ and } \left(\frac{\partial(n^1\circ f)}{\partial t}, \frac{\partial(n^2\circ f)}{\partial t}\right)$$

$$= \frac{\partial(n^1\circ f)}{\partial s}\frac{\partial(n^2\circ f)}{\partial t} - \frac{\partial(n^1\circ f)}{\partial t}\frac{\partial(n^2\circ f)}{\partial s} .$$

This is precisely the formula Gauss obtains at the bottom of page 12. If we use the formula

$$n(x,y,g(x,y)) = \text{normalized cross product}$$

$$\left(1,0,\frac{\partial g}{\partial x}\right) \times \left(0,1,\frac{\partial g}{\partial y}\right) ,$$

we obtain, after a little calculation,

$$(*) \qquad k(x,y,g(x,y)) = \frac{\dfrac{\partial^2 g}{\partial x^2}\cdot\dfrac{\partial^2 g}{\partial y^2} - \dfrac{\partial^2 g}{\partial x \partial y}}{\left(1 + \left(\dfrac{\partial g}{\partial x}\right)^2 + \left(\dfrac{\partial g}{\partial y}\right)^2\right)^2} .$$

We can compare this with the results of Chapter 2, in which we picked our coordinate system so that

$$p = (0,0,0) = (0,0,g(0,0))$$

$$\frac{\partial g}{\partial x}(0,0) = \frac{\partial g}{\partial y}(0,0) = 0$$

$$\frac{\partial^2 g}{\partial x \partial y}(0,0) = 0 \quad .$$

In this case, we obtained the result that the minimum K_1 and maximum K_2 of all curvatures cut out by normal planes through p are the minimum and maximum of $\partial^2 g/\partial x^2(0,0)$ and $\partial^2 g/\partial y^2(0,0)$. Now for our special choice of coordinates in $\mathbb{R}^3$, formula (*) becomes

$$k(p) = \frac{\partial^2 g}{\partial x^2} \frac{\partial^2 g}{\partial y^2} = K_1 \cdot K_2 \quad .$$

(Notice that, just as k does not depend on the orientation of M, neither does the <u>product</u> $K_1 \cdot K_2$, even though K_1 and K_2 individually do). We thus have the following result:

The Gaussian curvature $k(p)$ at any point $p \in M$ is the product of the extreme curvatures of the curves through p cut out by normal planes.

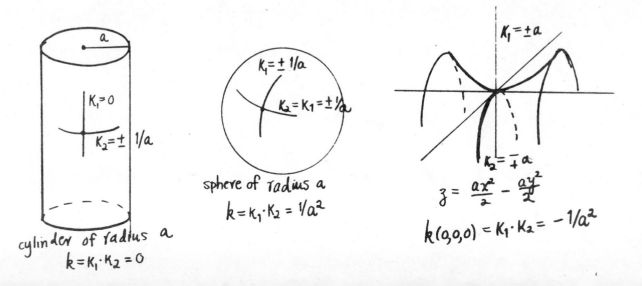

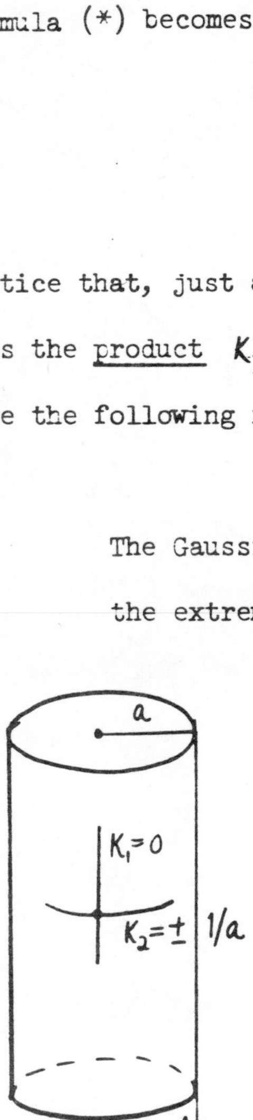

cylinder of radius a
$k = K_1 \cdot K_2 = 0$

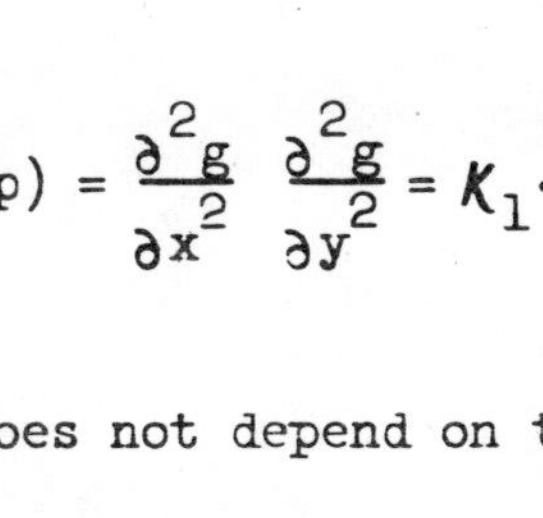

sphere of radius a
$k = K_1 \cdot K_2 = 1/a^2$

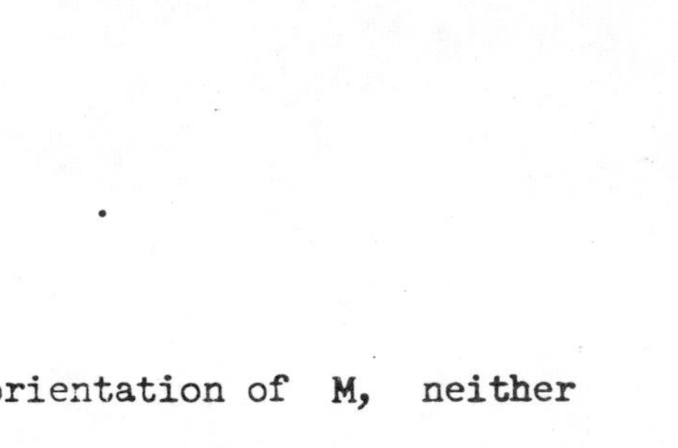

$z = \frac{ax^2}{2} - \frac{ay^2}{2}$
$k(0,0,0) = K_1 \cdot K_2 = -1/a^2$

To prove this result we have followed Gauss' exposition. In particular, the proof of Euler's Theorem which appeared in Chapter 2 is Gauss'. Undoubtedly, this proof is considerably simpler than Euler's, for Gauss remarks with pride "These conclusions contain almost all that the illustrious Euler was the first to prove on the curvature of curved surfaces." Nevertheless, later developments provided a nicer way of obtaining these results, which will consequently now be rederived.

The definition of curvature involves the map $n: M \longrightarrow S^2 \subset R^3$, but even more important, it involves the map $n_*: M_p \longrightarrow S^2{}_{n(p)}$. We can also think of n as an $\mathbb{R}^3$-valued function $n: M \longrightarrow \mathbb{R}^3$, and we then have a map

$$dn: M_p \longrightarrow \mathbb{R}^3 \quad ,$$

namely

$$\begin{aligned} dn(v_p) &= (dn^1(v_p),\ dn^2(v_p),\ dn^3(v_p)) \\ &= (v_p(n^1),\ v_p(n^2),\ v_p(n^3)) \quad . \end{aligned}$$

We claim that dn is essentially the same as n_*; to be precise,

$$n_*(v_p) = dn(v_p)_{n(p)} .$$

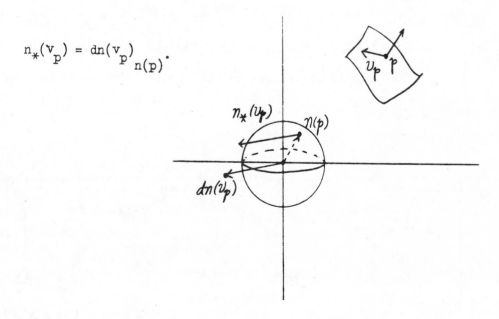

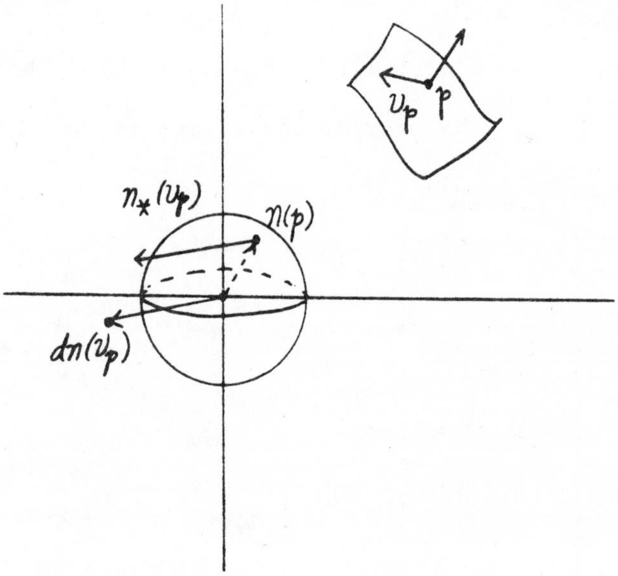

Probably the easiest way* to see this is to take a curve c in M with $c'(0) = v_p$. Then

$$n_*(v_p) = (n\circ c)'(0)$$

$$= \left(\left.\frac{d(n^1\circ c)}{dt}\right|_{t=0}, \left.\frac{d(n^2\circ c)}{dt}\right|_{t=0}, \left.\frac{d(n^3\circ c)}{dt}\right|_{t=0} \right)_{n(p)}$$

$$= (v_p(n^1), v_p(n^2), v_p(n^3))_{n(p)} \quad .$$

Taking advantage of the identification of $\mathbb{R}^3$ with $\mathbb{R}^3{}_p$, we can introduce one more confusion, and consider dn as a map

$$dn\colon M_p \to \mathbb{R}^3{}_p \quad .$$

Since $dn(v_p)$ is parallel to $n_*(v_p) \in S^2{}_{n(p)}$, and since $S^2{}_{n(p)}$ is parallel to M_p, we see that we actually have a map

$$dn\colon M_p \to M_p \quad .$$

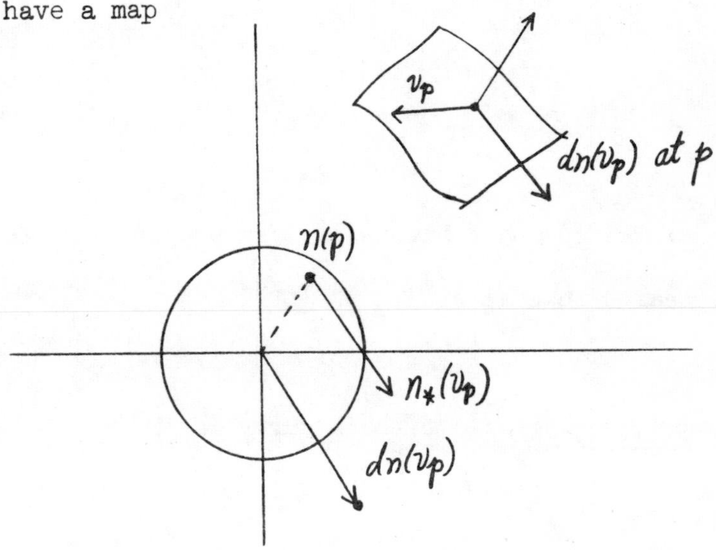

* A more rigorous way is to introduce the inclusion map $j\colon S^2 \to \mathbb{R}^3$, and observe that we are really trying to prove that

$$j_*n_*(v_p) = d(j\circ n)_{n(p)} \quad .$$

The result then follows from Problem I.4-3, applied to the component functions of n.

Despite the torturous process used to define this map, the net result can be described very simply: $dn(v_p)$ is just $n_*(v_p)$ moved back up to a parallel vector in M_p.

The map $dn\colon M_p \longrightarrow M_p$ is sometimes called the Weingarten map. Using it, we can define a tensor II on M which is covariant of order 2: for $v_p, w_p \in M_p$ we define

$$II(p)(v_p,w_p) = -\langle dn(v_p),w_p \rangle \quad .$$

Notice that II does depend on the choice of n, and hence on the orientation picked for M. This tensor II is called the second fundamental form of M. It is not alternating, as we shall shortly see, so the word "form" is a misnomer, a holdover from classical terminology. The same is true for another tensor that you may be worrying about. The first fundamental form I of M is just the induced Riemannian metric $i^*\langle\ ,\ \rangle$ on M (where $\langle\ ,\ \rangle$ is the usual Riemannian metric on $\mathbb{R}^3$). Thus

$$I(p)(v_p,w_p) = \langle v_p,w_p \rangle \quad [= \langle v,w \rangle] \quad .$$

The second fundamental form provides us with a name for a quantity that appeared in Chapter 2:

0. PROPOSITION. Let c be a curve in M which is parameterized by arclength. Let $c(0) = p$, and let $X = c'(0) \in M_p$. Then

$$\langle c''(0),n(p) \rangle = II(X,X) \qquad [\text{i.e. } II(p)(X,X)] \quad .$$

Consequently, $II(X,X)$ is the signed curvature κ_X of the curve cut out on M

by the normal plane through $n(p)$ and X (with $X, n(p)$ positively oriented). Moreover, if κ_φ is the curvature of the curve c_φ cut out by the plane which contains X and makes an angle of φ with the normal plane, then

$$\kappa_\varphi \cdot \cos\varphi = \kappa_X \quad .$$

Proof. Clearly

$$(1) \quad \left.\frac{dn(c(s))}{ds}\right|_{s=0} = dn(X) \quad .$$

Since

$$< c'(s), n(c(s)) > = 0 \qquad \text{for all } s,$$

differentiating yields, using (1),

$$\begin{aligned} < c''(0), n(p) > &= - < c'(0), dn(X) > \\ &= - < X, dn(X) > \\ &= II(X,X) \quad . \end{aligned}$$

This, of course, is precisely the result (proved in precisely the same way) which was derived in Chapter 2. The rest of the theorem is proved just as before. ∎

Unlike Meusnier's Theorem, which involves II in a trivial way, Euler's Theorem involves a crucial property of II:

1. THEOREM. The second fundamental form II of M is symmetric,

$$II(p)(X_p, Y_p) = II(p)(Y_p, X_p) \qquad \text{for } X_p, Y_p \in M_p \quad .$$

First Proof. Near p we can represent M as the image of a function $f: \mathbb{R}^2 \to \mathbb{R}^3$. Let $N = n \circ f$, so that N is "n, considered as a function on $\mathbb{R}^2$." Then

$$\begin{aligned} dn\left(\frac{\partial f}{\partial s}\right) &= dn\left(f_*\left(\frac{\partial}{\partial s}\right)\right) \\ &= f_*\left(\frac{\partial}{\partial s}\right)(n) \\ &= \frac{\partial}{\partial s}(n \circ f) \\ &= \frac{\partial N}{\partial s} \quad . \end{aligned}$$

Now we clearly have

$$< N , \frac{\partial f}{\partial t} > = 0 \quad .$$

Differentiating with respect to s gives

$$< N , \frac{\partial^2 f}{\partial s \partial t} > = - < \frac{\partial N}{\partial s} , \frac{\partial f}{\partial t} > = - < dn\left(\frac{\partial f}{\partial s}\right) , \frac{\partial f}{\partial t} > = II\left(\frac{\partial f}{\partial s} , \frac{\partial f}{\partial t}\right) \quad .$$

Exactly the same argument gives

$$\text{(a)} \quad < N, \frac{\partial^2 f}{\partial t \partial s} > = II\left(\frac{\partial f}{\partial t} , \frac{\partial f}{\partial s}\right) \quad .$$

Thus

$$II\left(\frac{\partial f}{\partial s} , \frac{\partial f}{\partial t}\right) = II\left(\frac{\partial f}{\partial t} , \frac{\partial f}{\partial s}\right) \quad .$$

Since $\partial f/\partial s$, $\partial f/\partial t$ are a basis for the tangent space of M at each point, II is symmetric. For later use we also note that similar arguments lead to the equations

(b) $$II\left(\frac{\partial f}{\partial s}, \frac{\partial f}{\partial s}\right) = \langle N, \frac{\partial^2 f}{\partial s^2} \rangle$$

(c) $$II\left(\frac{\partial f}{\partial t}, \frac{\partial f}{\partial t}\right) = \langle N, \frac{\partial^2 f}{\partial t^2} \rangle .$$

Equations (a) - (c) are sometimes called the Weingarten equations.

Second (coordinate free, fancy) Proof. Let Y be a vector field in a neighborhood $U \subset M$ of p whose value at p is Y_p, and such that $Y(q) \in M_q$ for all $q \in U$. Since

$$\langle n, Y \rangle = 0 \quad \text{on } U ,$$

we have

$$\begin{aligned} 0 = X_p(\langle n, Y \rangle) &= \langle X_p(n), Y_p \rangle + \langle n(p), X_p(Y) \rangle \\ &= \langle dn(X_p), Y_p \rangle + \langle n(p), X_p(Y) \rangle , \end{aligned}$$

where $X_p(Y)$ denotes the vector whose i^{th} component is $X_p(Y^i)$, for Y^i the ith component of Y. This shows that

$$II(p)(X_p, Y_p) = \langle n(p), X_p(Y) \rangle .$$

If a vector field X is picked similarly, then we have a corresponding equation, and it follows that

$$II(p)(X_p, Y_p) - II(p)(Y_p, X_p) = \langle n(p), X_p(Y) - Y_p(X) \rangle .$$

It is easy to check that in $\mathbb{R}^n$ we have $X_p(Y) - Y_p(X) = [X,Y](p)$, so that

$$II(p)(X_p, Y_p) - II(p)(Y_p, X_p) = \langle n(p), [X,Y](p) \rangle .$$

But the right side is 0, since $[X,Y](p) \in M_p$. ∎

The symmetry of II states an important property of the map $dn\colon M_p \longrightarrow M_p$; this map is self-adjoint,

$$< dn(X),Y > = < X,dn(Y) > \quad X,Y \in M_p \quad .$$

Recall that if V is a vector space with an inner product $< , >$, then a linear transformation $T\colon V \longrightarrow V$ is called <u>self-adjoint</u> (with respect to $< , >$) if

$$< Tv,w > = < v,Tw > \qquad \text{for all } v,w \in V \quad .$$

This is equivalent to saying that the matrix of T with respect to an <u>orthonormal</u> basis is symmetric. It is an elementary fact that eigenvectors v_1 and v_2 of T with distinct eigenvalues must be orthogonal, for if $Tv_i = \lambda_i v_i$ with $\lambda_1 \neq \lambda_2$, then

$$\lambda_1 < v_1,v_2 > = < Tv_1,v_2 > = < v_1,Tv_2 > = \lambda_2 < v_1,v_2 >. \quad .$$

The main theorem about self-adjoint transformations (the "spectral theorem") states that a self-adjoint $T\colon V \longrightarrow V$ has a basis of eigenvectors $v_1,\ldots,v_n$. We have seen that eigenvectors with distinct eigenvalues are orthogonal. If two or more eigenvectors have the same eigenvalue, then all vectors in the subspace they span are eigenvalues, so we can select an orthogonal collection spanning the subspace. If we also choose our eigenvectors to be of unit length, we thus obtain an orthonormal basis of eigenvalues. Applying this result to $dn\colon M_p \longrightarrow M_p$ we see that there is an orthonormal basis X_1,X_2 of M_p with

$$dn(X_i) = \lambda_i X_i \quad .$$

This fact is what is behind

<u>2. THEOREM (EULER)</u>. The curvatures κ_X have a minimum κ_1 in one direction and a maximum κ_2 in a perpendicular direction. For a direction X making an angle of θ with the first direction we have

$$\kappa_X = \kappa_1 \cos^2\theta + \kappa_2 \sin^2\theta \quad .$$

<u>Proof</u>. Let X_1 and X_2 be unit eigenvectors. Then by Proposition 0

$$\begin{aligned} \kappa_{X_i} &= II(X_i, X_i) = - \langle dn(X_i), X_i \rangle \\ &= - \langle \lambda_i \cdot X_i, X_i \rangle \\ &= - \lambda_i \quad . \end{aligned}$$

If we express any other unit vector $X \in M_p$ as

$$X = \cos\theta\, X_1 + \sin\theta\, X_2,$$

then

$$\begin{aligned} \kappa_X &= II(X,X) = - \langle dn(X), X \rangle \\ &= - \langle \lambda_1 \cos\theta\, X_1 + \lambda_2 \sin\theta\, X_2,\ \cos\theta\, X_1 + \sin\theta\, X_2 \rangle \\ &= - \lambda_1 \cos^2\theta - \lambda_2 \sin^2\theta \quad . \end{aligned}$$

As before, this completes the proof. ∎

(Notice that we have reduced Euler's Theorem to a fact about the eigenvalues $\lambda_1 \leq \lambda_2$ of a self-adjoint transformation $T: V \to V$ on a 2-dimensional vector space:

$$\lambda_1 = \min_{\|x\|=1} \langle Tv, v \rangle$$

$$\lambda_2 = \max_{\|v\|=1} \langle Tv, v \rangle \quad .$$

For higher dimensions there is a minimax definition of the various eigenvalues. See Courant, Über die Abhängigkeit ... , Nachrichten, Königlichen Gesellschaft der Wissenschaften zu Göttingen, Math. Phys. Klasse 1919, pp. 255-264.)

To connect the curvatures κ_1, κ_2 with k, we first note that k can be expressed very succinctly in terms of dn.

3. PROPOSITION. The Gaussian curvature $k(p)$ at $p \in M$ is

$$k(p) = \text{determinant of } dn\colon M_p \to M_p \quad .$$

Proof. If $Y_1, Y_2 \in M_p$ are linearly independent, then

$$k(p) = \frac{n^*(\sigma')(p)(Y_1, Y_2)}{dV(p),(Y_1, Y_2)}$$

$$= \frac{\sigma'(n(p))(n_*Y_1, n_*Y_2)}{dV(p)(Y_1, Y_2)} \quad .$$

Using the fact that $S^2{}_{n(p)}$ is parallel to M_p, and remembering that we are considering dn as a map into M_p, this can be written simply

$$k(p) = \frac{dV(p)(dn(Y_1), dn(Y_2))}{dV(p)(Y_1, Y_2)} .$$

Since dV is a 2-form, this ratio is indeed just det dn. ∎

4. COROLLARY. Let $Y_1, Y_2 \in M_p$ be orthonormal. Then

$$k(p) = \det(II(Y_i, Y_j)) = \det \begin{pmatrix} II(Y_1, Y_1) & II(Y_1, Y_2) \\ II(Y_2, Y_1) & II(Y_2, Y_2) \end{pmatrix} .$$

Proof. Since $II(Y_i, Y_j) = \langle -dn(Y_i), Y_j \rangle$, the matrix of dn with respect to Y_1, Y_2 is

$$\begin{pmatrix} -II(Y_1, Y_1) & -II(Y_2, Y_1) \\ -II(Y_1, Y_2) & -II(Y_2, Y_2) \end{pmatrix} ,$$

which has the same determinant as the matrix $(II(Y_i, Y_j))$. ∎

5. COROLLARY. Let $Y_1, Y_2 \in M_p$ be linearly independent. Then

$$k(p) = \frac{\det(II(Y_i, Y_j))}{\det(\langle Y_i, Y_j \rangle)} = \frac{\det(II(Y_i, Y_j))}{\det(I(Y_i, Y_j))} .$$

Proof. Corollary 4 proves the formula for orthonormal X_1, X_2. If $Y_i = \Sigma a_{ij} X_j$, then replacing X_1, X_2 by Y_1, Y_2 multiplies both numerator and denominator by $\det(a_{ij})$. ∎

6. COROLLARY. For any $p \in M$ we have

$$k(p) = \kappa_1 \cdot \kappa_2 .$$

<u>Proof</u>. Apply Corollary 4 when $Y_1, Y_2 \in M_p$ are the orthonormal basis of eigenvalues of dn, with eigenvectors $-K_1$, $-K_2$. Then

$$k(p) = \det \begin{pmatrix} -K_1 & 0 \\ 0 & -K_2 \end{pmatrix} = K_1 \cdot K_2 \qquad . \blacksquare$$

Corollary 5 allows us to develop an explicit formula for k, which involves some standard symbolism to be introduced first. If $\chi = (x,y)$ is a coordinate system on M, then we write the first fundamental form I as

$$I = E\,dx \otimes dx + F\,dx \otimes dy + F\,dy \otimes dx + G\,dy \otimes dy \qquad .$$

If $f\colon R^2 \longrightarrow R^3$ is the inverse of χ, so that

$$\frac{\partial}{\partial x} = \frac{\partial f}{\partial s}$$

$$\frac{\partial}{\partial y} = \frac{\partial f}{\partial t} \qquad ,$$

then

$$E = \left\langle \frac{\partial f}{\partial s}, \frac{\partial f}{\partial s} \right\rangle = \sum_{i=1}^{3} \left(\frac{\partial f^i}{\partial s} \right)^2$$

(*)

$$F = \sum_{i=1}^{3} \frac{\partial f^i}{\partial s} \frac{\partial f^i}{\partial t}$$

$$G = \sum_{i=1}^{3} \left(\frac{\partial f^i}{\partial t} \right)^2$$

[Notice that the left sides of these equations really mean $E(f(s,t))$, etc. (the functions E,F,G themselves are defined on M). This is just the form we want. For example, to compute $\partial E/\partial x$ at $q = f(\bar{s},\bar{t})$ we have

$$\frac{\partial E}{\partial x}(q) = D_1(E\circ\chi^{-1})(\chi(q))$$

$$= D_1(E\circ f)(\chi(q))$$

$$= \frac{\partial E}{\partial s}(\bar{s},\bar{t}) \quad ;$$

the E on the last line denotes the same function on R^2 which appears in the equations (*)].

The symbols E,F,G were introduced by Gauss himself (at the beginning of section 11 of his paper), and they have remained standard ever since. There are also standard symbols for the second fundamental form:

$$II = L\, dx \otimes dx + M\, dx \otimes dy + M\, dy \otimes dx + N\, dy \otimes dy \quad .$$

To obtain formulas for L,M,N, we look at the Weingarten equations in the first proof of Theorem 1, and note that

$$L = II\left(\frac{\partial f}{\partial s}, \frac{\partial f}{\partial s}\right) = \left\langle N, \frac{\partial^2 f}{\partial s^2}\right\rangle = \sum_{i=1}^{3} n^i(f(s,t))\, \frac{\partial^2 f^i}{\partial s^2}$$

$$(**) \qquad M = \sum_{i=1}^{3} n^i(f(s,t))\, \frac{\partial^2 f^i}{\partial s \partial t}$$

$$N = \sum_{i=1}^{3} n^i(f(s,t))\, \frac{\partial^2 f^i}{\partial t^2} \quad ;$$

formulas for $n^i(f(s,t))$ have already been given on page 3 . Using Corollary 5, we now have the classical formula

$$k(p) = \frac{LN - M^2}{EG - F^2}(p) \quad .$$

The symbols L,M,N do not appear in Gauss, who instead uses the symbols D,D',D" for certain quantities proportional to them. These symbols are introduced on page 18; it is easy to see that Gauss' formula for k on this page is equivalent to the one we have just derived, although Gauss obtained it in a different way, by beginning with the formula which we derived on page 10 .

Our next theorem probably requires an apology in advance. The result looks amazingly unappetizing, it's hard to see why anyone would want it even if he had it, and the proof is merely an involved calculation. Nevertheless, we will justify its existence soon after proving it. The calculation appearing in the proof may be a little easier to follow than Gauss', though it is essentially equivalent.

7. THEOREM. Let (x,y) be a coordinate system on a neighborhood of $p \in M \subset \mathbb{R}^3$, and let

$$I = i^* \langle\ ,\ \rangle = E\, dx \otimes dx + F\, dx \otimes dy + F\, dy \otimes dx + G\, dy \otimes dy \quad .$$

Then

$$\begin{aligned}
4(EG - F^2)^2 k = {} & E\left(\frac{\partial E}{\partial y}\frac{\partial G}{\partial y} - 2\frac{\partial F}{\partial x}\frac{\partial G}{\partial y} + \left(\frac{\partial G}{\partial x}\right)^2\right) \\
& + F\left(\frac{\partial E}{\partial x}\frac{\partial G}{\partial y} - \frac{\partial E}{\partial y}\frac{\partial G}{\partial x} - 2\frac{\partial E}{\partial y}\frac{\partial F}{\partial y} + 4\frac{\partial F}{\partial x}\frac{\partial F}{\partial y} - 2\frac{\partial F}{\partial x}\frac{\partial G}{\partial x}\right) \\
& + G\left(\frac{\partial E}{\partial x}\frac{\partial G}{\partial x} - 2\frac{\partial E}{\partial x}\frac{\partial F}{\partial y} + \left(\frac{\partial E}{\partial y}\right)^2\right) \\
& - 2(EG - F^2)\left(\frac{\partial^2 E}{\partial y^2} - 2\frac{\partial^2 F}{\partial x \partial y} + \frac{\partial^2 G}{\partial x^2}\right) \quad .
\end{aligned}$$

Proof. Consider the inverse $f: \mathbb{R}^2 \to \mathbb{R}^3$ of the coordinate system (x,y). To save space we will denote

$$\frac{\partial^2 f}{\partial s \partial t} = D_{12}f \qquad \text{simply by } f_{12} \quad ,$$

and similarly for other partial derivatives. We have

$$k = \frac{LN - M^2}{EG - F^2} \quad ,$$

where, by (**),

$$L = \langle f_{11}, N \rangle = \langle f_{11}, \frac{f_1 \times f_2}{\sqrt{EG - F^2}} \rangle$$

$$M = \langle f_{12}, N \rangle = \langle f_{12}, \frac{f_1 \times f_2}{\sqrt{EG - F^2}} \rangle$$

$$N = \langle f_{22}, N \rangle = \langle f_{22}, \frac{f_1 \times f_2}{\sqrt{EG - F^2}} \rangle \quad .$$

Thus

$$\begin{aligned} k(EG - F^2)^2 &= \langle f_{11}, f_1 \times f_2 \rangle \cdot \langle f_{22}, f_1 \times f_2 \rangle - \langle f_{12}, f_1 \times f_2 \rangle^2 \\ &= \det(f_{11}, f_1, f_2) \cdot \det(f_{22}, f_1, f_2) - \det(f_{12}, f_1, f_2) \cdot \det(f_{12}, f_1, f_2) \\ &= \det \begin{pmatrix} f_{11} \\ f_1 \\ f_2 \end{pmatrix} \cdot \det \begin{pmatrix} f_{22} \\ f_1 \\ f_2 \end{pmatrix} - \det \begin{pmatrix} f_{12} \\ f_1 \\ f_2 \end{pmatrix} \cdot \det \begin{pmatrix} f_{12} \\ f_1 \\ f_2 \end{pmatrix} \quad . \end{aligned}$$

In this equation each f_i and f_{ij} is considered as a row of the matrix. If we use f_i^T and f_{ij}^T to denote the columns with the same entries, then we also have

$$K(EG - F^2)^2 = \det\begin{pmatrix} f_{11} \\ f_1 \\ f_2 \end{pmatrix} \cdot \det(f_{22}{}^T, f_1{}^T, f_2{}^T) - \det\begin{pmatrix} f_{12} \\ f_1 \\ f_2 \end{pmatrix} \cdot \det(f_{12}{}^T, f_1{}^T, f_2{}^T)$$

$$= \det\left[\begin{pmatrix} f_{11} \\ f_1 \\ f_2 \end{pmatrix} \cdot (f_{22}{}^T, f_1{}^T, f_2{}^T)\right] - \det\left[\begin{pmatrix} f_{12} \\ f_1 \\ f_2 \end{pmatrix} \cdot (f_{12}{}^T, f_1{}^T, f_2{}^T)\right] ,$$

which is easily seen to be equal to

$$\det\begin{pmatrix} <f_{11}, f_{22}> & <f_{11}, f_1> & <f_{11}, f_2> \\ <f_1, f_{22}> & E & F \\ <f_2, f_{22}> & F & G \end{pmatrix} - \det\begin{pmatrix} <f_{12}, f_{12}> & <f_{12}, f_1> & <f_{12}, f_2> \\ <f_{12}, f_1> & E & F \\ <f_{12}, f_2> & F & G \end{pmatrix}.$$

We thus have

$$k(EG - F^2)^2 = (<f_{11}, f_{22}> - <f_{12}, f_{12}>) \det\begin{pmatrix} E & F \\ F & G \end{pmatrix}$$

$$+ \det\begin{pmatrix} 0 & <f_{11}, f_1> & <f_{11}, f_2> \\ <f_1, f_{22}> & E & F \\ <f_2, f_{22}> & F & G \end{pmatrix} - \det\begin{pmatrix} 0 & <f_{12}, f_1> & <f_{12}, f_2> \\ <f_{12}, f_1> & E & F \\ <f_{12}, f_2> & F & G \end{pmatrix}.$$

Now we also have, from (*),

$$< f_{11}, f_1 > = \frac{1}{2} E_1 \qquad (E_1 = \frac{\partial E}{\partial s})$$
$$< f_{12}, f_1 > = \frac{1}{2} E_2$$
$$< f_{22}, f_2 > = \frac{1}{2} G_2$$
$$< f_{12}, f_2 > = \frac{1}{2} G_1$$
$$< f_{11}, f_2 > = F_1 - \frac{1}{2} E_2$$
$$< f_{22}, f_1 > = F_2 - \frac{1}{2} G_1 \quad .$$

From the fourth and fifth equations we obtain

$$\frac{1}{2} G_{11} = \frac{\partial}{\partial s} < f_{12}, f_2 > = < f_{121}, f_2 > + f_{12}, f_{21} >$$
$$F_{12} - \frac{1}{2} E_{22} = \frac{\partial}{\partial t} < f_{11}, f_2 > = < f_{112}, f_2 > + < f_{11}, f_{22} > \quad .$$

Subtracting the first from the second, we then obtain

$$< f_{11}, f_{22} > - < f_{12}, f_{12} > = -\frac{1}{2} G_{11} + F_{12} - \frac{1}{2} E_{22} \quad .$$

Substituting all of these into the previous equation, we obtain the desired result! ∎

From Theorem 7 we can deduce an immediate Corollary, which appears at the very bottom of page 20 of Gauss' work:

"Thus the formula of the preceding article leads of itself to the remarkable

THEOREM. If a curved surface is developed upon any other surface whatever, the measure of curvature in each point remains unchanged."

The Latin word for remarkable has become part of the traditional name for this result:

8. COROLLARY (THEOREMA EGREGIUM). If $f, g: M \longrightarrow \mathbb{R}^3$ are two imbeddings (or even immersions) such that $f^* < , > = g^* < , >$, then the Gaussian curvature of $f(M) \subset \mathbb{R}^3$ at $f(p)$ equals the Gaussian curvature of $g(M)$ at $g(p)$.

The Theorema Egregium justifies the close attention which we have given to the Gaussian curvature k of a surface. Although defined in terms of the imbedding of the surface in $\mathbb{R}^3$, it turns out to depend only on the Riemannian metric induced by that imbedding. This shows why the Gaussian curvature of a cylinder must be 0 -- there is a (local) isometry from the plane to the cylinder. If the cylinder $Z \subset \mathbb{R}^3$ is

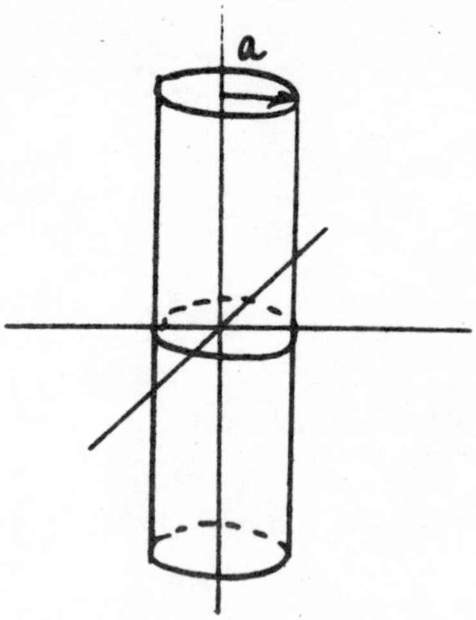

$$Z = \{(x,y,z): x^2 + y^2 = a^2\} \quad ,$$

then a local isometry $f\colon \mathbb{R}^2 \longrightarrow Z$ is given by

$$f(s,t) = (a \cos \frac{s}{a}, a \sin \frac{s}{a}, t).$$

It is easily checked that f_* is an isometry at any point, but the result should be clear without any calculations whatsoever. To prove that Z is locally isometric to $\mathbb{R}^2$, just take a piece of paper, and roll it up into a cylinder.

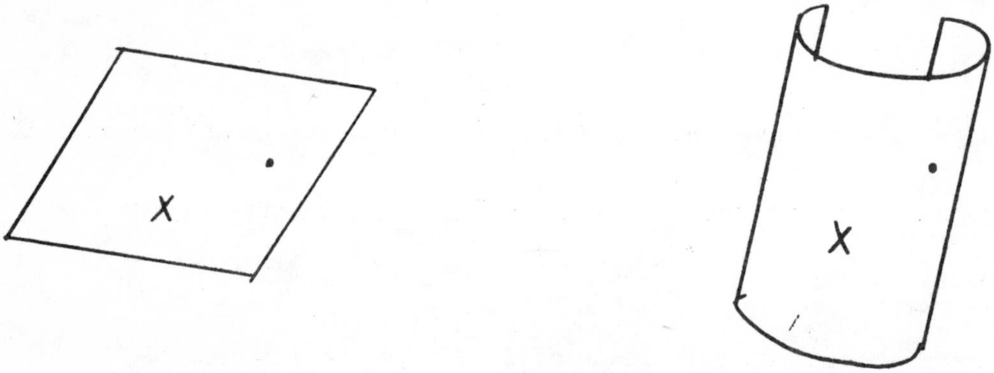

The map which takes a point on the flat piece of paper into the corresponding point on the rolled up piece of paper is an isometry. The isometric properties of this map are expressed by the everyday experience that paper cannot be stretched, but merely bent. The Theorema Egregium is often expressed by saying that Gaussian curvature is a "bending invariant." Anyone who has ever made a paper dunce hat knows (though he may not know that he knows) that the cone is also locally isometric to the plane, and hence has Gaussian curvature 0.

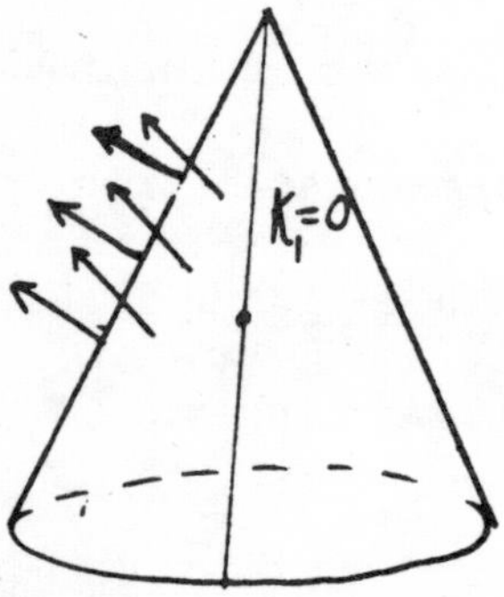

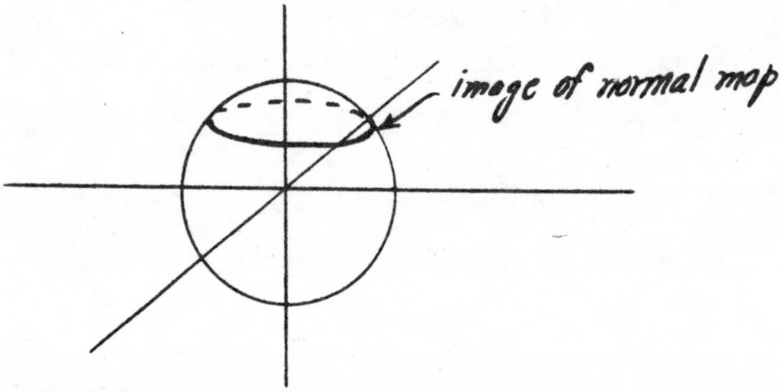

Map makers, and anyone who has had to wrap a spherical object, know that a piece of paper cannot be bent onto even a small portion of a sphere. A mathematical proof follows immediately from the Theorema Egregium, for a sphere has non-zero Gaussian curvature. The situation for surfaces is thus completely differently from that for curves. All 1-dimensional Riemannian manifolds are locally isometric to $\mathbb{R}^1$, for if we choose an immersed curve c in the manifold, then the arclength function of c is an isometry into $\mathbb{R}^1$. So there are no interesting bending invariants of a curve; all the interesting characteristics of a curve are invariants under the group of Euclidean motions. The Gaussian curvature is, of course, an invariant under the group of Euclidean motions, but it is also invariant under the much larger (but still important) group of maps which are merely defined on the surface, and are isometries there. (As a contrast, the mean curvature $\frac{1}{2}(\kappa_1 + \kappa_2)$ is invariant under the group of Euclidean motions, but it is not a bending invariant; for example, the plane has mean curvature 0, while a cylinder of radius a has mean curvature $a/2$.)

In the previous paragraph we appealed to the Theorema Egregium to prove that a sphere is not locally isometric to the plane. But it is also possible to give a much more elementary proof of this fact, that will convince some one who knows a little geometry. If there were an isometry from a neighborhood of $p \in S^2$

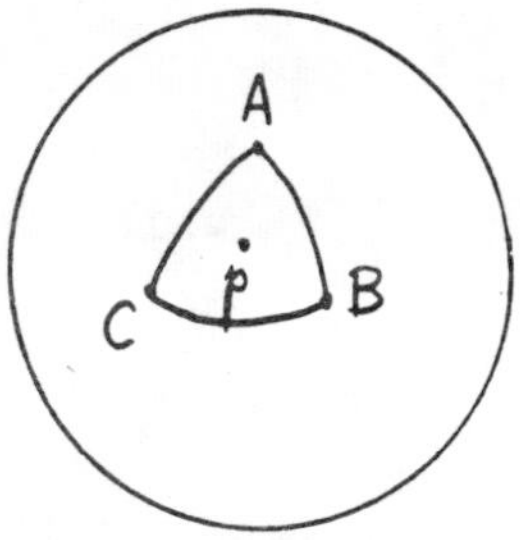

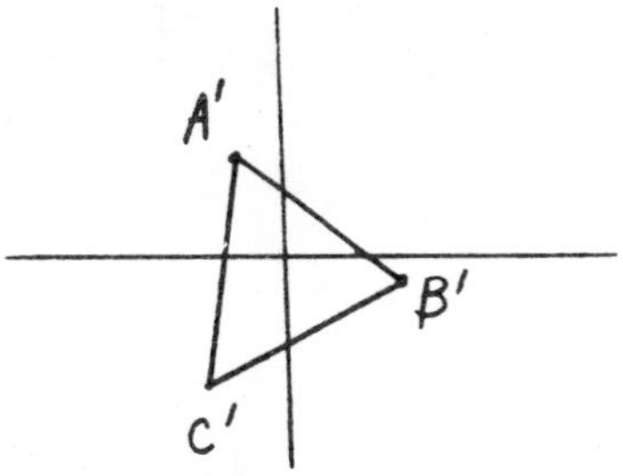

into the plane, then a small triangle ABC around p, with portions of great circles as sides, would have to be mapped into an ordinary triangle A'B'C' on the plane, since great circles are geodesics on the sphere. The angles at A,B, and C would also have to equal the angles A',B',C'. This is impossible, since $\angle A + \angle B + \angle C > \pi$, while $\angle A' + \angle B' + \angle C' = \pi$. This phenomenon turns out to have a generalization to arbitrary surfaces, a result that Gauss felt "ought to be counted among the most elegant in the theory of curved surfaces." In deriving this result we will essentially follow Gauss, but we will use some of our previous results about geodesics, and suppress some of the additional formulas which Gauss obtains along the way, so that the argument may appear somewhat simpler.

In expounding the theory of geodesics, Gauss derives two conditions, of entirely different natures, for a geodesic on a surface. Although the first of these conditions is not necessary for our final goal, it is an interesting exercise in the Calculus of Variations, as well as an interesting result in its own right. We consider a curve $c\colon [a,b] \longrightarrow M$ and a variation α keeping endpoints fixed. Looking at the energy function (Gauss looks at length instead), we have

$$\left.\frac{dE(\bar{\alpha}(u))}{du}\right|_{u=0} = \left.\frac{d}{du}\right|_{u=0} \frac{1}{2}\int_a^b < \frac{\partial\alpha}{\partial t}(u,t), \frac{\partial\alpha}{\partial t}(u,t) > dt$$

$$= \int_a^b < \gamma'(t), \frac{\partial^2\alpha}{\partial u\partial t}(0,t) > dt$$

$$= -\int_a^b < \gamma''(t), \frac{\partial\alpha}{\partial u}(0,t) > dt \quad , \qquad \text{using integration by parts.}$$

This result holds for <u>any</u> variation α keeping endpoints fixed. The final integral must therefore be zero whenever $\partial\alpha/\partial u(0,t) \in M_{\gamma(t)}$ for all t, since there is then a variation α <u>through curves on</u> M with these values of

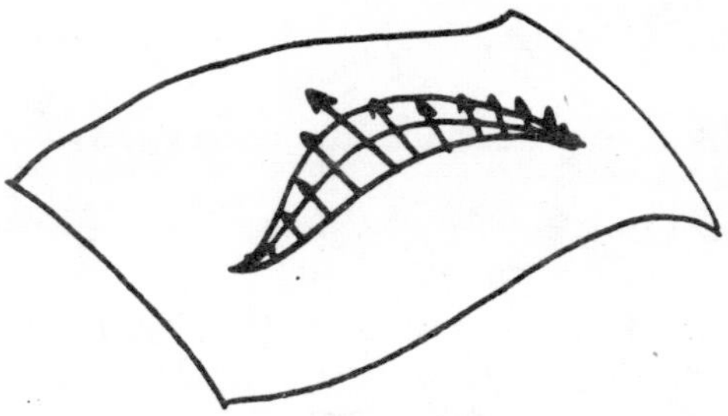

$\partial\alpha/\partial u(0,t)$. In other words, γ is a geodesic on M if and only if

$$\int_a^b < \gamma''(t), \eta(t) > = 0 \quad \text{for every } \eta \text{ satisfying } < \eta(t), n(\gamma(t)) > = 0 .$$

In particular, we can choose

$$\eta(t) = \varphi(t)\ [\gamma''(t) - < \gamma''(t), n(\gamma(t)) > n(\gamma(t))]$$

where φ is a C^∞ function on $[a,b]$ with $\varphi > 0$ on (a,b) and $\varphi(a) = \varphi(b) = 0$. We then obtain

$$0 = \int_a^b \varphi(t)[< \gamma''(t), \gamma''(t) > - < \gamma''(t), n(\gamma(t)) >^2]dt \quad .$$

Since $n(\gamma(t))$ has length 1, the Schwarz inequality (Theorem I.9-1(2)) shows that the term in brackets is always ≤ 0. Since $\varphi(t) > 0$ on (a,b), the term in brackets must actually be 0 everywhere. This implies that $\gamma''(t)$ and $n(\gamma(t))$ are everywhere <u>linearly dependent</u>. In other words

> The curve γ on M is a geodesic if and only if γ'' is always perpendicular to M.

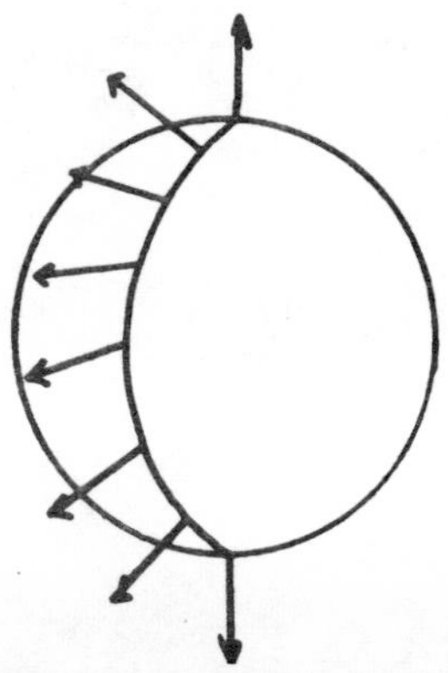

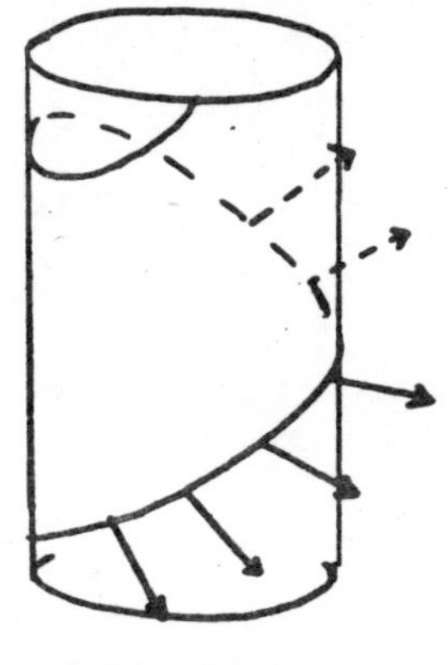

Notice that this condition makes sense only for a surface in $\mathbb{R}^3$; the vector γ'' has no meaning for an abstract Riemannian manifold.

We now pass to the equations for a geodesic on any 2-dimensional Riemannian manifold. However, we will work with a special coordinate system. Consider a neighborhood of $p \in M$ which is $\exp(U)$, where $U \subset M_p$ is a neighborhood of 0 on which $\exp$ is one-one. Identify M_p with $\mathbb{R}^2$ by choosing an orthonormal basis for M_p. Introducing polar coordinates $(\rho, \emptyset)$ on M_p (minus some ray), we obtain a coordinate system $(r, \varphi) = (\rho, \emptyset) \circ \exp^{-1}$ on $\exp(U)$ (minus some geodesic ray), where ρ is now chosen so that $\rho = 1$

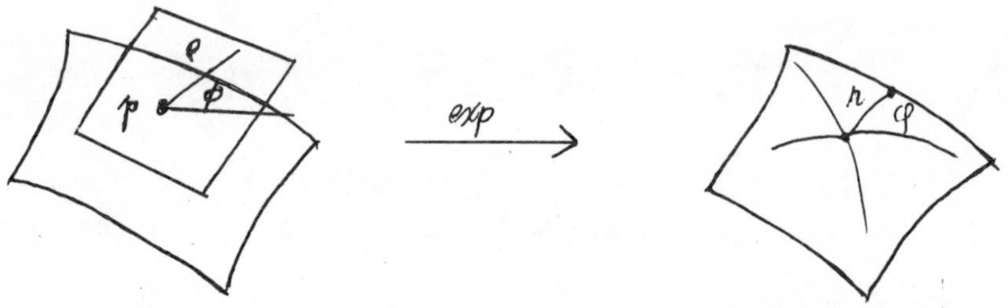

for vectors in M_p with norm 1. This implies that $\|\partial/\partial r\| = 1$. We also know that $\langle \partial/\partial r, \partial/\partial \varphi \rangle = 0$, by Gauss' Lemma. So we have

$$\langle\ ,\ \rangle = dr \otimes dr + G\, d\varphi \otimes d\varphi$$

for some function G. The function G is just $G(q) = \langle \partial/\partial \varphi\big|_q, \partial/\partial \varphi\big|_q \rangle$. Clearly, G can be considered as defined on $\exp(U) - \{p\}$, even though any particular coordinate system (r, φ) can be defined only on $\exp(U)$ minus some geodesic ray. In terms of the g_{ij} notation we have

$$\begin{aligned} g_{11} &= 1 & g^{11} &= 1 \\ g_{12} = g_{21} &= 0 & g^{12} = g^{21} &= 0 \\ g_{22} &= G & g^{22} &= \frac{1}{G}\,. \end{aligned}$$

An easy calculation from the definitions (pp. I.9-37 and I.9-39) then gives

$$[12,2] = [22,2] = \frac{1}{2}\frac{\partial G}{\partial r}$$

$$[22,1] = -\frac{1}{2}\frac{\partial G}{\partial r}$$

all other $[ij,k] = 0$;

$$\Gamma^2_{12} = \Gamma^2_{21} = \frac{1}{2G}\frac{\partial G}{\partial r}$$

$$\Gamma^1_{22} = -\frac{1}{2}\frac{\partial G}{\partial r}$$

all other $\Gamma^k_{ij} = 0$.

The equations for a geodesic (pg. I.9-40) thus give

$$(1)\quad \frac{d^2\gamma^1}{ds^2} = \frac{1}{2}\frac{\partial G}{\partial r}(\gamma(s))\left(\frac{d\gamma^2}{ds}\right)^2 .$$

Now suppose γ is some curve parameterized by arclength, and let $\theta(s)$ be the angle between $\gamma'(s)$ and $\partial/\partial r|_{\gamma(s)}$. Clearly

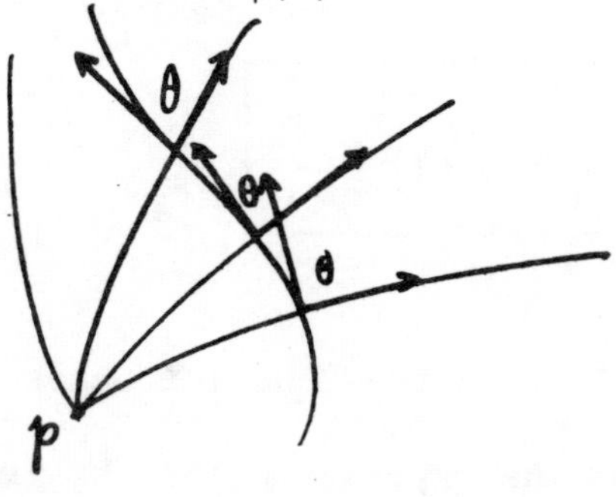

$$(2)\quad \cos\theta(s) = \left< \gamma'(s), \left.\frac{\partial}{\partial r}\right|_{\gamma(s)} \right>$$

$$= \left< \frac{d\gamma^1}{ds}\left.\frac{\partial}{\partial r}\right|_{\gamma(s)} + \frac{d\gamma^2}{ds}\left.\frac{\partial}{\partial\varphi}\right|_{\gamma(s)} , \left.\frac{\partial}{\partial r}\right|_{\gamma(s)} \right>$$

$$= \frac{d\gamma^1}{ds} .$$

For a geodesic γ we obtain from (1) and (2),

$$(3)\quad \frac{1}{2}\frac{\partial G}{\partial r}(\gamma(s))\left(\frac{d\gamma^2}{ds}\right)^2 = \frac{d\cos\theta(s)}{ds}$$

$$= -\sin\theta(s)\,\frac{d\theta(s)}{ds} .$$

Finally, note that the area of the parallelogram spanned by the unit vectors $\gamma'(s)$ and $\partial/\partial r\big|_{\gamma(s)}$ equals

$$(4)\quad \sin\theta(s) \quad \text{and also equals} \quad dV\left(\frac{\partial}{\partial r}\bigg|_{\gamma(s)}, \gamma'(s)\right)$$

$$= \sqrt{G}(\gamma(s))dr \wedge d\varphi\left(\frac{\partial}{\partial r}\bigg|_{\gamma(s)}, \frac{d\gamma^1}{ds}\frac{\partial}{\partial r}\bigg|_{\gamma(s)} + \frac{d\gamma^2}{ds}\frac{\partial}{\partial\varphi}\bigg|_{\gamma(s)}\right)$$

$$= \sqrt{G}(\gamma(s))\,\frac{d\gamma^2}{ds} \quad .$$

So we obtain

$$\frac{1}{2}\frac{\partial G}{\partial r}(\gamma(s))\left(\frac{d\gamma^2}{ds}\right)^2 = -\sqrt{G}(\gamma(s))\,\frac{d\gamma^2}{ds}\,\frac{d\theta}{ds} \quad ,$$

$$\frac{d\theta}{ds} = -\frac{1}{2}\,\frac{\frac{\partial G}{\partial r}}{\sqrt{G}}(\gamma(s))\,\frac{d\gamma^2}{ds} \quad ,$$

$$\boxed{(*)\quad \frac{d\theta}{ds} = \frac{\partial\sqrt{G}}{\partial r}(\gamma(s))\,\frac{d\gamma^2}{ds}}$$

for any geodesic γ. This is the equation Gauss finally obtains on page 28. .

Gauss also obtains the expression for k in this special coordinate system. Theorem 7 now takes the much simpler form

$$4G^2k = \left(\frac{\partial G}{\partial r}\right)^2 - 2G\,\frac{\partial^2 G}{\partial r^2} \quad ,$$

which gives

$$\boxed{(**)\; k = -\frac{1}{\sqrt{G}}\,\frac{\partial^2\sqrt{G}}{\partial r^2}} \quad .$$

It will now be necessary to obtain some further information about the function $\sqrt{G}$, for which Gauss gives very brief arguments. We will find it convenient to express G "as a function of ρ and ϕ"; that is, we consider

$$g = G \circ \exp \circ P^{-1} \quad ,$$

where $P\colon M_p - \text{ray} \to (0,\delta] \times (0,2\pi)$ is $P = (\rho,\phi)$, so that $g(\rho_0,\phi_0)$ is $G(\exp v)$, where $v \in M_p$ has polar coordinates (ρ_0,ϕ_0). At times it will also be convenient to use (ρ,ϕ) to stand for a point in M_p, as well as standing for the coordinate functions themselves.

The function g can be considered as defined on $(0,\delta] \times [0,2\pi]$ (with the same values at $(\rho,0)$ as at $(\rho,2\pi)$). What we want to examine is the behavior of g near $(0,\phi)$. We do this by comparing distances on M_p with those on M. On the vector space M_p we have an inner product $\langle\ ,\ \rangle_p$; since the tangent space $(M_p)_v$ of M_p at $v \in M_p$ can be identified with M_p, we can use $\langle\ ,\ \rangle_p$ to obtain a Riemannian metric on M_p. To keep things straight, we will use v,w to denote elements of M_p, and X_v, Y_v to denote tangent vectors in the tangent space $(M_p)_v$. For each $X_v \in (M_p)_v$ we thus have a certain norm, which we will denote by $|||X_v|||$. Now recall that $\exp_*\colon (M_p)_0 \to M_p$ is the identity map (when we identify $(M_p)_0$ with M_p), so that we certainly have

$$|||X_0||| = \|\exp_*(X_0)\|_p , \quad X_0 \in (M_p)_0 \quad .$$

If $\varepsilon > 0$, it follows that for $v \in M_p$ sufficiently close to 0 and $X_v \in (M_p)_v$ of unit norm $|||\ \ |||$ we have

$$\Big|\, |||X_v||| - \|\exp_*(X_0)\|_{\exp(v)} \Big| < \varepsilon \quad ,$$

so that for any $Y_v \in (M_p)_v$ we have

$$\Big| \, |||Y_v||| - \|\exp_*(Y_v)\|_{\exp(v)} \Big| < \varepsilon \cdot |||Y_v||| \ .$$

Noting that $\exp_*(\partial/\partial\emptyset|_v) = \partial/\partial\varphi|_{\exp(v)}$, we have

$$(1) \qquad \left| \, \left|\left|\left| \frac{\partial}{\partial\emptyset}\right|_v \right|\right|\right| - \left\| \frac{\partial}{\partial\varphi}\right|_{\exp(v)} \right\| \, \right| < \varepsilon \cdot \left|\left|\left| \frac{\partial}{\partial\emptyset}\right|_v \right|\right|\right| ,$$

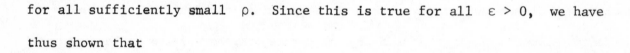

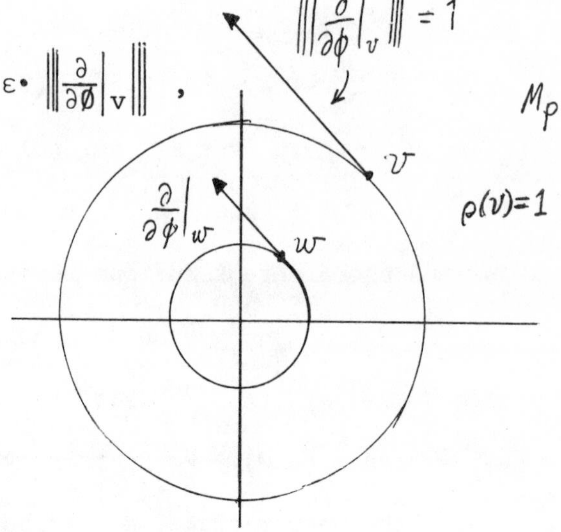

for sufficiently small v, while clearly

$$\left|\left|\left| \frac{\partial}{\partial\emptyset}\right|_v \right|\right|\right| = \|v\|_p = \rho(v) \ .$$

Dividing all terms of (1) by ρ yields

$$\left| 1 - \frac{\sqrt{g}(\rho,\emptyset)}{\rho} \right| < \varepsilon ,$$

for all sufficiently small ρ. Since this is true for all $\varepsilon > 0$, we have thus shown that

$$(2) \qquad \sqrt{g}(\rho,\emptyset) = \rho + o(\rho) ,$$

where $o(\rho)$ denotes a function on $(0,\delta] \times [0,2\pi]$ such that

$$\lim_{\rho\to 0} \frac{o(\rho)}{\rho} = 0 \quad \text{(uniformly in } \emptyset) \ .$$

Clearly $\sqrt{g}$ remains continuous on $[0,\delta] \times [0,2\pi]$ if we define

$$(***)\ \sqrt{g}(0,\phi) = 0\ .$$

Notice, moreover, that equation (2) now immediately implies that

$$(****)\ \frac{\partial\sqrt{g}}{\partial\rho}(0,\phi) = 1$$

(where $\partial\sqrt{g}/\partial\rho(0,\phi)$ really denotes a right hand derivative). However, we want to know that $\partial\sqrt{g}/\partial\rho$ is actually continuous on $[0,\delta] \times [0,2\pi]$; the argument for this will require another step.

From equation (**) we have, on $(0,\delta] \times [0,2\pi]$,

$$\frac{\partial^2\sqrt{g}}{\partial\rho^2}(\rho,\phi) = -\sqrt{g}(\rho,\phi)\cdot k(\exp(\rho,\phi))$$ [where $\exp(\rho,\phi)$ really means $\exp(v)$, where $v \in M_p$ has polar coordinates (ρ,ϕ)] ,

which shows that $\partial^2\sqrt{g}/\partial\rho^2(\rho,\phi) \longrightarrow 0$ as $\rho \longrightarrow 0$. It follows, in particular, that $\partial^2\sqrt{g}/\partial\rho^2$ is bounded. For $\rho > 0$ we have

$$\frac{\partial\sqrt{g}}{\partial\rho}(\rho,\phi) = \frac{\partial\sqrt{g}}{\partial\rho}(\delta,\phi) - \int_\rho^\delta \frac{\partial^2\sqrt{g}}{\partial\rho^2}(t,\phi)dt \quad ,$$

which immediately implies that

$$\lim_{\rho\to 0} \frac{\partial\sqrt{g}}{\partial\rho}(\rho,\phi) \quad \text{exists.}$$

By a standard theorem of calculus (see e.g., p. 178 of the author's Calculus), the limit must be $\partial\sqrt{g}/\partial\rho(0,\phi)$, which equals 1, by (****). It follows, in particular, that

$$(*****)\qquad \int_0^{\rho_0} -\frac{\partial^2\sqrt{g}}{\partial\rho^2}(\rho,\phi)\,d\rho = \lim_{\varepsilon\to 0}\int_\varepsilon^{\rho_0} -\frac{\partial^2\sqrt{g}}{\partial\rho^2}(\rho,\phi)\,d\rho$$

$$= \lim_{\varepsilon\to 0} \frac{\partial\sqrt{g}}{\partial\rho}(\varepsilon,\phi) - \frac{\partial\sqrt{g}}{\partial\rho}(\rho_0,\phi)$$

$$= 1 - \frac{\partial\sqrt{g}}{\partial\rho}(\rho_0,\phi) \quad .$$

We are ready to prove a theorem.

9. THEOREM. Let $A, B, C,$ be three points of $\exp(U)$, where U is a convex neighborhood of $0 \in M_A$, on which $\exp$ is a diffeomorphism, and let α be a geodesic in $\exp(U)$ between B and C. Denote the geodesic segment from A to C by β and the geodesic segment from A to B by γ, and let $\triangle ABC$ be the "geodesic triangle" bounded by α, β, γ. Also let $\angle A$ denote the angle between β and γ, etc. Then

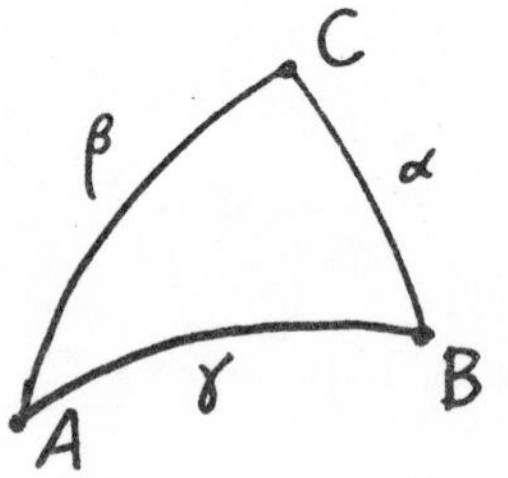

$$\int_{\triangle ABC} k\,dV = \angle A + \angle B + \angle C - \pi \quad .$$

Proof. Let $b, c \in M_p$ be the vectors with $\exp(b) = B$ and $\exp(c) = C$. Choose the polar coordinates ρ, ϕ on M_p so that $\phi(b) = 0$. Then $\phi(c)$ is just $\angle A$.

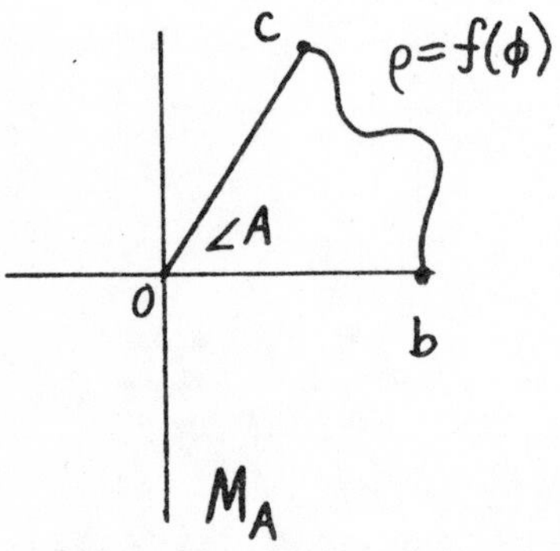

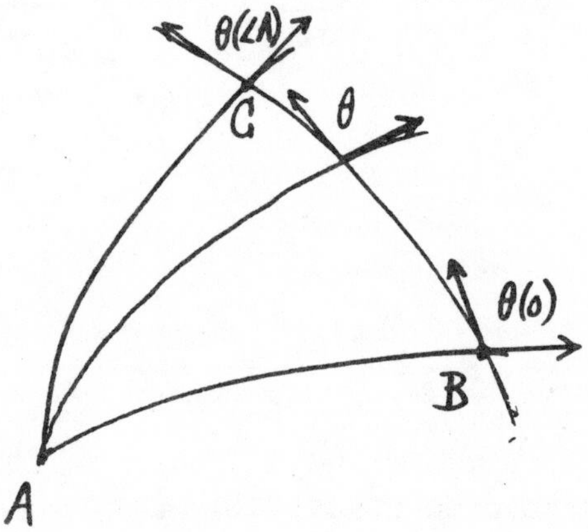

It is easy to see that α cannot intersect the same geodesic ray through A twice, so it must be the image under exp of a curve in M_A which is the graph $\rho = f(\phi)$ of some function in the polar coordinates ρ, ϕ. Let $\theta(\phi)$ be the angle between $\partial/\partial r|_q$ and $\alpha'(q)$, where q is exp of the vector with polar coordinates $(f(\phi), \phi)$.

We clearly have (see the figure)

$$\theta(0) = \pi - \angle B$$
$$\theta(\angle A) = \angle C \qquad .$$

Consequently,

$$\int_{\Delta ABC} k\, dV = \int_{\exp^{-1}(\Delta ABC)} \exp^*(k\, dV)$$

$$= \int_{\exp^{-1}(\Delta ABC)} -\frac{1}{\sqrt{g}} \frac{\partial^2 \sqrt{g}}{\partial \rho^2} \cdot \sqrt{g}\, d\rho \wedge d\phi \qquad \text{by } (**)$$

$$= \int_0^{\angle A} \left(\int_0^{f(\phi)} -\frac{\partial^2 \sqrt{g}}{\partial \rho^2}(\rho, \phi)\, d\rho \right) d\phi$$

$$= \int_0^{\angle A} \left(1 - \frac{\partial \sqrt{g}}{\partial \rho}(f(\phi), \phi) \right) d\phi \qquad \text{by } (*****)$$

$$= \int_0^{\angle A} \left(1 + \frac{d\theta}{d\phi}(\phi)\right) d\phi \qquad \text{by } (*)$$

$$= \angle A + \theta(\angle A) - \theta(0)$$

$$= \angle A + \angle C + \angle B - \pi \quad . \blacksquare$$

According to Theorem 9, on a surface with everywhere positive curvature the sum of the angles of a triangle with geodesic sides is always $> \pi$, while on a surface with everywhere negative curvature the sum of the angles is always $< \pi$. This certainly looks like the case in pictures, and one can even see that the bigger the triangle, the bigger the difference between $\angle A + \angle B + \angle C$ and π.

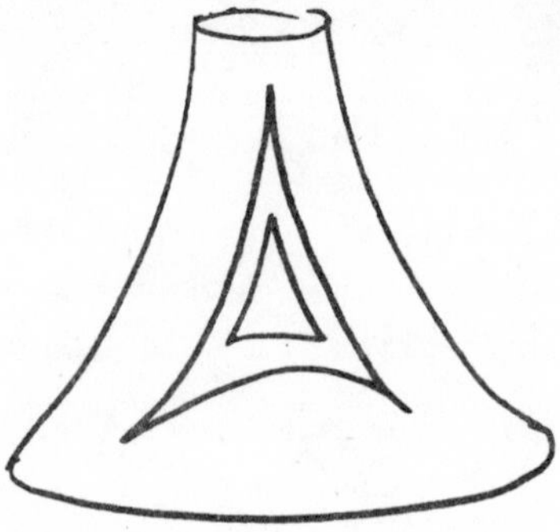

Notice that in the proof of Theorem 9 we are essentially converting an integral over the region $\triangle ABC$ into an integral over (one of) its sides — this looks suspiciously like Stokes' Theorem. Later on, we will indeed be able to present a much nicer proof of Theorem 9, which does not depend on a special coordinate system, and which makes explicit the role of Stokes' Theorem. Moreover, we will be able to derive other important consequences of the same results. However, for the moment, we are more interested in two questions. How did Gauss think of the Theorema Egregium? and What does it really mean?

The paper which we have just examined was based on an earlier paper which Gauss did not publish; the Raven Press has included a translation of this paper together with the paper of 1827. From the earlier paper it appears that Gauss first proved the result in Theorem 9. Notice that this result gives the Theorema Egregium as a corollary, for it implies that

$$k(p) = \lim_{\triangle ABC \to p} \frac{\angle A + \angle B + \angle C - \pi}{\text{area } \triangle ABC},$$

and this limit is defined totally in terms of the Riemannian metric on the surface. After realizing this, Gauss probably argued that since k depends only on the metric it should be possible to show this by a direct computation.

To answer the second question — What does the Theorema Egregium really mean? — requires a more serious effort. The original definition of k seemed perfectly satisfactory — it has immediate geometric appeal and is even fairly easy to compute. The only defect of the definition is that the concept being defined turns out to be too good; it turns out to be invariant under isometries, while its definition is not.

One of the dogmas of modern mathematics is that for any object that is invariant in any sort of way, a definition must be found that exhibits this invariance directly (even if such a definition is harder to understand than the original!). The definition of determinants is a good example. An elementary treatment of determinants usually begins by defining the determinant of a <u>matrix</u>, either by writing down a messy formula, or in an inductive way that really amounts to the messy formula. It is then shown that $\det(A \cdot B) = \det A \cdot \det B$; from this it follows that one can define the determinant of a <u>linear transformation</u> to be the determinant of its matrix with respect to any basis. This naturally leads one to seek a definition of the determinant of a linear transformation $T\colon V \to V$ which does not require a choice of basis. After one has defined $\Omega^k(V)$, it is possible to define $\det T$ to be the constant such that $T^*\colon \Omega^n(V) \to \Omega^n(V)$ is $\det T$ times the identity. This definition is indeed independent of a choice of basis, but when one looks a little harder at it, one sees that it isn't all that different from the messy formula defining the determinant of a matrix. Indeed, one proves that $\dim \Omega^n(V) \neq 0$ by writing down an explicit non-zero element of it (in terms of a basis!) which involves permutations in exactly the same way as the original definition of determinants. Finally, if one can tolerate even more complicated constructions, the "exterior algebra" of V can be used to produce a definition of $\det T$ which is completely independent of bases, and does not even mention permutations. In fact, the sign of a permutation σ can then be <u>defined</u> as the determinant of the linear transformation $T\colon \mathbb{R}^n \to \mathbb{R}^n$ with $T(e) = e_{\sigma(i)}$. This approach is expounded in Chevalley's book, <u>The Construction and Study of Certain</u>

Important Algebras.

The definition of curvature follows a similar course. We first defined the curvature of a surface in a way which depends on an imbedding in $\mathbb{R}^3$; our final goal is a definition of curvature which depends only on the Riemannian metric on the surface. It should be noted that we already have a candidate — the formula in the Theorema Egregium may be used as a definition of curvature! It must still be checked that this definition does not depend on the coordinate system (an uninviting task), but at least the definition involves only a coordinate system on the surface, not an imbedding of the surface in $\mathbb{R}^3$. Presumably, few mathematicians would accept this definition as a reasonable one. As we proceed to frame more acceptable definitions of curvature, we will rely, just as in the case of determinants, on more complicated and abstract constructions, so it is important that we follow the historical evolution of the definition of curvature, in order not to lose sight of its geometric significance.

One further comparison with the case of determinants will point out the direction which our investigations will have to take. It seems safe to assert that the modern invariant definitions of determinants would never have come into being if mathematicians had always considered only determinants of 2×2 matrices — our perception of the structure is complicated too much by the simplicity of this special case. Similarly for curvature. The essence of curvature is revealed more fully only when one transcends the limits of intuition and considers manifolds of arbitrary dimensions, conceived as existing in their own right, not as subsets of a Euclidean space.

Addendum. The Formula of Bertrand and Puiseux; Diquet's Formula

Consider the function $\sqrt{g}$ which was used in the proof of Theorem 9. This function is defined on some $[0,\delta] \times [0,2\pi]$ and

(1) $\sqrt{g}(\rho,0) = \sqrt{g}(\rho,2\pi)$ for $\rho \in [0,\delta]$

(2) $\sqrt{g}(0,\phi) = 0$

(3) $\dfrac{\partial\sqrt{g}}{\partial\rho}(0,\phi) = \lim\limits_{\rho\to 0} \dfrac{\partial\sqrt{g}}{\partial\rho}(\rho,\phi) = 1$.

We have also noted that on $(0,\delta] \times [0,2\pi]$ we have

(4) $$\frac{\partial^2\sqrt{g}}{\partial\rho^2}(\rho,\phi) = -\sqrt{g}(\rho,\phi)\cdot k(\exp(\rho,\phi))$$

$$= -\sqrt{g}(\rho,\phi)\cdot K(\rho,\phi) \quad , \quad \text{say.}$$

This shows (appealing once again to the standard theorem of calculus used before) that

(5) $$\frac{\partial^2\sqrt{g}}{\partial\rho^2}(0,\phi) = \lim_{\rho\to 0} \frac{\partial^2\sqrt{g}}{\partial\rho^2}(\rho,\phi) = 0 \quad .$$

Differentiating equation (4) yields

(6) $$\frac{\partial^3\sqrt{g}}{\partial\rho^3}(\rho,\phi) = -\frac{\partial\sqrt{g}}{\partial\rho}(\rho,\phi)\cdot K(\rho,\phi) - \sqrt{g}(\rho,\phi)\,\frac{\partial K}{\partial\rho}(\rho,\phi) \quad .$$

We note that the term $\partial K/\partial\rho(\rho,\phi)$ makes sense even for $\rho = 0$; it is just a directional derivative of K, which is a C^∞ function on M_p. Consequently, $\partial K/\partial\rho(\rho,\phi)$ approaches a limit as $\rho \to 0$. From (6) we thus obtain

(7) $$\frac{\partial^3\sqrt{g}}{\partial\rho^3}(0,\phi) = \lim_{\rho\to 0} \frac{\partial^3\sqrt{g}}{\partial\rho^3}(\rho,\phi)$$

$$= -k(p) \ ,$$ using (2) and (3), and noting that $K(0,\phi) = k(p)$.

Using (2), (3), (5), (7), we now have the Taylor polynomial expansion

$$\boxed{\sqrt{g}(\rho,\phi) = \rho - \frac{k(p)\rho^3}{6} + o(\rho^3)} \quad .$$

[There is still a technical detail which must be taken care of. We actually want to know that the remainder

$$R(\rho) = \sqrt{g}(\rho,\phi) - \rho - \frac{k(p)\rho^3}{6}$$

satisfies

$$\lim_{\rho\to 0} \frac{R(\rho)}{\rho^3} \longrightarrow 0 \qquad \underline{\text{uniformly}}\ \underline{\text{in}}\ \phi.$$

This can be seen from the proof that the Taylor polynomial approximates the function; the proof involves L'Hôpital's Theorem, which in turn depends on the Cauchy mean value theorem (<u>Calculus</u>, pg. 178), whose role will be made explicit. We have

$$\lim_{\rho\to 0} \frac{\sqrt{g}(\rho,\phi) - \rho - k(p)\rho^3/6}{\rho^3}$$

$$= \lim_{\rho\to 0} \frac{\frac{\partial\sqrt{g}}{\partial\rho}(\bar{\rho},\phi) - 1 - k(p)\bar{\rho}^2/2}{3\bar{\rho}^2} \qquad 0 < \bar{\rho} < \rho, \text{ by the Cauchy mean value theorem}$$

$$= \lim_{\rho\to 0} \frac{\frac{\partial^2\sqrt{g}}{\partial\rho^2}(\bar{\bar{\rho}},\phi) - k(p)\bar{\bar{\rho}}}{6\bar{\bar{\rho}}} \qquad 0 < \bar{\bar{\rho}} < \bar{\rho}$$

$$= \lim_{\rho\to 0} \frac{\frac{\partial^3\sqrt{g}}{\partial\rho^3}(\bar{\bar{\bar{\rho}}},\phi) - k(p)}{6} \quad . \qquad 0 < \bar{\bar{\bar{\rho}}} < \bar{\bar{\rho}}$$

$= 0$, by (7).]

10. PROPOSITION (BERTRAND AND PUISEUX; 1848). Let $C(\rho)$ be the circumference of the "geodesic circle" of radius ρ around $p \in M$, consisting of the end-points of geodesic segments of length ρ which start at p. Then

$$k(p) = \lim_{\rho \to 0} 3 \cdot \frac{2\pi\rho - C(\rho)}{\pi\rho^3} .$$

Proof. Clearly

$$C(\rho) = \int_0^{2\pi} \left\| \frac{\partial}{\partial\varphi}\bigg|_{\exp(\rho,\phi)} \right\| d\phi = \int_0^{2\pi} \sqrt{g}(\rho,\phi)\, d\phi = \int_0^{2\pi} \left(\rho - \frac{k(p)\rho^3}{6}\right) d\phi + \int_0^{2\pi} o(\rho^3)\, d\phi$$

$$= 2\pi\left(\rho - \frac{k(p)\rho^3}{6}\right) + o(\rho^3) .$$

So

$$k(p) = 3 \cdot \frac{2\pi\rho - C(\rho)}{\pi\rho^3} + \frac{o(\rho^3)}{\rho^3} . \blacksquare$$

According to Proposition 10, on a surface of positive curvature, like a sphere, geodesic circles are always "too small", while on surfaces of negative curvature, they are always "too large." Notice that Proposition 10 gives

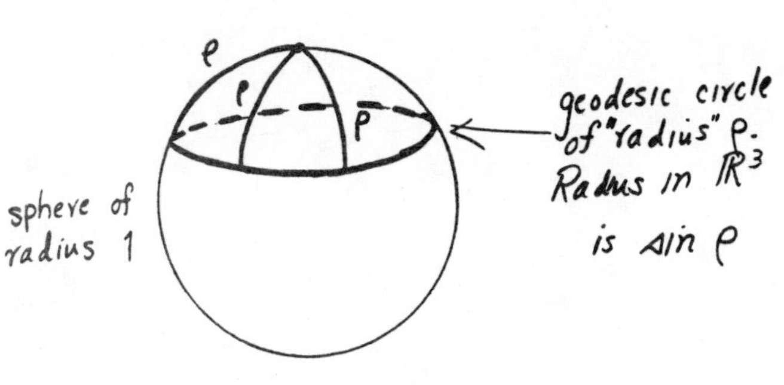

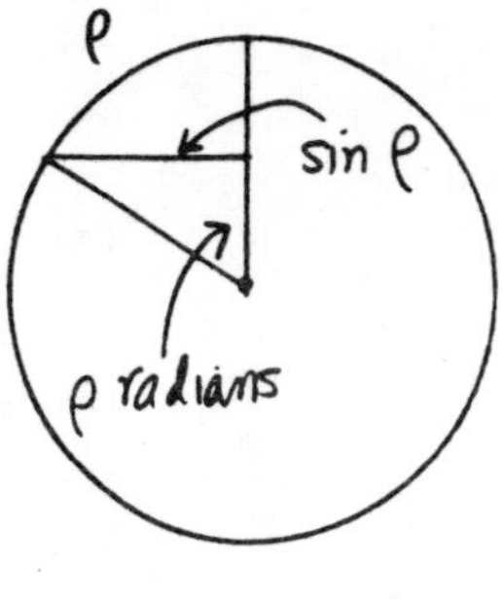

another interpretation of curvature totally in terms of the Riemannian metric on M. There is yet another formula of the same type.

11. PROPOSITION (DIQUET; 1848). Let $A(\rho)$ be the area enclosed by the geodesic circle of radius ρ around $p \in M$. Then

$$k(p) = \lim_{\rho \to 0} 12 \frac{\pi\rho^2 - A(\rho)}{\pi\rho^4}$$

Proof. Clearly

$$A(\rho) = \int_{\substack{\text{inclosed} \\ \text{region}}} dV = \int_0^{2\pi}\int_0^{\rho} \sqrt{g}(\rho,\phi)\,d\rho\,d\phi$$

$$= \int_0^{2\pi}\int_0^{\rho} \left(\rho - \frac{k(p)\rho^3}{6}\right) d\rho\,d\phi + \int_0^{2\pi}\int_0^{\rho} o(\rho^3)\,d\rho\,d\phi$$

$$= 2\pi\left(\frac{\rho^2}{2} - \frac{k(p)\rho^4}{24}\right) + o(\rho^4) \quad ,$$

which easily yields the desired result. ∎

Chapter 4A. An Inaugural Lecture

On June 10, 1854 the faculty of Göttingen University heard a lecture entitled Über die Hypothesen, welche der Geometrie zu Grunde liegen (On the Hypotheses which lie at the Foundations of Geometry). This lecture was delivered by Georg Friedrich Bernhard Riemann, who had been born just a year before Gauss' paper of 1827. Although the lecture was not published until 1866, the ideas contained within it proved to be the most influential in the entire history of differential geometry. To be sure, mathematicians had not neglected the study of surfaces in the meantime; in fact, Gauss' work had inspired a tremendous amount of work along these lines. But the results obtained in those years can all be proved with much greater ease after we have followed the long series of developments initiated by the turning point in differential geometry which Riemann's lecture provided.

A short account of the life and character of Riemann can be found in the biography by Dedekind* which is included in Riemann's collected works (published by Dover). His interest in many fields of mathematical physics, together with a demand for perfection in all he did, delayed until 1851 the submission of his doctoral dissertation Grundlagen für eine allegemeine Theorie der Functionen einer veränderlichen complexen Grösse (Foundations for a general theory of functions of a complex variable). Gauss' official report to the Philosophical Faculty of the University of Göttingen stated "The dissertation submitted by Herr Riemann offers convincing evidence of the author's thorough and penetrating investigations in those parts of the subject treated in the dissertation, of a creative, active

---

* Even for those who can only plod through German, this is preferable to the account in E. T. Bell's Men of Mathematics, which is hardly more than a translation of Dedekind, written in a racy style and interladen with supercilious remarks of questionable taste.

truly mathematical mind, and of a gloriously fertile originality."

Riemann was now qualified to seek the position of Privatdocent (a lecturer who received no salary, but was merely forwarded fees paid by those students who elected to attend his lectures). To attain this position he first had to submit an "inaugural paper" (Habilitationsschrift). Again there were delays, and it was not until the end of 1853 that Riemann submitted the Habilitationsschrift, Über die Darstellbarkeit einer Function durch eine trigonometrische Reihe (On the representability of a function by a trigonometric series). Now Riemann still had to give a probationary inaugural lecture on a topic chosen by the faculty, from a list of three proposed by the candidate. The first two topics which Riemann submitted were ones on which he had already worked, and he had every reason to expect that one of these two would be picked; for the third topic he chose the foundations of geometry. Contrary to all traditions, Gauss passed over the first, and picked instead the third, in which he had been interested for years. At this time Riemann was also investigating the connection between electricity, magnetism, light, and gravitation, in addition to acting as an assistant in a seminar on mathematical physics. The strain of carrying out another major investigation, aggravated perhaps by the hardships of poverty, brought on a temporary breakdown. However, Riemann soon recovered, disposed of some other work which had to be completed, and then finished his inaugural lecture in about seven more weeks.

Riemann hoped to make his lecture intelligible even to those members of the faculty who knew little mathematics. Consequently, hardly any formulas appear and the analytic investigations are completely suppressed. Although Dedekind describes the lecture as a masterpiece of exposition, it is questionable how many of the faculty comprehended it. In making the following translation*, I was

---

* The original is contained, of course, in Riemann's collected works. Two English translations are readily available, one in Volume 2 of Smith's Source Book in Mathematics (Dover), and one in Clifford's Mathematical Papers (Chelsea).

aided by the fact that I already had some idea what the mathematical results were supposed to be. The uninitiated reader will probably experience a great deal of difficulty merely understanding what Riemann is trying to say (the proofs of Riemann's assertions are spread out over the next several Chapters). We can be sure, however, that one member of the faculty appreciated Riemann's work. Dedekind tells us that Gauss sat at the lecture "which surpassed all his expectations, in the greatest astonishment, and on the way back from the faculty meeting he spoke to Wilhelm Weber, with the greatest appreciation, and with an excitement rare for him, about the depth of the ideas presented by Riemann."

RIEMANN

ON THE HYPOTHESES WHICH LIE AT THE FOUNDATIONS OF GEOMETRY

Plan of the Investigation

As is well known, geometry presupposes the concept of space, as well as assuming the basic principles for constructions in space. It gives only nominal definitions of these things, while their essential specifications appear in the form of axioms. The relationship between these presuppositions [the concept of space, and the basic properties of space] is left in the dark; we do not see whether, or to what extent, any connection between them is necessary, or *a priori* whether any connection between them is even possible.

From Euclid to the most famous of the modern reformers of geometry, Legendre, this darkness has been dispelled neither by the mathematicians nor by the philosophers who have concerned themselves with it. This is undoubtedly because the general concept of multiply extended quantities, which includes spatial quantities, remains completely unexplored. I have therefore first set myself the task of constructing the concept of a multiply extended quantity from general notions of quantity. It will be shown that a multiply extended quantity is susceptible of various metric relations, so that Space constitutes only a special case of a triply extended quantity. From this however it is a necessary consequence that the theorems of geometry cannot be deduced from general notions of quantity, but that those properties which distinguish Space from other conceivable triply extended quantities can only be deduced from experience. Thus arises the problem of seeking out the simplest data from which the metric relations of Space can be determined, a problem which by its very nature is not completely determined, for there may be several systems of simple data which suffice to determine the metric relations of Space; for the present purposes, the most important system is that laid down as a foundation of geometry by Euclid. These data are — like all data — not logically necessary, but only of empirical certainty, they are hypotheses; one can therefore investigate their likelihood, which is certainly very great within the bounds of observation, and afterwards decide upon the legitimacy of extending them beyond the bounds of observation, both in the direction of the immeasurably large, and in the direction of the immeasurably small.

I. Concept of an n-fold extended quantity

In proceeding to attempt the solution of the first of these problems, the development of the concept of multiply extended quantity, I feel particularly entitled to request an indulgent criticism, as I am little trained in these tasks of a philosophical nature where the difficulties lie more in the concepts than in the construction, and because I could not make use of any previous studies, except for some very brief hints on the subject which Privy Councillor Gauss has given in his second memoir on Biquadratic Residues, in the Göttingen Gelehrte Anzeige and in his Jubilee-book, and some philosophical researches of Herbart.

1.

Notions of quantity are possible only when there already exists a general concept which admits particular instances. These instances form either a continuous or a discrete manifold, depending on whether or not a continuous transition of instances can be found between any two of them; individual instances are called points in the first case and elements of the manifold in the second. Concepts whose particular instances form a discrete manifold are so numerous that some concept can always be found, at least in the more highly developed languages, under which any given collection of things can be comprehended (and consequently, in the study of discrete quantities, mathematicians could unhesitatingly proceed from the principle that given objects are to be regarded as all of one kind).

On the other hand, opportunities for creating concepts whose instances form a continuous manifold occur so seldom in everyday life that color and the position of sensible objects are perhaps the only simple concepts whose instances form a multiply extended manifold. More frequent opportunities for creating and developing these concepts first occur in higher mathematics.

Particular portions of a manifold, distinguished by a mark or by a boundary, are called quanta. Their quantitative comparison is effected in the case of discrete quantities by counting, in the case of continuous quantities by measurement. Measuring involves the superposition of the quantities to be compared; it therefore requires a means of transporting one quantity to be used as a standard for the others. Otherwise, one can compare two quantities only when one is a part of the other, and then only as to "more" or "less", not as to "how much". The investigations which can be carried out in this case form a general divison of the science of quantity, independent of measurement, where quantities are regarded, not as existing independent of position and not as expressible in terms of a unit, but as regions in a manifold. Such investigations have become a necessity for several parts of mathematics, e.g., for the treatment of many-valued analytic functions, and the dearth of such studies is one of the principal reasons why the celebrated theorem of Abel and the contributions of Lagrange, Pfaff and Jacobi to the general theory of differential equations have remained unfruitful for so long. From this portion of the science of extended quantity, a portion which proceeds without any further assumptions, it suffices for the present purposes to emphasize two points, which will make clear the essential characteristic of an n-fold extension. The first of these concerns the generation of the concept of a multiply extended manifold, the second involves reducing position fixing in a given manifold to numerical determinations.

2.

In a concept whose instances form a continuous manifold, if one passes from one instance to another in a well-determined way, the instances through which one has passed form a simply extended manifold, whose essential characteristic is, that from any point in it a continuous movement is possible in only two directions, forwards and backwards. If one now imagines that this manifold passes to another, completely different one, and once again in a well-determined way, that is, so that every point passes to a well-determined point of the other, then the instances form, similarly, a double extended manifold. In a similar way, one obtains a triply extended manifold when one imagines that a doubly extended one passes in a well-determined way to a completely different one, and it is easy to see how one can continue this construction. If one considers the process as one in which the objects vary, instead of regarding the concept as fixed, then this construction can be characterized as a synthesis of a variability of n + 1 dimensions from a variability of n dimensions and a variability of one dimension.

3.

I will now show, conversely, how one can break up a variability, whose boundary is given, into a variability of one dimension and a variability of lower dimension. One considers a piece of a manifold of one dimension — with a fixed origin, so that points of it may be compared with one another — varying so that for every point of the given manifold it has a definite value, continuously changing with this point. In other words, we take within the given manifold a continuous function of position, which, moreover, is not constant on any part of the manifold. Every system of points where the function has a constant value then forms a continuous manifold of fewer dimensions than the given one. These manifolds pass continuously from one to another as the function changes; one can therefore assume that they all emanate from one of them, and generally speaking this will occur in such a way that every point of the first passes to a definite point of any other; the exceptional cases, whose investigation is important, need not be considered here. In this way, the determination of position in the given manifold is reduced to a numerical determination and to the determination of position in a manifold of fewer dimensions. It is now easy to show that this manifold has n-1 dimensions, if the given manifold is an n-fold extension. By an n-time repetition of this process, the determination of position in an n-fold extended manifold is reduced to n numerical determinations, and therefore the determination of position in a given manifold is reduced, whenever this is possible, to a finite number of numerical determinations. There are, however, also manifolds in which the fixing of position requires not a finite number, but either an infinite sequence or a continuous manifold of numerical measurements. Such manifolds form e.g., the possibilities for a function in a given region, the possible shapes of a solid figure, etc.

II. Metric Relations of which a manifold of n dimensions is susceptible, on the assumption that lines have a length independent of their configuration, so that every line can be measured by every other.

Now that the concept of an n-fold extended manifold has been constructed, and its essential characteristic has been found in the fact that position fixing in the manifold can be reduced to n numerical determinations, there follows, as the second of the problems proposed above, an investigation of the metric relations of which such a manifold is susceptible, and of the conditions which suffice to determine them. These metric relations can be investigated only in abstract terms, and their interdependence exhibited only through formulas. Under certain assumptions, however, one can resolve them into relations which are individually capable of geometric representation, and in this way it becomes possible to express the results of calculation geometrically. Thus, although an abstract investigation with formulas certainly cannot be avoided, the results can be presented in geometric garb. The foundations of both parts of the question are contained in the celebrated treatise of Privy Councillor Gauss on curved surfaces.

1.

Measurement requires an independence of quantity from position, which can occur in more than one way. The hypothesis which first presents itself, and which I shall develop here, is just this, that the length of lines is independent of their configuration, so that every line can be measured by every other. If position-fixing is reduced to numerical determinations, so that the position of a point in the given n-fold extended manifold is expressed by n varying quantities x_1, x_2, x_3 and so forth up to x_n, then specifying a line amounts to giving the quantities x as functions of one variable. The problem then is, to set up a mathematical expression for the length of a line, for which purpose the quantities x must be thought of as expressible in units. I will treat this problem only under certain restrictions, and I first limit myself to lines in which the ratios of the quantities dx — the increments in the quantities x — vary continuously. One can then regard the lines as broken up into elements within which the ratios of the quantities dx may be considered to be constant. The problem then reduces to setting up a general expression for the line element ds at every point, an expression which will involve the quantities x and the quantities dx. I assume, secondly, that the length of the line element remains unchanged, up to first order, when all the points of this line element suffer the same infinitesimal displacement, whereby I simply mean that if all the quantities dx increase in the same ratio, the line element changes by the same ratio. Under these assumptions, the line element can be an arbitrary homogeneous function of the first degree in the quantities dx which remains the same when all the quantities dx change sign, and in which the arbitrary constants are functions of the quantities x. To find the simplest cases, I first seek an expression for the (n-1)-fold extended manifolds which are everywhere equidistant from the origin of the line element, i.e., I seek a continuous function of position which distinguishes them from one another. This must either decrease or increase in all directions from the origin; I will assume that it increases in all directions and therefore has a minimum at the origin. Then if its first and second differential quotients are finite, the first order differential must vanish and the second order differential cannot be negative; I assume that it is always positive. This differential expression of the second order remains constant if ds remains constant and increases quadratically when the quantities dx, and thus also ds, all increase in the same ratio; it is therefore equal to a constant times ds^2, and consequently ds equals the squareroot of an everywhere positive homogeneous function of the second degree in the quantities dx, in which the coefficients are continuous functions of the quantities x. In Space, if one expresses the location of a point by rectilinear coordinates, then $ds = \sqrt{\Sigma(dx)^2}$; Space is therefore included in this simplest case. The next simplest case would perhaps include the manifolds in which the line element can be expressed as the fourth root of a differential expression of the fourth degree. Investigation of this more general class would actually require no essentially different principles, but it would be rather time consuming and throw proportionally little new light on the study of Space, especially since the results cannot be expressed geometrically;

I consequently restrict myself to those manifolds where the line element can be expressed by the square root of a differential expression of the second degree. One can transform such an expression into another similar one by substituting for the n independent variables, functions of n new independent variables. However, one cannot transform any expression into any other in this way; for the expression contains $n \frac{n+1}{2}$ coefficients which are arbitrary functions of the independent variables; by the introduction of new variables one can satisfy only n conditions, and can therefore make only n of the coefficients equal to given quantities. There remain $n \frac{n-1}{2}$ others, already completely determined by the nature of the manifold to be represented, and consequently $n \frac{n-1}{2}$ functions of position are required to determine its metric relations. Manifolds, like the Plane and Space, in which the line element can be brought into the form $\sqrt{\Sigma dx^2}$ thus constitute only a special case of the manifolds to be investigated here; they clearly deserve a special name, and consequently, these manifolds, in which the square of the lines element can be expressed as the sum of the squares of complete differentials, I propose to call flat. In order to survey the essential differences of the manifolds representable in the assumed form, it is necessary to eliminate the features depending on the mode of presentation, which is accomplished by choosing the variable quantities according to a definite principle.

2.

For this purpose, one constructs the system of shortest lines emanating from a given point; the position of an arbitrary point can then be determined by the initial direction of the shortest line in which it lies, and its distance, in this line, from the initial point. It can therefore be expressed by the ratios of the quantities dx^0, i.e., the quantities dx at the origin of this shortest line, and by the length s of this line. In place of the dx^0 one now introduces linear expressions $d\alpha$ formed from them in such a way that the initial value of the square of the line element will be equal to the sum of the squares of these expressions, so that the independent variables are: the quantity s and the ratio of the quantities $d\alpha$. Finally, in place of the $d\alpha$ choose quantities $x_1, x_2, \ldots, x_n$ proportional to them, but such that the sum of their squares equals s^2. If one introduces these quantities, then for infinitely small values of x the square of the line element equals Σdx^2, but the next order term in its expansion equals a homogeneous expression of the second degree in the $n\frac{n-1}{2}$ quantities $(x_1 dx_2 - x_2 dx_1)$, $(x_1 dx_3 - x_3\, dx_1), \ldots,$ and is consequently an infinitely small quantity of the fourth order, so that one obtains a finite quantity if one divides it by the square of the infinitely small triangle at whose vertices the variables have the values $(0,0,0,\ldots)$, $(x_1, x_2, x_3, \ldots)$, $(dx_1, dx_2, dx_3, \ldots)$. This quantity remains the same as long as the quantities x and dx are contained in the same binary linear forms, or as long as the two shortest lines from the initial point to x and from the initial point to dx remain in the same surface element, and therefore depends only on the position and direction of that element. It obviously equals zero if the manifold in question is flat, i.e., if the square of the line element is reducible to Σdx^2, and can therefore be regarded as the measure of deviation from flatness in this surface direction at this point. When multiplied by $-3/4$ it becomes equal to the quantity which Privy Councillor Gauss has called the curvature of a surface. Previously, $n\frac{n-1}{2}$ functions of position were found necessary in order to determine the metric relations of an n-fold extended manifold representable in the assumed form; hence if the curvature is given in $n\frac{n-1}{2}$ surface directions at every point, then the metric relations of the manifold may be determined, provided only that no identical relations can be found between these values, and indeed in general this does not occur. The metric relations of these manifolds, in which the line element can be represented as the squareroot of a differential expression of the second degree, can thus be expressed in a way completely independent of the choice of the varying quantities. A similar path to the same goal could also be taken in those manifolds in which the line element is expressed in a less simple way, e.g., by the fourth root of a differential expression of the fourth degree. The line element in this more general case would not be reducible to the square root of a quadratic sum of differential expressions, and therefore in the expression for the square of the line element the deviation from flatness would be an infinitely small quantity of the second dimension, whereas for the other manifolds it was an infinitely small quantity of the fourth dimension. This peculiarity of the latter manifolds therefore might well be called planeness in the smallest parts. For present purposes, however, the most important peculiarity of these manifolds, on whose account alone they have been examined here, is this, that the metric relations of the doubly extended ones can be represented geometrically by surfaces and those of the multiply extended ones can be reduced to those of the surfaces contained within them, which still requires a brief discussion.

3.

In the conception of surfaces, the inner metric relations, which involve only the lengths of paths within them, are always bound up with the way the surfaces are situated with respect to points outside them. We may, however, abstract from external relations by considering deformations which leave the lengths of lines within the surfaces unaltered, i.e., by considering arbitrary bendings — without stretching — of such surfaces, and by regarding all surfaces obtained from one another in this way as equivalent. Thus, for example, arbitrary cylindrical or conical surfaces count as equivalent to a plane, since they can be formed from a plane by mere bending, under which the inner metric relations remain the same; and all theorems about the plane — hence all of planimetry — retain their validity. On the other hand, they count as essentially different from the sphere, which cannot be transformed into the plane without stretching. According to the previous investigations, the inner metric relations at every point of a doubly extended quantity, if its line element can be expressed as the square root of a differential expression of the second degree, which is the case with surfaces, is characterized by the curvature. For surfaces, this quantity can be given a visual interpretation as the product of the two curvatures of the surface at this point, or by the fact that its product with an infinitely small triangle formed from shortest lines is, in proportion to the radius, half the excess of the sum of its angles over two right angles [that is, equal to the excess of the sum over π, when the angles are measured in radians]. The first definition would presuppose the theorem that the product of the two radii of curvatures is unaltered by mere bendings of a surface, the second, that at each point the excess over two right angles of the sum of the angles of any infinitely small triangle is proportional to its area. To give a tangible meaning to the curvature of an n-fold extended manifold at a given point, and in a given surface direction through it, we first mention that a shortest line emanating from a point is completely determined if its initial direction is given. Consequently we obtain a certain surface if we prolong all the initial directions from the given point which lie in the given surface element, into shortest lines; and this surface has a definite curvature at the given point, which is equal to the curvature of the n-fold extended manifold at the given point, in the given surface direction.

4.

Before applying these results to Space, it is still necessary to make some general considerations about flat manifolds, i.e. about manifolds in which the square of the line element can be represented as the sum of squares of complete differentials.

In a flat n-fold extended manifold the curvature in every direction, at every point, is zero; but according to the preceding investigation, in order to determine the metric relations it suffices to know that at each point the curvature is zero in $\frac{1}{2}n(n-1)$ independent surface-direction. The manifolds whose curvature is everywhere 0 can be considered as a special case of those manifolds whose curvature is everywhere constant. The common character of those manifolds whose curvature is constant may be expressed as follows: figures can be moved in them without stretching. For obviously figures could not be freely shifted and rotated in them if the curvature were not the same in all directions, at all points. On the other hand, the metric properties of the manifold are completely determined by the curvature; they are therefore exactly the same in all the directions around any one point as in the directions around any other, and thus the same constructions can be effected starting from either; consequently, in the manifolds with constant curvature figures may be given any arbitrary position. The metric relations of these manifolds depend only on the value of the curvature, and it may be mentioned, as regards the analytic presentation, that if one denotes this value by α, then the expression for the line element can be put in the form

$$\frac{1}{1+\frac{\alpha}{4}\Sigma x^2}\sqrt{\Sigma dx^2} \quad .$$

5.

The consideration of surfaces with constant curvature may serve for a geometric illustration. It is easy to see that the surfaces whose curvature is positive can always be rolled onto a sphere whose radius is the reciprocal of the curvature; but in order to survey the multiplicity of these surfaces, let one of them be given the shape of a sphere, and the others the shape of surfaces of rotation which touch it along the equator. The surfaces with greater curvature than the sphere will then touch the sphere from inside and take a form like the portion of the surface of a ring, which is situated away from the axis; they could be rolled upon zones of spheres with smaller radii, but would go round more than once. Surfaces with smaller positive curvature are obtained from spheres of larger radii by cutting out a portion bounded by two great semi-circles, and bringing together the cut-lines. The surface of curvature zero will be a cylinder standing on the equator; the surfaces with negative curvature will touch this cylinder from outside and be formed like the part of the surface of a ring which is situated near the axis. If one regards these surfaces as possible positions for pieces of surface moving in them, as Space is for bodies, then pieces of surface can be moved in all these surfaces without stretching. The surfaces with positive curvature can always be so formed that pieces of surface can even be moved arbitrarily without bending, namely as spherical surfaces, but those with negative curvature cannot. Aside from this independence of position for surface pieces, in surfaces with zero curvature there is also an independence of position for directions, which does not hold in the other surfaces.

III. Applications to Space

1.

Following these investigations into the determination of the metric relations of an n-fold extended quantity, the conditions may be given which are sufficient and necessary for determining the metric relations of Space, if we assume beforehand the independence of lines from configuration and the possibility of expressing the line element as the square root of a second order differential expression, and thus flatness in the smallest parts.

First, these conditions may be expressed by saying that the curvature at every point equals zero in three surface directions, and thus the metric relations of Space are implied if the sum of the angles of a triangle always equals two right angles.

But secondly, if one assumes with Euclid not only the existence of lines independently of configuration, but also of bodies, then it follows that the curvature is everywhere constant, and the angle sum in all triangles is determined if it is known in one.

In the third place, finally, instead of assuming the length of lines to be independent of place and direction, one might assume that their length and direction is independent of place. According to this conception, changes or differences in position are complex quantities expressible in three independent units.

2.

In the course of the previous considerations, the relations of extension or regionality were first distinguished from the metric relations, and it was found that different metric relations were conceivable along with the same relations of extension; then systems of simple metric specifications were sought by means of which the metric relations of Space are completely determined, and from which all theorems about it are a necessary consequence. It remains now to discuss the question how, to what degree, and to what extent these assumptions are borne out by experience. In this connection there is an essential difference between mere relations of extension and metric relations, in that among the first, where the possible cases form a discrete manifold, the declarations of experience are to sure never completely certain, but they are not inexact, while for the second, where the possible cases form a continuous manifold, every determination from experience always remains inexact — be the probability ever so great that it is nearly exact. This circumstance becomes important when these empirical determinations are extended beyond the limits of observation into the immeasurably large and the immeasurably small; for the latter may obviously become ever more inexact beyond the boundary of observation, but not so the former.

When constructions in Space are extended into the immeasurably large, unboundedness is to be distinguished from infinitude; one belongs to relations of extension, the other to metric relations. That Space is an unbounded triply extended manifold is an assumption which is employed for every apprehension of the external world, by which at every moment the domain of actual perception is supplemented, and by which the possible locations of a sought for object are constructed; and in these applications it is continually confirmed. The unboundedness of space consequently has a greater empirical certainty than any experience of the external world. But its infinitude does not in any way follow from this; quite to the contrary, Space would necessarily be finite if one assumed independence of bodies from position, and thus ascribed to it a constant curvature, as long as this curvature had ever so small a positive value. If one prolonged the initial directions lying in a surface direction into shortest lines, one would obtain an unbounded surface with constant positive curvature, and thus a surface which in a flat triply extended manifold would take the form of a sphere, and consequently be finite.

3.

Questions about the immeasurably large are idle questions for the explanation of Nature. But the situation is quite different with questions about the immeasurably small. Upon the exactness with which we pursue phenomenon into the infinitely small, does our knowledge of their causal connections essentially depend. The progress of recent centuries in understanding the mechanisms of Nature depends almost entirely on the exactness of construction which has become possible through the invention of the analysis of the infinite and through the simple principles discovered by Archimedes, Galileo and Newton, which modern physics makes use of. By contrast, in the natural sciences where the simple principles for such constructions are still lacking, to discover causal connections one pursues phenomenon into the spatially small, just so far as the microscope permits. Questions about the metric relations of Space in the immeasurably small are thus not idle ones.

If one assumes that bodies exist independently of position, then the curvature is everywhere constant, and it then follows from astronomical measurements that it cannot be different from zero; or at any rate its reciprocal must be an area in comparison with which the range of our telescopes can be neglected. But if such an independence of bodies from position does not exist, then one cannot draw conclusions about metric relations in the infinitely small from those in the large; at every point the curvature can have arbitrary values in three directions, provided only that the total curvature of every measurable portion of Space is not perceptibly different from zero. Still more complicated relations can occur if the line element cannot be represented, as was presupposed, by the square root of a differential expression of the second degree. Now it seems that the empirical notions on which the metric determinations of Space are based, the concept of a solid body and that of a light ray, lose their validity in the infinitely small; it is therefore quite definitely conceivable that the metric relations of Space in the infinitely small do not conform to the hypotheses of geometry; and in fact one ought to assume this as soon as it permits a simpler way of explaining phenomena.

The question of the validity of the hypotheses of geometry in the infinitely small is connected with the question of the basis for the metric relations of space. In connection with this question, which may indeed still be ranked as part of the study of Space, the above remark is applicable, that in a discrete manifold the principle of metric relations is already contained in the concept of the manifold, but in a continuous one it must come from something else. Therefore, either the reality underlying Space must form a discrete manifold, or the basis for the metric relations must be sought outside it, in binding forces acting upon it.

An answer to these questions can be found only by starting from that conception of phenomena which has hitherto been approved by experience, for which Newton laid the foundation, and gradually modifying it under the compulsion of facts which cannot be explained by it. Investigations like the one just made, which begin from general concepts, can serve only to insure that this work is not hindered by too restricted concepts, and that progress in comprehending the connection of things is not obstructed by traditional prejudices.

This leads us away into the domain of another science, the realm of physics, into which the nature of the present occasion does not allow us to enter.

Chapter 4B. What did Riemann say?

Upon a first reading, Riemann's lecture may appear to have almost no mathematical content. But this is only because the analytic investigations, which occur in Part II, have been drastically condensed, while Part I explains, in general philosophical terms, important mathematical concepts which succeeding generations of investigators were eventually able to express with mathematical preciseness; finally, Part III of the lecture deals with applications of the mathematical discoveries to questions in physics, a process which is perhaps not yet complete.

In this commentary on Riemann's lecture, we will follow closely the order of Riemann's exposition, referring often to the various sections (1,2, etc.) within each part (I,II,III). It should not be expected that all details will be cleared up, even in the remaining portions of this Chapter, for the complete consideration of Riemann's ideas will occupy several of the succeeding Chapters. Consequently, the remaining parts of Chapter 4 may be the hardest reading encountered in either of the two volumes of these notes. Nevertheless, we hope that in the end a clear view of all these ideas will be obtained.

In the "Plan of the Investigation", Riemann begins by accounting for the confusion over the status of non-Euclidean geometry, which at this time was still not completely accepted. In 1829, Lobachevsky and Bolyai had independently constructed a system of geometry which began by assuming that through a point not on a line there was more than one line parallel to it (as opposed to the assumption that there is only one parallel line, which is equivalent to Euclid's Fifth Postulate); but it was still supposed by some that contradictions in this system would eventually be found. Riemann attributes the difficulties encountered in the study of non-Euclidean geometry to the fact that geometers had never separated what we would call the topological properties of space from its metric properties; in the axiomatic development of geometry, even the notion of space itself is undefined, and its properties are developed through the axioms . Riemann proposes to distinguish the metric properties from the topological properties, and promises that we will discover how different metric structures can be put on the triply extended quantity which constitutes Space, so that one cannot possibly expect to deduce the parallel postulate of Euclid from topological considerations alone. This implies that experimental data must be used to determine what metric properties Space actually has, and raises the question which data we should seek, and what we can expect to say about the regions of space too distant, or too small, to be investigated experimentally.

In Part I, "Concept of an n-fold extended quantity", Riemann is clearly trying to define a manifold. It is impossible to tell from this lecture, intended for non-mathematicians, how far Riemann had advanced toward the precise solution of this problem, and whether he had any way of expressing concretely the notion of a metric or topological space, which is essentially prerequisite to the definition of a manifold. However, it is quite obvious that the notion was thoroughly clear in his own mind and that he recognized that manifolds were characterized by the fact that they are locally like n-dimensional Euclidean space. It is also clear that he understood the importance of infinite dimensional spaces, such as the set of all real-valued functions on a space (it is interesting that quite recently some of these infinite dimensional spaces have been given the structure of "infinite dimensional manifolds", and differential geometric methods have been applied to them with great success).

Part II contains nearly all the mathematical results, and the discussion of this Part will take up most of the present Chapter. The difficulties in

Part II begin right with the title, "Metric relations of which a manifold of n dimensions is susceptible, on the assumption that lines have a length independent of their configuration, so that every line can be measured by every other." To understand what Riemann means, it is necessary to recall the process by which lengths are assigned to curves in the plane or 3-dimensional space of analytic geometry. In this case, we begin with the notion of distance between pairs of points, which amounts to saying that we first assign a length to straight lines; the length of other lines is then defined as the least upper bound of inscribed curves made up of straight lines, a process which can be reduced to integration. In this method of assigning lengths to curves, it may be said that all curves are measured by means of straight lines. By contrast, Riemann proposes to consider a uniform method of assigning lengths to all curves in a manifold, a method which does not depend on first distinguishing a particular class of curves. This is to be done by measuring the lengths of tangent vectors, so that the lengths of curves can be defined by an integral (a restriction to C^1 curves is first indicated). Riemann assumes that this "length" function is continuous on each tangent space and also positive homogeneous — the length of λv is $|\lambda|$ times the length of v. Now there are many kinds of positive homogeneous functions on a finite dimensional vector space; any subset of the vector space which is symmetric with respect to the origin, and intersects each ray through 0 just once, can be used as the set of vectors of length 1. By a process that I do not at all under-

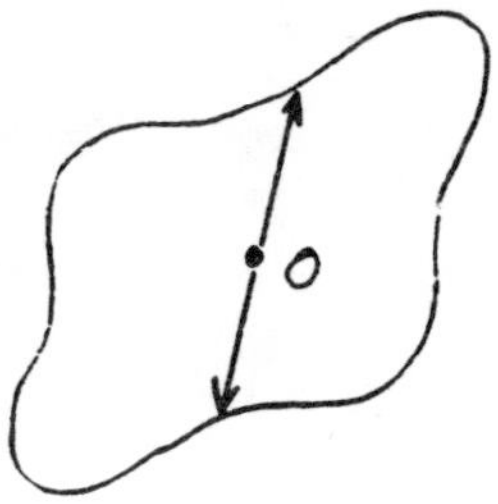

stand, from among all these length functions Riemann selects as the simplest, those which can be expressed as $\sqrt{\Sigma g_{ij}(p)dx^i \cdot dx^j}$ for certain numbers $g_{ij}(p)$. An assignment to each tangent space M_p of such a norm, or more precisely the inner product from which it comes, is, of course, what we now call a Riemannian metric on the manifold M. Riemann points out that it is merely to save time, and to allow geometric descriptions of the results, that he restricts his attention to the special case. Certain more general cases, though not the most general of all, was investigated by Finsler in his thesis (1918), and are now known as Finsler metrics; it seems clear, however that Riemann must have already known the basic facts about these more general metrics (some information on Finsler metrics is given in the Addendum). Having restricted his attention to "Riemannian manifolds", Riemann now asks the crucial question: when does the introduction of a new coordinate system change the metric $\Sigma g_{ij} dy^i \otimes dy^j$ into some given metric $\Sigma a_{ij} dx^i \otimes dx^j$; in other words, when are two Riemannian manifolds locally isometric? Riemann here presents one of his famous "counting arguments", which enabled him to guess results that in some cases were not rigorously proved until a hundred years later. Riemann argues that the expression $\Sigma g_{ij} dx^i \otimes dx^j$ contains $n\frac{n+1}{2}$ functions (not n^2, for $g_{ij} = g_{ji}$) while a new coordinate system involves only n functions, so that we can change only n of the g_{ij}, leaving $n\frac{n-1}{2}$ other functions which depend on the metric; consequently, Riemann argues, there should be some set of $n\frac{n-1}{2}$ functions which will determine the metric completely.

In section 2 of Part II, Riemann indicates how such functions are to be found. We are going to apply a standard technique for the study of differentiable functions — we examine the Taylor polynomials approximating the functions g_{ij}. If x is a coordinate system on M, with $x(p) = 0$, and the Riemannian metric is given by $\langle\ ,\ \rangle = \Sigma g_{ij} dx^i \otimes dx^j$, then for the Taylor expansion of

$g_{ij} \circ x^{-1} \colon \mathbb{R}^n \longrightarrow \mathbb{R}$ we have

$$g_{ij} \circ x^{-1}(t) = (g_{ij} \circ x^{-1})(0) + \sum_{k=1}^{n} D_k(g_{ij} \circ x^{-1})(0)t^k$$

$$+ \frac{1}{2} \sum_{k,\ell=1}^{n} D_{k,\ell}(g_{ij} \circ x^{-1})(0)t^k t^\ell + o(|t|^2) \quad .$$

Hence on M we have

$$(*) \qquad g_{ij} = g_{ij}(p) + \sum_{k=1}^{n} \frac{\partial g_{ij}}{\partial x^k}(p)x^k + \frac{1}{2} \sum_{k,\ell=1}^{n} \frac{\partial g_{ij}}{\partial x^k \partial x^\ell}(p)x^k x^\ell$$

$$+ o(|x|^2) \quad ,$$

where $o(|x|^2)$ denotes a function f on M such that

$$\lim_{q \to p} \frac{f(q)}{|x(q)|^2} = 0 \quad .$$

However, and this is the important device Riemann introduces, we will select a very special coordinate system around each point $p \in M$. We choose an orthonormal basis $X_1, \ldots, X_n \in M_p$, and define a coordinate system $\chi \colon M_p \longrightarrow \mathbb{R}^n$ on M_p by $\chi(\Sigma a^i X_i) = (a^1, \ldots, a^n)$. Then we let x be the coordinate system $\chi \circ \exp^{-1}$. (This coordinate system is introduced at the very beginning of section 2, but it takes a little work to decipher Riemann's description of it.) The coordinate system x is not uniquely determined, for it depends on the choice of the orthonormal basis $X_1, \ldots, X_n \in M_p$; but any two differ by an element $O(n)$, so it will not be hard to take into account the way any of our results depend on this choice. These coordinate systems are called <u>Riemannian normal coordinates</u> at p. Notice that since $\exp_* \colon (M_p)_0 \longrightarrow M_p$ is the identity (upon identifying

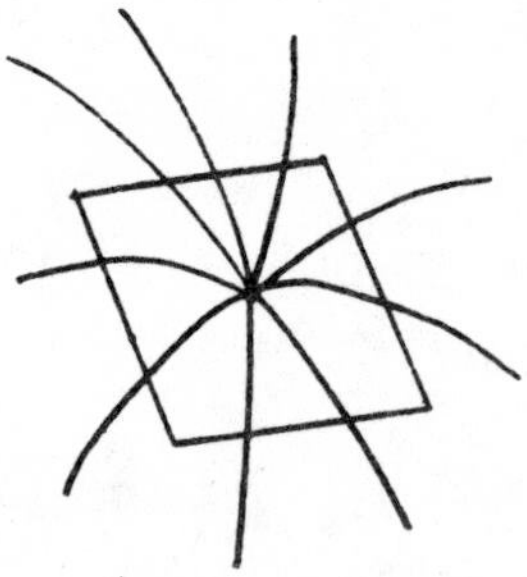

$(M_p)_0$ with M_p), we have

$$\left.\frac{\partial}{\partial x^i}\right|_p = \exp_* X_i = X_i \in M_p \quad .$$

We can quickly give some information about the first 2 terms in the expansion (*) of g_{ij}:

<u>1. PROPOSITION</u>. In a Riemannian normal coordinate system x at p we have

$$g_{ij}(p) = \delta_{ij}$$

$$\frac{\partial g_{ij}}{\partial x^k}(p) = 0 \quad .$$

<u>Proof</u>. The first set of equations is clear, for

$$g_{ij}(p) = \left\langle \left.\frac{\partial}{\partial x^i}\right|_p , \left.\frac{\partial}{\partial x^j}\right|_p \right\rangle = \langle X_i, X_j \rangle = \delta_{ij} \quad .$$

To prove the second set of equations, we recall the equations for a geodesic γ:

$$\frac{d^2\gamma^k}{dt^2} + \sum_{i,j=1}^{n} \Gamma^k_{ij}(\gamma(t)) \frac{d\gamma^i}{dt} \frac{d\gamma^j}{dt} = 0 \quad .$$

In Riemannian normal coordinates the geodesics through p are just $\exp \circ c$, where c is a straight line in M_p. This means that for all n-tuples $(\xi^1,\ldots,\xi^n)$, the geodesics through p are the curves γ with $\gamma^k(t) = \xi^k t$. Hence

$$\sum_{i,j=1}^{n} \Gamma^k_{ij}(\gamma(t))\xi^i\xi^j = 0 \qquad \text{for the geodesic } \gamma^k(t) = \xi^k t \quad .$$

In particular, since $p = \gamma(0)$ is on all these geodesics, we have

$$\sum_{i,j=1}^{n} \Gamma^k_{ij}(p)\xi^i\xi^j = 0 \qquad \text{for } \underline{\text{all}} \text{ n-tuples } (\xi^1,\ldots,\xi^n) \quad .$$

This shows that all $\Gamma^k_{ij}(p)$ are 0 (choosing all $\xi^\alpha = 0$ except $\xi^i = 1$ gives $\Gamma^k_{ii} = 0$; then choosing all $\xi^\alpha = 0$ except $\xi^i = \xi^j = 1$ gives

$$0 = \Gamma^k_{ij}(p) + \Gamma^k_{ji}(p) + \Gamma^k_{ii}(p) + \Gamma^k_{jj}(p) = 2\Gamma^k_{ij}(p).)$$

It follows that

$$[ij,k] = \sum_{\alpha=1}^{n} g_{\alpha \ell}\Gamma^{\ell}_{ij} = 0 \qquad .$$

Making use of the equation on page I.9-41, we have finally

$$\frac{\partial g_{ij}}{\partial x^k} = [ik,j] + [jk,i] = 0 \qquad . \blacksquare$$

In view of Proposition 1, we can now use (*) to expand the squared norm $\| \ \|^2$ as

$$\| \ \|^2 = \sum_{i,j=1}^{n} g_{ij} dx^i dx^j$$

$$= \sum_{i=1}^{n} dx^i dx^i + \frac{1}{2} \sum_{i,j;k,\ell} \frac{\partial g_{ij}}{\partial x^k \partial x^\ell}(p) x^k x^\ell dx^i dx^j + o(|x|^2) \ .$$

(This is an equation for tangent vectors near p, and $o(|x|^2)$ now denotes a function f on tangent vectors; in order to have

$$\lim_{q \to p} \frac{f(v_q)}{|x|^2} = 0 \quad ,$$

we must rectrict v_q to be of some bounded length.) Riemann's main assertion involves the term

$$\frac{1}{2} \sum_{i,j;k,\ell} \frac{\partial g_{ij}}{\partial x^k \partial x^\ell}(p) x^k x^\ell dx^i dx^j = \sum_{i,j;k,\ell} c_{ij,k\ell} x^k x^\ell dx^i dx^j, \quad \text{say.}$$

Riemann asserts that there are numbers $C_{ij,k\ell}$ such that we can write

$$\sum_{i,j;k,\ell} c_{ij,k\ell} x^k x^\ell dx^i dx^j = \sum_{i,j;k,\ell} C_{ij,k\ell} (x^k dx^i - x^i dx^k) \cdot (x^\ell dx^j - x^j dx^\ell) \ .$$

This assertion immediately suggests three questions — Why did Riemann suspect this was true? How did he prove it? What is its significance?

We will begin by giving a partial answer to the third of these questions. Notice that the equation in question is supposed to hold for all tangent vectors v at all points q in a neighborhood of p. Consequently, the numbers $dx^i(v)$ [and $x^i(q)$] can take on all [sufficiently small] values. The coordinate system x and the Riemannian metric $\langle \ , \ \rangle$ are used to obtain the n^4 numbers $c_{ij,k\ell} = \frac{1}{2} \partial g_{ij} / \partial x^k \partial x^\ell (p)$; but beyond this, the above equation has nothing to do with the manifold at all. If we define a quadratic polynomial Q in $2n$ variables by

$$Q(X,Y) = Q(X_1,\ldots,X_n,Y_1,\ldots,Y_n) = \sum_{ij;k\ell} c_{ij,k\ell} X_i X_j Y_k Y_\ell \quad ,$$

then Riemann is asserting that this quadratic polynomial can be written as

$$Q(X,Y) = \sum_{i,j;k,\ell} c_{ij,k\ell}(X_i Y_k - X_k Y_i)\cdot(X_j Y_\ell - X_\ell Y_j) \quad .$$

To obtain the geometric consequences of this fact, we observe what it says when we select two vectors $v_p, w_p \in M_p$ and let $X_i = dx^i(v_p)$ and $Y_i = dx^i(w_p)$; denoting $Q(X,Y)$ by $Q(v_p,w_p)$, we have

$$Q(v_p,w_p) = \sum_{i,j;k,\ell} c_{ij,k\ell} dx^i(v_p) dx^j(v_p)\cdot dx^k(w_p) dx^\ell(w_p)$$

$$= \sum_{i,j;k,\ell} c_{ij,k\ell} [(dx^i \wedge dx^k)(v_p,w_p)]\cdot[(dx^j \wedge dx^\ell)(v_p,w_p)] \quad .$$

(We can also write simply

$$Q = \sum_{i,j;k,\ell} c_{ij,k\ell} dx^i dx^j \otimes dx^k dx^\ell = \sum_{i,j;k,\ell} c_{ij,k\ell}(dx^i \wedge dx^k)\cdot(dx^j \wedge dx^\ell) \quad .)$$

Now suppose $v'_p, w'_p \in M_p$ span the same subspace as v_p, w_p, so that we can write

$$\begin{aligned} v'_p &= a_{11} v_p + a_{21} w_p \\ w'_p &= a_{12} v_p + a_{22} w_p \end{aligned} \qquad \det(a_{ij}) \neq 0 \quad .$$

The right side of the above equation for $Q(v_p,w_p)$ shows that

$$Q(v'_p,w'_p) = [\det(a_{ij})]^2 \cdot Q(v_p,w_p) \quad ,$$

since each $dx^{\alpha} \wedge dx^{\beta}$ is multiplied by the factor $\det(a_{ij})$. If we use $\|v_p, w_p\|$ to denote the area of the parallelogram spanned by v_p and w_p, then we also have

$$\|v'_p, w'_p\|^2 = [\det(a_{ij})]^2 \cdot \|v_p, w_p\|^2 \quad .$$

Consequently,

$$\frac{Q(v'_p, w'_p)}{\|v'_p, w'_p\|^2} = \frac{Q(v_p, w_p)}{\|v_p, w_p\|^2} \quad .$$

We therefore have a way of assigning a number to every 2-dimensional subspace of the tangent space at p. (Riemann sticks to the original quadratic function of the x^i and dx^i, which puts him in the position of having to divide by the squared area of a very strange triangle, with one vertex at x^i, and one at dx^i.)

It is easy to see that if we pick a different Riemannian normal coordinate system at p, then the resulting function on the 2-dimensional subspaces of M_p will be the same, for $Q(v_p, w_p) = Q(dx^1(v_p), \ldots, dx^n(v_p), dx^1(w_p), \ldots, dx^n(w_p))$ will change by $(\det B)^2$, where $B \in O(n)$, so that $\det B = \pm 1$. We will examine later the significance of this new function on 2-dimensional subspaces of the tangent space. For the present we take up the other questions — Why did Riemann think it was true, and how did he prove it?

Of course, an answer to the first question is not only doomed to be mere conjecture, but is always foolhardy to put forth, for there is no accounting for genius. The best suggestion I can offer is that the dependence of $Q(v_p, w_p)$ on the span of v_p and w_p alone is certainly an attractive one, and as we shall see later, in one special case which Riemann may have investigated first, the result appears in a rather natural way. It is also impossible to say for sure how Riemann proved the result, for his own investigations were never published. I have used the remarks by H. Weber in Riemann's collected works (pp. 405-409), as well as the commentary given by Hermann Weyl in a special edition of Riemann's lecture. There are two parts to the proof, a purely algebraic one about quadratic functions, which determines what relations the numbers $c_{ij,k\ell}$ ought to satisfy, and an analytic one which establishes these relations.

For the algebraic part, we will be considering a quadratic function Q of 2n variables

$$Q(X,Y) = Q(X_1,\ldots,X_n,Y_1,\ldots,Y_n) = \sum_{i,j;k,\ell} c_{ij,k\ell} X_i X_j Y_k Y_\ell \quad .$$

Note that for our Q we have

$$c_{ij,k\ell} = c_{ji,k\ell} = c_{ij,\ell k} \quad ,$$

(using $g_{ij} = g_{ji}$, and $\partial^2/\partial x^k \partial x^\ell = \partial^2/\partial x^\ell \partial x^k$). If $A = (a_{ij})$ is a 2×2 matrix, we will use $A(X,Y)$ to denote the 2n tuple

$$\begin{aligned} A(X,Y) &= (a_{11}X + a_{21}Y, a_{12}X + a_{22}Y) \\ &= (a_{11}X_1 + a_{21}Y_1,\ldots,a_{11}X_n + a_{21}Y_n, a_{12}X_1 + a_{22}Y_1,\ldots,a_{12}X_n + a_{22}Y_n) \quad . \end{aligned}$$

2. PROPOSITION. Let Q be a quadratic function of 2n variables,

$$Q(X,Y) = \sum_{i,j;k,\ell} c_{ij,k\ell} X_i X_j Y_k Y_\ell \quad ,$$

where

(1) $$c_{ij,k\ell} = c_{ji,k\ell} = c_{ij,\ell k} \quad .$$

Then

$$Q(A(X,Y)) = (\det A)^2 Q(X,Y)$$

for all 2 x 2 matrices A if and only if:

(2) $$c_{ij,k\ell} = c_{k\ell,ij}$$

(3) $$c_{\ell i,jk} + c_{\ell j,ki} + c_{\ell k,ij} = 0 \quad .$$

Proof. It is clear, first of all, that the equation $Q(A(X,Y)) = (\det A)^2 Q(X,Y)$ holds for all 2 x 2 matrices A if and only if it holds for the non-singular ones, since both sides of the equation are continuous functions of A, and the non-singular matrices are dense.

Now it is well known that all non-singular 2 x 2 matrices can be written as a product of the matrices

$$\begin{pmatrix} a & 0 \\ 0 & 1 \end{pmatrix}, \begin{pmatrix} 1 & 0 \\ 0 & a \end{pmatrix}, \begin{pmatrix} 0 & 1 \\ 1 & 0 \end{pmatrix}, \begin{pmatrix} 1 & 0 \\ 1 & 1 \end{pmatrix}, \begin{pmatrix} 1 & 1 \\ 0 & 1 \end{pmatrix} .$$

[This comes from the fact that any non-singular matrix can be obtained from the identity by a sequence of elementary row operations, and every row operation may be accomplished by multiplying by one of the above matrices.] So our condition holds for all A if and only if it holds for the above matrices. We can disregard the last matrix, since

$$\begin{pmatrix}1&1\\0&1\end{pmatrix}=\begin{pmatrix}0&1\\1&0\end{pmatrix}\begin{pmatrix}1&0\\1&1\end{pmatrix}\begin{pmatrix}0&-1\\-1&0\end{pmatrix}\left[-\begin{pmatrix}0&1\\1&0\end{pmatrix}\begin{pmatrix}1&0\\1&1\end{pmatrix}\begin{pmatrix}0&1\\1&0\end{pmatrix}\right]^{-1}.$$

For the matrix $A=\begin{pmatrix}a&0\\0&1\end{pmatrix}$, the condition $Q(A(X,Y)) = (\det A)^2 Q(X,Y)$ becomes simply

$$Q(aX,Y) = a^2 Q(X,Y) \qquad ,$$

which is automatically true. The same result holds for the second matrix on our list, so all the conditions finally come down to

$$\text{(a)} \qquad Q(Y,X) = Q(X,Y) \qquad A=\begin{pmatrix}0&1\\1&0\end{pmatrix}$$

$$\text{(b)} \qquad Q(X+Y,Y) = Q(X,Y) \qquad A=\begin{pmatrix}1&0\\1&1\end{pmatrix}.$$

Now equation (a) becomes

$$\sum_{i,j;k,\ell} c_{ij,k\ell} X_i X_j Y_k Y_\ell = \sum_{i,j;k,\ell} c_{ij,k\ell} Y_i Y_j X_k X_\ell \qquad .$$

Since this must be a polynomial identity, we obtain (2) immediately, by looking at the coefficient of $X_i X_j Y_k Y_\ell$ on both sides.

Equation (b) becomes

$$\sum_{i,j;k,\ell} c_{ij,k\ell}(X_i + Y_i)(X_j + Y_j)Y_k Y_\ell = \sum_{i,j;k,\ell} c_{ij,k\ell} X_i X_j Y_k Y_\ell \quad ,$$

or

$$\sum_{i,j;k,\ell} c_{ij,k\ell}[X_i Y_j Y_k Y_\ell + X_j Y_i Y_k Y_\ell + Y_i Y_j Y_k Y_\ell] = 0 \qquad .$$

Letting $X = 0$, we obtain

$$\text{(b1)} \qquad \sum_{i,j;k,\ell} c_{ij,k\ell} Y_i Y_j Y_k Y_\ell = 0$$

and then, in consequence,

$$\text{(b2)} \qquad \sum_{i,j;k,\ell} c_{ij,k\ell}[X_iY_jY_kY_\ell + X_jY_iY_kY_\ell] = 0 \ .$$

On the other hand, (b2) implies (b1), so (b) is equivalent to (b2) alone. Finally, since $c_{ij,k\ell} = c_{ji,k\ell}$, equation (b2) is equivalent to

$$\text{(b3)} \qquad \sum_{i,j;k,\ell} c_{ij,k\ell} X_iY_jY_kY_\ell = 0 \ .$$

Looking at the coefficient of a particular $X_iY_jY_kY_\ell$ we obtain

$$c_{ij,k\ell} + c_{ij,\ell k} + c_{ik,j\ell} + c_{ik,\ell j} + c_{i\ell,jk} + c_{i\ell,kj} = 0 \ .$$

Using the symmetry with respect to the last two indices, this is equivalent to equation (3). ∎

3. PROPOSITION. A quadratic function

$$Q(X,Y) = \sum_{i,j;k,\ell} c_{ij,k\ell} X_iX_jY_kY_\ell \ ,$$

with

$$\text{(1)} \qquad c_{ij,k\ell} = c_{ji,k\ell} = c_{ij,\ell k}$$

satisfies the two equivalent conditions of Proposition 2 if and only if it can be written as

$$Q(X,Y) = \sum_{i,j;k,\ell} c_{ij,k\ell}(X_iY_k - X_kY_i)\cdot(X_jY_\ell - X_\ell Y_j) \ .$$

Proof. If Q can be written this way, then clearly $Q(A(X,Y)) = (\det A)^2 Q(X,Y)$ for all A. Conversely, suppose this holds for all A, so that we also have

$$(2) \qquad c_{ij,k\ell} = c_{k\ell,ij}$$

$$(3) \qquad c_{\ell i,jk} + c_{\ell j,ki} + c_{\ell k,ij} = 0 \quad .$$

We begin by writing 4 equivalent expressions for Q:

$$\begin{aligned} Q(X,Y) &= \sum c_{ij,k\ell} X_i X_j Y_k Y_\ell \\ &= \sum c_{jk,i\ell} X_j X_k Y_i Y_\ell \\ &= \sum c_{i\ell,jk} X_i X_\ell Y_j Y_k \\ &= \sum c_{k\ell,ij} X_k X_\ell Y_i Y_j \quad . \end{aligned}$$

Now, by (3) we have

$$c_{jk,i\ell} = - c_{ji,\ell k} - c_{j\ell,ki} \quad ,$$

so

$$\sum_{i,j,k,\ell} c_{jk,i\ell} X_j X_k Y_i Y_\ell =$$

$$= - \sum_{i,j,k,\ell} c_{ji,\ell k} X_j X_k Y_i Y_\ell - \sum_{i,j,k,\ell} c_{j\ell,ki} X_j X_k Y_i Y_\ell$$

$$= - \sum_{i,j,k,\ell} c_{ji,\ell k} X_j X_k Y_i Y_\ell - \sum_{i,j,k,\ell} c_{k\ell,ji} X_j X_k Y_i Y_\ell$$

(interchanging j and k in the second sum)

$$= - 2 \sum_{i,j,k,\ell} c_{ij,k\ell} X_j X_k Y_i Y_\ell \quad , \quad \text{using (1) and (2)} \quad .$$

If we apply a similar process to the third expression for Q, use (2) on the fourth, and leave the first unaltered, we obtain

$$\begin{aligned} Q(X,Y) &= \Sigma\, c_{ij,k\ell} X_i X_j Y_k Y_\ell \\ \tfrac{1}{2} Q(X,Y) &= -\, \Sigma\, c_{ij,k\ell} X_j X_k Y_i Y_\ell \\ \tfrac{1}{2} Q(X,Y) &= -\, \Sigma\, c_{ij,k\ell} X_i X_\ell Y_j Y_k \\ Q(X,Y) &= \Sigma\, c_{ij,k\ell} X_k X_\ell Y_i Y_j \;. \end{aligned}$$

Adding, we obtain the desired result,

$$3Q(X,Y) = \sum_{i,j;k,\ell} c_{ij,k\ell}(X_i Y_k - X_k Y_i)(X_j Y_\ell - X_\ell Y_j) \;. \quad \blacksquare$$

[We note in passing that these results can easily be generalized to quadratic functions in mn variables,

$$Q(X^{(1)},\ldots,X^{(m)}) = \sum_{i_\alpha, j_\alpha = 1}^{n} c_{i_1 j_1,\ldots,i_m j_m} X^{(1)}_{i_1} X^{(1)}_{j_1} \cdots X^{(m)}_{i_m} X^{(m)}_{j_m}$$

for which the c's are symmetric in each pair of indices. In order to have

$$Q(A(X^{(1)},\ldots,X^{(m)})) = (\det A)^m Q(X^{(1)},\ldots,X^{(m)})$$

for all $m \times m$ matrices A, it is necessary and sufficient for this to be true for

(1)
$$A = \begin{bmatrix} 1 & & & & \bigcirc \\ & \ddots & & & \\ & & a & & \\ & & & \ddots & \\ \bigcirc & & & & 1 \end{bmatrix},$$

(2) A any permutation matrix,

and

$$(3) \qquad A = \begin{pmatrix} 1 & 0 & 0 \\ 1 & 1 & 0 \\ & 0 & \ddots & \\ & & & 1 \end{pmatrix} \quad ;$$

all other matrices with just one 1 off the diagonal are conjugates of this one by some permutation matrix. From this we easily derive the necessary and sufficient conditions

$$c_{i_1j_1,\ldots,i_\alpha j_\alpha,\ldots,i_\beta j_\beta,\ldots,i_m j_m} = c_{i_1j_1,\ldots,i_\beta j_\beta,\ldots,i_\alpha j_\alpha,\ldots,i_m j_m}$$

$$c_{\ell i,jk,i_3j_3,\ldots,i_mj_m} + c_{\ell j,ki,i_3j_3,\ldots,i_mj_m} + c_{\ell k,ij,i_3j_3,\ldots,i_mj_m} = 0 \quad .$$

Using these equations, it can then be shown that Q can be written as a quadratic function in the $X_i^{(\alpha)}Y_k^{(\beta)} - X_i^{(\beta)}Y_k^{(\alpha)}$. There surely ought to be a more elegant way to state and prove these results, but I have not been able to find out what it is.]

We now proceed with the hardest part of the investigation, a hairy calculation indeed.

4. PROPOSITION. In a Riemannian normal coordinate system x at p, the numbers

$$c_{ij,k\ell} = \frac{1}{2}\frac{\partial g_{ij}}{\partial x^k \partial x^\ell}(p)$$

satisfy

$$c_{ij,k\ell} = c_{k\ell,ij}$$

$$c_{\ell i,jk} + c_{\ell j,ki} + c_{\ell k,ij} = 0 \quad .$$

Proof. We begin with an equation derived in the proof of Proposition 1. For the geodesic $\gamma^k(t) = \xi^k t$ we have

$$\sum_{i,j=1}^{n} \Gamma_{ij}^k(\gamma(t))\xi^i\xi^j = 0 \quad ;$$

multiplying by t^2, we have

$$\sum_{i,j=1}^{n} \Gamma_{ij}^k(\gamma(t))x^i(\gamma(t))x^j(\gamma(t)) = 0 \quad .$$

Since these geodesics go through all points in a neighborhood of p, we have the following relation between the functions Γ_{ij}^k and x^i:

$$(1) \qquad \sum_{i,j=1}^{n} \Gamma_{ij}^k x^i x^j = 0 \quad .$$

Since the tangent vector to the geodesic $\gamma^k(t) = \xi^k t$ has constant length, we also obtain

$$\left\langle \frac{d\gamma}{dt}, \frac{d\gamma}{dt} \right\rangle = \sum_{i=1}^{n} (\xi^i)^2 \quad ,$$

which leads, in the same way, to the equation

$$(2) \qquad \sum_{i,j=1}^{n} g_{ij}x^i x^j = \sum_{i=1}^{n} (x^i)^2 \quad .$$

Now equation (1) leads to

$$\sum_{i,j=1}^{n} [ij,k]x^i x^j = 0 \quad ,$$

i.e., to

$$\sum_{i,j=1}^{n} \frac{1}{2}\left(\frac{\partial g_{ik}}{\partial x^j} + \frac{\partial g_{jk}}{\partial x^i} - \frac{\partial g_{ij}}{\partial x^k}\right) x^i x^j = 0 \quad .$$

Interchanging the indices i and j in the second term, we can write

$$(3) \qquad \sum_{i,j=1}^{n} \left(\frac{\partial g_{ik}}{\partial x^j} - \frac{1}{2}\frac{\partial g_{ij}}{\partial x^k}\right) x^i x^j = 0 \quad .$$

Our penultimate goal is to break this equation up into two sums, each of which is individually 0; the conditions on the c's, which are our ultimate goal, will then follow fairly easily. To achieve this, our antepenultimate goal is to prove that $x^\beta = \sum_\alpha g_{\beta\alpha} x^\alpha$; these equations are at least reasonable, for they imply (2). To prove these relations, we first introduce the functions $\overline{x}^\beta$ defined by

$$\overline{x}^\beta = \sum_{\alpha=1}^{n} g_{\beta\alpha} x^\alpha \quad .$$

Note that

$$\frac{\partial \overline{x}^\beta}{\partial x^\gamma} = \sum_{\alpha=1}^{n} \frac{\partial g_{\beta\alpha}}{\partial x^\gamma} x^\alpha + g_{\beta\gamma} \quad .$$

Substituting in (3), we obtain

$$0 = \sum_{j=1}^{n}\left(\sum_{i=1}^{n} \frac{\partial g_{ik}}{\partial x^j} x^i\right)x^j - \frac{1}{2}\sum_{i=1}^{n}\left(\sum_{j=1}^{n} \frac{\partial g_{ij}}{\partial x^k} x^j\right)x^i$$

$$= \sum_{j=1}^{n}\left(\frac{\partial \bar{x}^k}{\partial x^j} - g_{kj}\right)x^j - \frac{1}{2}\sum_{i=1}^{n}\left(\frac{\partial \bar{x}^i}{\partial x^k} - g_{ik}\right)x^i$$

$$= \sum_{j=1}^{n} \frac{\partial \bar{x}^k}{\partial x^j} x^j - \bar{x}^k - \frac{1}{2}\left(\sum_{i=1}^{n} \frac{\partial \bar{x}^i}{\partial x^k} x^i - \bar{x}^k\right)$$

$$= \sum_{j=1}^{n} \frac{\partial \bar{x}^k}{\partial x^j} x^j - \frac{1}{2}\left(\sum_{i=1}^{n} \frac{\partial \bar{x}^i}{\partial x^k} x^i + \bar{x}^k\right)$$

$$= \sum_{j=1}^{n} \frac{\partial \bar{x}^k}{\partial x^j} x^j - \frac{1}{2}\cdot \frac{\partial\left(\sum_{i=1}^{n} x^i\bar{x}^i\right)}{\partial x^k} .$$

Now by (2) and the definition of $\bar{x}^i$, we have

$$\sum_{i=1}^{n} x^i\bar{x}^i = \sum_{i=1}^{n} (x^i)^2 ,$$

so we obtain

$$0 = \sum_{j=1}^{n} \frac{\partial \bar{x}^k}{\partial x^j} x^j - x^k$$

$$= \sum_{j=1}^{n} \frac{\partial(\bar{x}^k - x^k)}{\partial x^j} x^j .$$

This equation shows that along any geodesic $\gamma(t) = \xi^i t$ we have

$$\frac{d[\bar{x}^k - x^k](\gamma(t))}{dt} = 0 ,$$

so that $\bar{x}^k - x^k$ is constant along the geodesic. Since $g_{ij}(p) = \delta_{ij}$, we clearly have $\bar{x}^k(p) = x^k(p)$. Moreover, these geodesics pass through all points in a neighborhood of p. Thus $\bar{x}^k = x^k$ in a neighborhood of p, so that we finally obtain the desired equations

$$(4) \qquad \sum_{\alpha=1}^{n} g_{k\alpha} x^\alpha = x^k .$$

Now we differentiate (4) to obtain

$$\sum_{\alpha=1}^{n} \frac{\partial g_{k\alpha}}{\partial x^\ell} x^\alpha + g_{k\ell} = \delta_{k\ell} ;$$

multiplying by x^ℓ and summing, we obtain

$$\sum_{\alpha,\ell=1}^{n} \frac{\partial g_{k\alpha}}{\partial x^\ell} x^\alpha x^\ell = \sum_{\ell=1}^{n} - g_{k\ell} x^\ell + \delta_{k\ell} x^\ell ,$$

which, together with (4) gives

$$\sum_{\alpha,\ell=1}^{n} \frac{\partial g_{k\alpha}}{\partial x^\ell} x^\alpha x^\ell = - x^k + x^k = 0 ,$$

and we have thus obtained the first part of our penultimate goal,

$$(5) \qquad \sum_{i,j=1}^{n} \frac{\partial g_{ik}}{\partial x^j} x^i x^j = 0 .$$

Together with (3), it implies the other part,

$$(6) \qquad \sum_{i,j=1}^{n} \frac{\partial g_{ij}}{\partial x^k} x^i x^j = 0 .$$

We now obtain the desired equations as follows. Along the geodesic $\gamma^k(t) = \xi^k t$ we have, by (6),

$$(7) \qquad \sum_{i,j=1}^{n} \frac{\partial g_{ij}}{\partial x^k}(\gamma(t))\xi^i\xi^j t^2 = 0 \quad .$$

This implies that

$$(8) \qquad \sum_{i,j=1}^{n} \frac{\partial g_{ij}}{\partial x^k}(\gamma(t))\xi^i\xi^j = 0$$

for $t \neq 0$, and hence even for $t = 0$, by continuity. Differentiating (7) with respect to t gives

$$0 = \sum_{i,j=1}^{n} \frac{\partial g_{ij}}{\partial x^k}(\gamma(t))\xi^i\xi^j \cdot 2t + \sum_{i,j,\ell=1}^{n} \frac{\partial g_{ij}}{\partial x^k \partial x^\ell}(\gamma(t))\xi^\ell\xi^i\xi^j t^2$$

$$= \sum_{i,j,\ell=1}^{n} \frac{\partial g_{ij}}{\partial x^k \partial x^\ell}(\gamma(t))\xi^\ell\xi^i\xi^j t^2 \ , \quad \text{by (8);}$$

consequently,

$$0 = \sum_{i,j,\ell=1}^{n} \frac{\partial g_{ij}}{\partial x^k \partial x^\ell}(\gamma(t))\xi^i\xi^j\xi^\ell$$

for all $t \neq 0$, and hence also for $t = 0$. Setting $t = 0$, we obtain

$$\sum_{i,j,\ell=1}^{n} \frac{\partial g_{ij}}{\partial x^k \partial x^\ell}(p)\xi^i\xi^j\xi^\ell = 0 \quad .$$

This equation holds for <u>all</u> n-tuples $\xi^1,\ldots,\xi^n$. From this we easily derive

$$(A) \qquad c_{ij,k\ell} + c_{i\ell,jk} + c_{j\ell,ik} = 0 \quad .$$

Applying the same process to (5), we obtain

$$\text{(B)} \qquad c_{ki,j\ell} + c_{kj,\ell i} + c_{k\ell,ij} = 0 \ .$$

In (B) we interchange k and ℓ, to obtain

$$c_{\ell i,jk} + c_{\ell j,ki} + c_{k\ell,ij} = 0 \ .$$

Comparing this equation with (A), we obtain the first of the desired relations,

$$c_{ij,k\ell} = c_{k\ell,ij} \ .$$

Moreover, using this relation with either (A) or (B), we obtain the second of the desired relations,

$$c_{\ell i,jk} + c_{\ell j,ki} + c_{\ell k,ij} = 0.$$

And thus we are done! ∎

When we put all these results together we see that the quadratic function

$$Q(v_p,w_p) = \frac{1}{2} \sum_{i,j;k,\ell} \frac{\partial^2 g_{ij}}{\partial x^k \partial x^\ell} dx^i(v_p)dx^j(v_p)dx^k(w_p)dx^\ell(w_p)$$

can be written

$$\boxed{Q(v_p,w_p) = \frac{1}{3} \sum_{i,j;k,\ell} c_{ij,k\ell}(dx^i \wedge dx^k)\cdot(dx^j \wedge dx^\ell)(v_p,w_p) \ .}$$

We thus see that the quadratic function Q, obtained from the Taylor expansion of $\| \ \|^2$ in Riemannian normal coordinates, has special properties which allow us to define, for any 2-dimensional subspace $W \subset M_p$, a number

$$Q(W) = \frac{Q(v_p, w_p)}{\|v_p, w_p\|^2} \qquad v_p, w_p \text{ any basis for } W.$$

The work of the last four Propositions, which establishes this fact, is completely suppressed in Riemann's account, where the final result is merely stated, at the beginning of section 2 of Part II. Riemann then makes some remarkable claims. First, Riemann interprets Q for a surface:

(1) If M is 2-dimensional and $W = M_p$, then $-3Q(W)$ is just the Gaussian curvature $k(p)$ given by Theorem 3B-7; we thus have an intrinsic definition of k, obtained by picking a special class of coordinate systems determined by the metric. (Riemann needs the factor $-3/4$ because he divides $Q(v_p, w_p)$ by the square of the area of the <u>triangle</u> spanned by v_p and w_p.)

At the end of section 3, Riemann interprets Q for an n-manifold:

(2) If M is n-dimensional, $W \subset M_p$ is a 2-dimensional subspace, and $\mathcal{O} \subset W$ is a neighborhood of $0 \in W$ on which exp is a diffeomorphism, then $-3Q(W)$ is the Gaussian curvature at p of the surface $\exp(\mathcal{O})$, with the metric it inherits as a submanifold of M.

But the most important claim is made in section 2. In an n-dimensional vector space there are $n\frac{n-1}{2}$ "independent" 2-dimensional subspaces: if $v_1, \ldots, v_n$ is a basis, we can choose the subspaces spanned by v_i and v_j, for $i < j$. Riemann claims that the metric $\langle\ ,\ \rangle$ is determined if $Q(W)$ is known for $n\frac{n-1}{2}$ independent 2-dimensional subspaces $W \subset M_q$ at each point q, for example, if Q is known for the subspaces spanned by each $\partial/\partial x^i|_q$ and $\partial/\partial x^j|_q$ $(i < j)$.

A very special case of this general claim is the following, which we will henceforth call the <u>Test Case</u>:

> (3) If M is n-dimensional and $Q = 0$ for $n\frac{n-1}{2}$ independent 2-dimensional subspaces of each M_q, then M is **flat**, that is, M is locally isometric to $\mathbb{R}^n$ with its usual inner product.

In connection with the Test Case, it should be pointed out that a local isometry with $\mathbb{R}^n$ is the best we can hope for, since there are Riemannian manifolds which are not homeomorphic to $\mathbb{R}^n$, but which are locally isometric to $\mathbb{R}^n$, and hence have $Q = 0$ everywhere. The simplest example of such a manifold, the "flat torus", is constructed as follows. The torus T can be obtained from $\mathbb{R}^2$ by identifying (x,y) with (x',y') if and only if

$$y' - y,\ x' - x \in \mathbb{Z}$$

(compare pg. I.10-2). The map $\pi\colon \mathbb{R}^2 \to T$, defined by taking (x,y) to its equivalence class, is locally a diffeomorphism, and there is clearly a unique metric $\langle\ ,\ \rangle$ on T such that $\pi^*\langle\ ,\ \rangle$ is the usual Riemannian metric on $\mathbb{R}^2$; consequently $(T, \langle\ ,\ \rangle)$ is locally isometric to $\mathbb{R}^2$ with its usual Riemannian metric. Notice that the usual torus in $\mathbb{R}^3$, with the induced Riemannian metric, is <u>not</u> flat; it has positive Gaussian curvature on the part

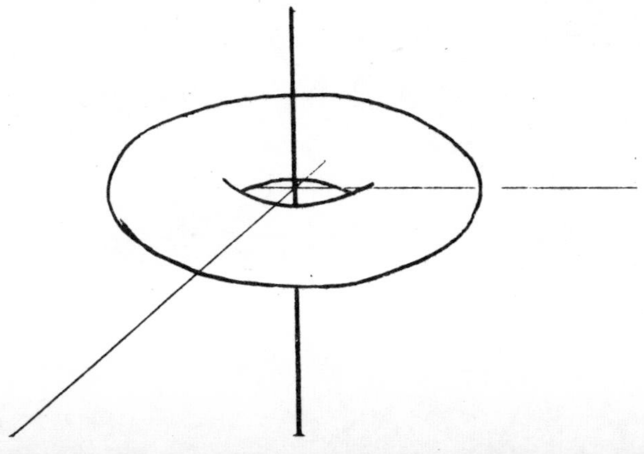

furthest from the axis, and negative Gaussian curvature on the part nearest the axis. However, if we consider $S^1 \subset \mathbb{R}^2$, then it is easy to see that $S^1 \times S^1 \subset \mathbb{R}^2 \times \mathbb{R}^2$, with the induced Riemannian metric, is flat.

One other remark should probably be made about the Test Case. At first sight, the Test Case might seem to be little more than a theorem about functions whose second partial derivatives are everywhere zero. However, it is actually quite different than this simple sort of result, since the value of Q at different points is defined in terms of different coordinate systems, each chosen specifically for one point.

Now our aim in the rest of this Chapter is to prove assertions (1), (2), and (3). (The general claim that Q determines the metric will be considered later, as will the information given in sections 4 and 5 of Part II). However, we will defer the proofs of assertions (1), (2), and (3) to another section of this Chapter, not only in order to provide ourselves with a brief respite, but also to allow Riemann to add one or two more brilliant ideas.

Addendum. Finsler Metrics

A Minkowski metric on a vector space V is a function $F: V \longrightarrow \mathbb{R}$ such that

$$F(v) > 0 \quad \text{for all} \quad v \neq 0$$

$$F(\lambda v) = |\lambda| F(v) \quad .$$

Clearly F is completely determined by its "unit sphere" $\{v: F(v) = 1\}$; the unit sphere is symmetric with respect to $0 \in V$, and intersects every ray through 0 exactly once. Moreover, any such set is clearly the unit sphere

for some F. The function F is never C^∞ at 0, but F^2 may be, in which case we simply say that F is C^∞. The general metric which Riemann mentions is essentially an assignment of a C^∞ Minkowski metric F_p to each tangent space M_p, in such a way that F_p varies smoothly with p. We will call such an assignment simply a "metric" on M.

If $c: [a,b] \longrightarrow M$ is a curve in a manifold M with such a metric, then we can define the length of c to be

$$\int_a^b F_{c(t)}(c'(t))\,dt \quad .$$

If $p: [a',b'] \to [a,b]$ is an increasing diffeomorphism, and we denote $F_{c(t)}(c'(t))$ by $g(t)$, then

$$\begin{aligned}\text{length of } c\circ p &= \int_{a'}^{b'} F_{c(p(t))}((c\circ p)'(t))dt \\ &= \int_{a'}^{b'} F_{c(p(t))}(p'(t)\cdot c'(p(t)))dt \\ &= \int_{p^{-1}(a)}^{p^{-1}(b)} p'(t)g(p(t))dt \\ &= \int_a^b g(t)dt = \text{length of } c.\end{aligned}$$

The same result clearly holds if p is decreasing; thus, the length of a curve is independent of parameterization. It is to insure this result that we require our metric to satisfy $F(\lambda v) = |\lambda|\cdot F(v)$.

Although a C^∞ metric F on a manifold M is not a tensor, it can be used to construct a tensor on the tangent bundle TM. To do this, we first consider a C^∞ function $f: V \to \mathbb{R}$. For any two vectors $v, w \in V$ we can form the second derivative

$$f_{**}(v)(w) = \left.\frac{d^2}{dt^2}\right|_{t=0} f(v + tw) \quad ;$$

this is a sort of second order directional derivative at v. If $v_1,\ldots,v_n$ is a basis for V, and $\varphi: \mathbb{R}^n \to \mathbb{R}$ is defined by

$$\varphi(a^1,\ldots,a^n) = f(\sum_{i=1}^n a^i v_i) \quad ,$$

then

$$f_{**}\left(\sum_{i=1}^{n} b^i v_i\right)\left(\sum_{i=1}^{n} c^i v_i\right) = \sum_{i,j} \frac{\partial^2 \varphi}{\partial x^i \partial x^j}(b)\cdot c^i c^j \quad .$$

When f is a Minkowski metric, it is clear from the definition that

$$(f^2)_{**}(v)(v) = 2[f(v)]^2 \quad .$$

The map $f_{**}(v)\colon V \longrightarrow \mathbb{R}$ is called the Hessian of f at $v \in V$. [It may be compared with the map f_{**}, also called the Hessian of f, which is defined in Problem I.5-17 for a function $f\colon M \longrightarrow \mathbb{R}$, at a point $p \in M$ where $f_{*p} = 0$. The latter is a bilinear function on M_p, whereas the present $f_{**}(v)$ is a quadratic function on V. The associated bilinear function is easily seen to be

$$(w_1, w_2) \longmapsto \left.\frac{\partial^2}{\partial s \partial t}\right|_{(s,t)=0} f(v + sw_1 + tw_2) \quad ,$$

and in terms of a basis it is given by

$$\left(\sum_{i=1}^{n} c^i v_i, \sum_{i=1}^{n} d^i v_i\right) \longmapsto \sum_{i,j} \frac{\partial^2 \varphi}{\partial x^i \partial x^j}(b)\cdot c^i d^j \quad ,$$

the same formula which occurs in Problem I.5-17. Our Hessian is defined even at points where $f_* \neq 0$ because we are working with a vector space, and identifying it with its tangent space at v; this amounts to saying that we are considering only linear changes of coordinates, all of which leave the quantity defined by this formula invariant.]

Now if M is a manifold, with a C^∞ metric F, and $v \in M_p$, then the tangent space $(TM)_v$ of TM at v can be identified with M_p; in accordance with this identification, we denote a vector in $(TM)_v$ by w_v. We can now define a tensor $\mathcal{F}$ on TM by

$$\mathcal{F}(w_v) = \tfrac{1}{2}(F_p{}^2)_{**}(v)(w) \quad .$$

If F is the norm $\| \ \|$ associated with a Riemannian metric $< , >$, then it is easy to see that

$$\mathcal{F}(w_v) = < v, w >_p \quad ,$$

and in general we always have

$$[F(v_p)]^2 = \mathcal{F}(v_v) \quad .$$

If x is a coordinate system on M, and $(x\circ\pi, \dot{x})$ is the corresponding coordinate system on TM (defined on page I.3-32), then

$$\mathcal{F} = \sum_{i,j=1} g_{ij}\, d\dot{x}^i \cdot d\dot{x}^j \quad ,$$

where

$$g_{ij}(v_p) = \frac{1}{2} \frac{\partial^2 (F_p{}^2)}{\partial \dot{x}^i \partial \dot{x}^j}(v_p) \quad .$$

Classically, one dealt only with the functions g_{ij}, defined by this formula, and checked that the function

$$\sum_{i=1}^{n} b^i \left.\frac{\partial}{\partial x^i}\right|_p \longmapsto \sum_{i,j=1}^{n} g_{ij} b^i b^j$$

was independent of the coordinate system x.

A Finsler metric is now defined to be a metric F such that the quadratic function $\mathcal{F}: (TM)_v \longrightarrow \mathbb{R}$ is positive definite for all $v \in TM$. In terms of a coordinate system x, this means that

$$(g_{ij}(v_p)) = \frac{1}{2}\left(\frac{\partial^2(F_p{}^2)}{\partial\dot{x}^i\partial\dot{x}^j}(v_p)\right)$$

is a positive definite matrix. To see what this really means, we consider once again a function $f: V \longrightarrow \mathbb{R}$. Suppose that $f_{**}(v)(w) > 0$ for all v, and all $w \neq 0$. This means that for all v, each function $\bar{f}$ obtained by restricting

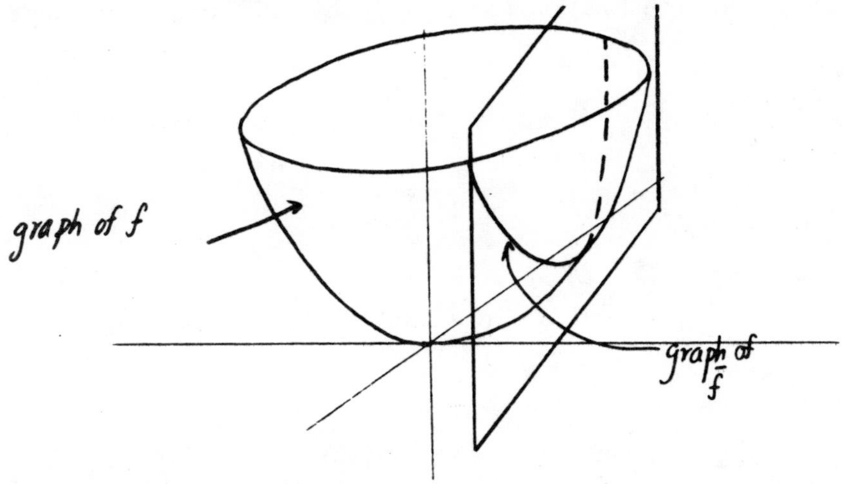

f to any line in V through v has an everywhere positive second derivative. By a standard theorem (Calculus, pg. 195) this means that the function $\bar{f}$ is convex (the set of points above or on its graph is a convex subset of the plane). It is easy to see that consequently the function f itself must be convex (the points on or above its graph is a convex subset of $V \times \mathbb{R}$).

When we apply this to a Finsler metric, we see that each function $F_p{}^2$ is convex on M_p. If $w_1, w_2 \in M_p$ are in the unit sphere of F_p, so that $F_p(w_1) = F_p(w_2) = 1$, then a point w on the line segment between w_1 and w_2 must have values of $F_p{}^2$ which are less than 1, and hence w lies inside the

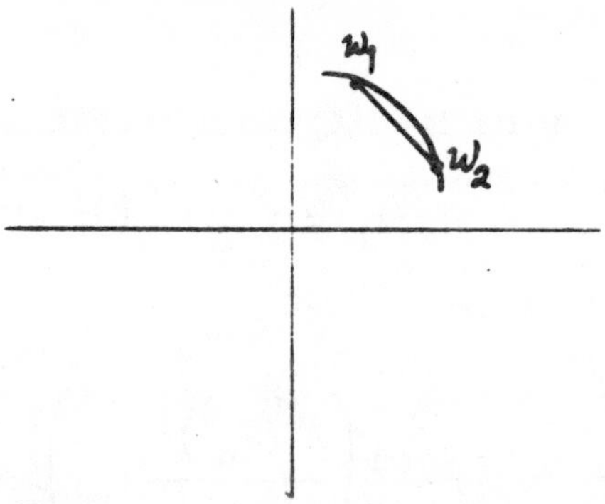

unit sphere. Thus the "unit ball" $\{w: F_p(w) \leq 1\}$ is convex. Conversely, if the unit ball is convex, then F_p is easily seen to be convex. Since $F_p > 0$, this implies that $F_p{}^2$ is convex (if $g > 0$ and $g'' > 0$, then clearly $(g^2)'' > 0$). Finally, we note that convexity of the unit ball is equivalent to the "triangle inequality"

$$F_p(w_1 + w_2) \leq F_p(w_1) + F_p(w_2) \quad .$$

In general, a function $\| \; \|: V \longrightarrow \mathbb{R}$ on a finite dimensional space V is called a <u>Banach space norm</u> if

$$\|v\| > 0 \quad \text{for} \quad v \neq 0$$
$$\|\lambda v\| = |\lambda| \cdot \|v\|$$
$$\|v + w\| \leq \|v\| + \|w\|$$

(if V is infinite dimensional, the definition is more involved). A Finsler metric on M is thus a C^∞ Banach space norm on each M_p, varying smoothly with p.

Although we will not develop the theory of Finsler, or more general, metrics here, we will mention a few facts. In the case of a Finsler metric, since the matrix $(g_{ij}(v_p))$ is non-degenerate, we can define $g^{ij}(v_p)$ so that

$$\sum_{j=1}^{n} g^{ij}(v_p)\cdot g_{jk}(v_p) = \delta^i_k \quad .$$

The reader may seek an invariant description of the $g^{ij}(v_p)$. We can also define the symbols

$$[ij,k] = \frac{1}{2}\left(\frac{\partial g_{ik}}{\partial x^j} + \frac{\partial g_{jk}}{\partial x^i} - \frac{\partial g_{ij}}{\partial x^k}\right)$$

$$\Gamma^{\ k}_{ij} = \sum_{\ell=1}^{n} g^{k\ell}[ij,\ell] \quad ,$$

as before, except that they are now functions on TM. It turns out that the critical paths for length are curves $c\colon [a,b] \longrightarrow M$ which, when parameterized by arclength, satisfy

$$\frac{d^2x^k(c(s))}{ds^2} + \sum_{i,j=1}^{n} \Gamma^{\ k}_{ij}(c'(s))\cdot\frac{d\dot{x}^i(c(s))}{ds}\,\frac{d\dot{x}^j(c(s))}{ds} = 0 \quad .$$

It also turns out, just as in the Riemannian case, that sufficiently small pieces of these critical paths are actually paths of shortest length. However, this is false for more general Minkowski metrics.

We shall not pursue the subject of Finsler metrics much further, but we will add some remarks about Minkowski metrics F on a vector space V. A Minkowski metric F can be used to define a "distance" function on $V \times V$, by $(v,w) \longmapsto F(w - v)$ (however, it is easy to see that this distance function satisfies the triangle inequality, and is consequently a metric, if and only if F is a Banach space norm on V). This is just the procedure by which, in analytic geometry, we define the distance between two points in $\mathbb{R}^n$; in this case, we choose $F(x) = (\Sigma(x^i)^2)^{1/2}$, motivated, of course, by the Pythagorean Theorem. After the Pythagorean Theorem has been incorporated into our definition of distance in this way, it is interesting to ask what content, if any, remains to this theorem. The answer is, that the Pythagorean Theorem has been declared true only for right triangles with sides parallel to the axes, but remains true for all right triangles. This is because, when $F(x) = (\Sigma(x^i)^2)^{1/2}$, the isometries of $\mathbb{R}^n$ are transitive on the unit sphere. That is to say, if p and q are in the unit sphere, then there is a linear transformation $\varphi\colon \mathbb{R}^n \longrightarrow \mathbb{R}^n$ with $\varphi(p) = q$ and $F(\varphi(x)) = F(x)$ for all x.

This same transitivity property holds for any F which is the norm $\| \ \|$ associated to a positive definite inner product $\langle\ ,\ \rangle$ on an n-dimensional vector space V, since there is an isomorphism $f\colon \mathbb{R}^n \longrightarrow V$ such that $f^*\langle\ ,\ \rangle$ is the usual inner product on $\mathbb{R}^n$ (Theorem I.9-3). It turns out that this property actually characterizes the Minskowski metrics F which arise from inner products. To prove this, we need an auxiliary concept, and a result from linear algebra.

An <u>ellipsoid</u> on a vector space V is a set of the from $\{v \in V\colon \langle v,v\rangle \leq 1\}$ for some positive definite inner product $\langle\ ,\ \rangle$. In particular, consider such an inner product $\langle\ ,\ \rangle$ on $\mathbb{R}^n$, which also has its standard inner product $\boldsymbol{\langle}\ ,\ \boldsymbol{\rangle}$. The ellipsoid $\{v \in V\colon \langle v,v\rangle \leq 1\}$ really looks like an ellipsoid,

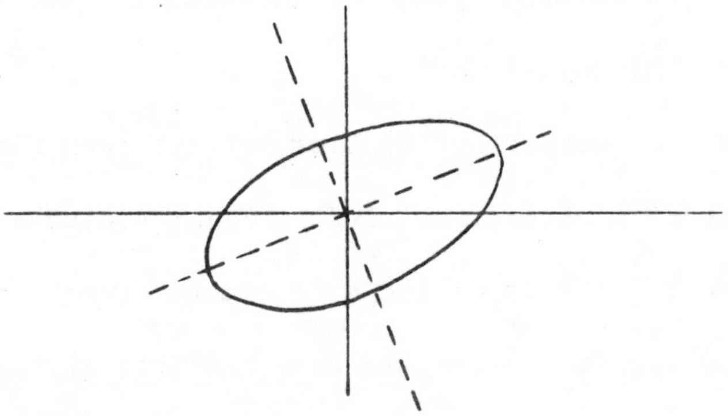

because of the following "well-known" result:

5. PROPOSITION (EXISTENCE OF PRINCIPAL AXES). If $\langle\ ,\ \rangle$ is any positive definite inner product on $\mathbb{R}^n$, then there is a basis for $\mathbb{R}^n$ which is ortho-

normal for $\langle\ ,\ \rangle$ and also <u>orthogonal</u> with respect to $\pmb{\langle}\ ,\ \pmb{\rangle}$.

<u>Proof</u>. For each $x \in \mathbb{R}^n$, the map $y \longmapsto \langle x,y \rangle$ is a linear functional, so there is a unique $Tx \in \mathbb{R}^n$ such that

$$\pmb{\langle} Tx,y \pmb{\rangle} = \langle x,y \rangle \qquad \text{for all } y\ .$$

It is easy to see that $x \longmapsto Tx$ is a linear transformation $T\colon \mathbb{R}^n \longrightarrow \mathbb{R}^n$. Moreover,

$$\pmb{\langle} Tx,y \pmb{\rangle} = \langle x,y \rangle = \langle y,x \rangle = \pmb{\langle} Ty,x \pmb{\rangle} = \pmb{\langle} x,Ty \pmb{\rangle}\ ,$$

so T is self-adjoint with respect to $\pmb{\langle}\ ,\ \pmb{\rangle}$. Thus T has a basis $x_1,\ldots,x_n$ of eigenvalues, $Tx_i = \lambda_i x_i$, and the x_i can be picked orthogonal with respect to $\pmb{\langle}\ ,\ \pmb{\rangle}$. Now

$$\begin{aligned}\lambda_j \langle x_i,x_j \rangle = \lambda_j \pmb{\langle} Tx_i,x_j \pmb{\rangle} &= \lambda_i\lambda_j \pmb{\langle} x_i,x_j \pmb{\rangle}\\ &= \lambda_i \pmb{\langle} x_i,Tx_j \pmb{\rangle}\\ &= \lambda_i \langle x_i,x_j \rangle\ .\end{aligned}$$

So $\langle x_i,x_j \rangle = 0$ if $\lambda_i \neq \lambda_j$. On the other hand, if two or more x_i have the same λ_i, then in the m-dimensional subspace which they span we can pick m eigenvectors which are orthogonal with respect to $\langle\ ,\ \rangle$. So we can assume that $\langle x_i,x_j \rangle = 0$ for $i \neq j$. Now we just normalize each x_i with respect to $\langle\ ,\ \rangle$. ∎

We now use this result to prove the basic lemma for our main assertion.

6. LEMMA. Let B be a bounded neighborhood of 0 in an n-dimensional vector space V. Then among all ellipsoids containing B there exists a unique one of smallest volume. (We assign a volume to the ellipsoids by choosing an isomorphism of V with $\mathbb{R}^n$. Choosing a different isomorphism clearly does not change the property of having the "smallest volume".)

Proof. We might as well assume that B is compact, since an ellipsoid containing B also contains $\bar{B}$. Every basis $b = (v_1,\ldots,v_n)$ of V determines an inner product (the one which makes $v_1,\ldots,v_n$ an orthonormal basis), and thus an ellipsoid $E(b)$ (different bases may determine the same ellipsoid). Clearly, $\{b\colon B \subset E(b) \subset 2B\}$ is a compact subset of the n-fold product $V \times \ldots \times V$. Hence $b \longmapsto$ volume of $E(b)$ takes on its minimum on this set. This proves existence.

Now consider two different ellipsoids containing B, with the same volume. Choose an isomorphism of V with $\mathbb{R}^n$ which makes the first of these ellipsoids correspond to the ordinary unit ball $\{x \in \mathbb{R}^n\colon f_1(x) \le 1\}$, where $f_1(x) = \Sigma(x^i)^2$. Proposition 5 shows that after a rotation of the axes, the second of the ellipsoids corresponds to $\{x \in \mathbb{R}^n\colon f_2(x) \le 1\}$, where

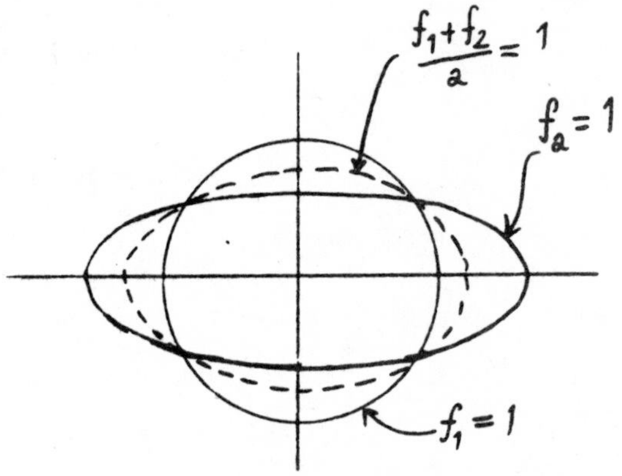

$$f_2(x) = a_1(x^1)^2 + \ldots + a_n(x^n)^2.$$

The volume of this ellipsoid is $\prod 1/\sqrt{a_i}$ times the volume of the unit ball. Since the two ellipsoids are assumed to have the same volume, this means that

$$\prod_{i=1}^{n} a_i = 1 \quad .$$

Now consider the ellipsoid

$$E = \{x \in \mathbb{R}^n : \frac{f_1 + f_2}{2}(x) \leq 1\} \quad .$$

Clearly E also contains B. Now the semi-axes of E have length

$$\frac{1}{\sqrt{\frac{1 + a_i}{2}}} \, ,$$

so the volume of E is the volume of the unit ball times

$$\prod_{i=1}^{n} \frac{1}{\sqrt{\frac{1 + a_i}{2}}} \quad .$$

Recall that for $a, b > 0$ we have

$$ab \leq \frac{a + b}{2} \qquad \text{with equality if and only if } a = b.$$

Consequently,

$$\prod_{i=1}^{n} \frac{1}{\sqrt{\frac{1 + a_i}{2}}} \leq \prod_{i=1}^{n} \frac{1}{\sqrt[4]{a_i}}$$

$$= \frac{1}{\sqrt[4]{\prod_{i=1}^{n} a_i}} = 1 \, ,$$

and strict equality holds if some $a_i \neq 1$, i.e. if the original two ellipsoids are different. This means that if two different ellipsoids containing B have the same volume, then there is another ellipsoid containing B with smaller volume. This clearly proves uniqueness of the ellipsoid with smallest volume. ∎

7. THEOREM. Let $F: V \longrightarrow \mathbb{R}$ be a continuous Minkowski metric on an n-dimensional vector space V. Suppose that for all p and q in the unit sphere $\{v \in V: F(x) = 1\}$, there is a linear transformation $\varphi: V \longrightarrow V$ such that $\varphi(p) = q$ and $F(\varphi(v)) = F(v)$ for all $v \in V$. Then F is the norm determined by some positive definite inner product.

Proof. Let $B = \{v: F(v) \leq 1\}$, and let E be the unique ellipsoid containing B of smallest volume. Clearly, there must be some point p with $F(p) = 1$ and $p \in$ boundary E. Let q be any other point with $F(q) = 1$, and

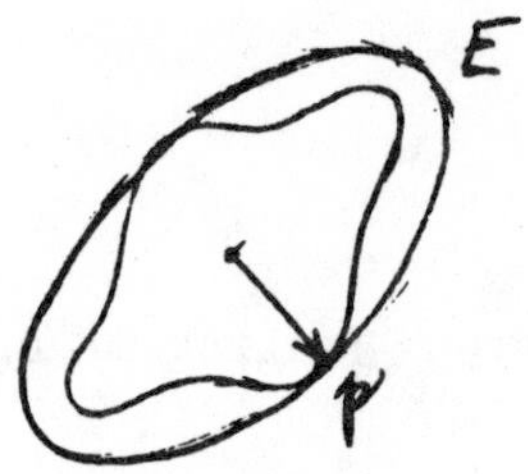

$\varphi: V \longrightarrow V$ a linear transformation with $\varphi(p) = q$ such that $F(\varphi(v)) = F(v)$ for all $v \in V$. It follows easily from the latter property that $\varphi(E) \supset B$. Moreover, $\varphi(B) = B$, so φ is volume preserving. By uniqueness of the ellipsoid E, it follows that $\varphi(E) = E$. Consequently, $q = \varphi(p) \in$ boundary E. In other words, every point q with $F(q) = 1$ is in boundary E. This means that $E = B$. ∎

Chapter 4C. A Prize Essay

The second edition of Riemann's collected works includes an unpublished paper, in Latin, which was submitted to the Paris Academy in 1861, to compete for a prize on a question involving heat conduction. In 1868, ten years after it had been offered, the prize was finally withdrawn. Because the way of obtaining the results of his essay were not fully explained, the prize was not awarded to Riemann, whose health prevented the more detailed handling of the subject which he had intended.

An extract from this paper is given below. We will be interested in the part which ends at the bottom of page 4C-3. It should not be very hard to read, but the significance of the equations obtained there is only suggested by Riemann's remarks in the last paragraph of page 4C-5; in the next part of this Chapter we will have a great deal more to add. Before these final remarks, Riemann inserts something which will present greater difficulties of interpretation than any thing else we have read; this material, on pages 4C-4 and 4C-5, will be discussed in the Addendum to Chapter 4D. In the translation I have made some minor changes of notation.

An Extract From Riemann's Paper of 1861

Second Part

On the transformation of the expression $\sum_{i,j} g_{ij}dy^i dy^j$ into the given form $\sum_{i,j} a_{ij}dx^i dx^j$.

When the inquiry of the third Academy is restricted to homogeneous bodies, in which the resulting conductivities are constants, we develop the first condition that the expression $\sum_{i,j} g_{ij}dy^i dy^j$, in which the y^i are functions of the x^i, can be transformed into the form $\sum_{i,j} a_{ij}dx^i dx^j$ with given constant coefficients a_{ij}. .

The expression $\sum_{i,j} a_{ij}dx^i dx^j$, if it is, as we shall suppose, a positive form in the dx^i, can always be put in the simplified form $\sum_i (dx^i)^2$. Thus if

$\sum_{i,j} g_{ij} dy^i dy^j$ can be transformed into the form $\sum_{i,j} a_{ij} dx^i dx^j$, it can likewise be reduced to the form $\sum_i (dx^i)^2$ and vice versa. We therefore ask whether it can be put in the form $\sum_i (dx^i)^2$.

Let $G = \det(g_{ij})$ and let γ_{ij} be the cofactor; in this way $\sum_i g_{ij}\gamma_{ij} = G$ and $\sum_i g_{ij}\gamma_{ik} = 0$ if $j \gtrless k$.

If $\sum_{i,j} g_{ij} dy^i dy^j = \sum_i (dx^i)^2$ for arbitrary values of the dx^i, substituting $d + \delta$ for d leads also to $\sum_{i,j} g_{ij} dy^i \delta y^j = \sum_i dx^i \delta x^i$ for arbitrary values of the dx^i and δx^i.

Consequently, if the dy^i are expressed in terms of the dx^i and the δx^i in terms of the δy^i, it follows that

$$(1) \qquad \frac{\partial x^\beta}{\partial y^\alpha} = \sum_i g_{\alpha i} \frac{\partial y^i}{\partial x^\beta}$$

and consequently

$$(2) \qquad \frac{\partial y^i}{\partial x^\beta} = \sum_\alpha \frac{\gamma_{\alpha i}}{G} \frac{\partial x^\beta}{\partial y^\alpha} \quad .$$

Thus we further deduce, seeing that

$$\sum_\alpha \frac{\partial y^i}{\partial x^\alpha} \frac{\partial x^\alpha}{\partial y^i} = 1 \quad \text{and} \quad \sum_\alpha \frac{\partial y^i}{\partial x^\alpha} \frac{\partial x^\alpha}{\partial y^j} = 0 \text{ if } i \gtrless j \ ,$$

$$(3) \qquad \sum_\alpha \frac{\partial x^\alpha}{\partial y^i} \frac{\partial x^\alpha}{\partial y^j} = g_{ij}, \qquad (4) \qquad \sum_\alpha \frac{\partial y^i}{\partial x^\alpha} \frac{\partial y^j}{\partial x^\alpha} = \frac{\gamma_{ij}}{G}$$

and differentiating formula (3),

$$\sum_\alpha \frac{\partial^2 x^\alpha}{\partial y^i \partial y^k} \frac{\partial x^\alpha}{\partial y^j} + \sum_\alpha \frac{\partial^2 x^\alpha}{\partial y^j \partial y^k} \frac{\partial x^\alpha}{\partial y^i} = \frac{\partial g_{ij}}{\partial y^k} \quad .$$

Now from these expressions for

$$\frac{\partial g_{ij}}{\partial y^k}, \frac{\partial g_{ik}}{\partial y^j}, \frac{\partial g_{jk}}{\partial y^i}$$

we can write

$$(5) \quad 2 \sum_{\alpha} \frac{\partial^2 x^\alpha}{\partial y^j \partial y^k} \frac{\partial x^\alpha}{\partial y^i} = \frac{\partial g_{ij}}{\partial y^k} + \frac{\partial g_{ik}}{\partial y^j} - \frac{\partial g_{jk}}{\partial y^i}$$

and if these quantities are designated by p_{ijk}, then

$$(6) \qquad 2 \frac{\partial^2 x^\alpha}{\partial y^j \partial y^k} = \sum_i \frac{\partial y^i}{\partial x^\alpha} p_{ijk} \quad .$$

Differentiating the quantities p_{ijk} again yields

$$\frac{\partial p_{ijk}}{\partial y^\ell} - \frac{\partial p_{ij\ell}}{\partial y^k} = 2 \sum_\nu \frac{\partial^2 x^\nu}{\partial y^j \partial y^k} \frac{\partial^2 x^\nu}{\partial y^i \partial y^\ell} - 2 \sum_\nu \frac{\partial^2 x^\nu}{\partial y^j \partial y^\ell} \frac{\partial^2 x^\nu}{\partial y^i \partial y^k} ,$$

whence finally, substituting the values found in (6) and (4),

$$(\mathrm{I}) \qquad \frac{\partial^2 g_{ik}}{\partial y^j \partial y^\ell} + \frac{\partial^2 g_{j\ell}}{\partial y^i \partial y^k} - \frac{\partial^2 g_{i\ell}}{\partial y^j \partial y^k} - \frac{\partial^2 g_{jk}}{\partial y^i \partial y^\ell}$$

$$+ \frac{1}{2} \sum_{\alpha,\beta} (p_{\alpha j\ell} p_{\beta ik} - p_{\alpha i\ell} p_{\beta jk}) \frac{\gamma_{\alpha\beta}}{G} = 0 \quad .$$

The functions g_{ij} must necessarily satisfy these equations whenever $\sum_{i,j} g_{ij} dy^i dy^j$ can be transformed into the form $\sum_i (dx^i)^2$: we denote the left side of this equation by

$$(ij, k\ell) \quad .$$

To make the nature of this equation more transparent, we form the expression

$$\delta\delta \sum_{i,j} g_{ij} dy^i dy^j - 2d\delta \sum_{i,j} g_{ij} dy^i \delta y^j + dd \sum_{i,j} g_{ij} \delta y^i \delta y^j \quad ,$$

the variations of the second order $d^2, d\delta, \delta^2$ being so determined that

$$\delta' \sum_{i,j} g_{ij} dy^i \delta y^j - \delta \sum_{i,j} g_{ij} dy^i \delta' y^j - d \sum_{i,j} g_{ij} \delta y^i \delta' y^j = 0$$

$$\delta' \sum_{i,j} g_{ij} dy^i dy^j - 2d \sum_{i,j} g_{ij} dy^i \delta' y^j = 0$$

$$\delta' \sum_{i,j} g_{ij} \delta y^i \delta y^j - 2\delta \sum_{i,j} g_{ij} \delta y^i \delta' y^j = 0 \quad ,$$

δ' denoting an arbitrary variation. In this way the above expression becomes equal to

$$\text{(II)} \qquad \Sigma(ij,k\ell)(dy^i \delta y^j - dy^j \delta y^i)(dy^k \delta y^\ell - dy^\ell \delta y^k) \quad .$$

Now from the formation of this expression it is immediately evident that a change of the independent variables changes it into a new form depending in the same way on the $\sum_{i,j} g_{ij} dx^i dx^j$. And if the quantities g_{ij} are constant, all coefficients of the expression (II) turn out to be equal to 0. Thus if $\sum_{i,j} g_{ij} dy^i dy^j$ can be transformed into a similar expression with constant coefficients, it is necessary that expression (II) vanishes identically.

In the same way it turns out that if the expression (II) does not vanish, the expression

$$\text{(III)} \quad -\frac{1}{2}\frac{\Sigma(ij,k\ell)(dy^i\delta y^j - dy^j\delta y^i)(dy^k\delta y^\ell - dy^\ell\delta y^k)}{\Sigma g_{ij}dy^i dy^j \, \Sigma g_{ij}dy^i\delta y^j - (\Sigma g_{ij}dy^i\delta y^j)^2}$$

does not change if the independent variables are changed, and moreover remains unchanged if in place of the variations $dy^i, \delta y^i$, arbitrary independent linear expressions of them, $\alpha dy^i + \beta\delta y^i, \gamma dy^i + \delta y^i$ are substituted. Moreover, the maximum and minimum values of the function (III) of the same $dy^i, \delta y^i$ depend neither on the form of the expression $\sum_{i,j} g_{ij}dy^i dy^j$ nor on the values of the variations $dy^i, \delta y^i$, whence from these values it can be determined when two expressions of this kind can be transformed into each other.

These interpretations can be illustrated by what one might call a geometrical interpretation, which, although it depends on unusual conceptions, it will never-the less help, as one goes along, to have touched upon.

The expression $\sqrt{\sum_{i,j} g_{ij}dy^i dy^j}$ can be regarded as just the line element in a generalized space of n dimensions transcending our intuition. If in this space at the point $(y^1,\ldots,y^n)$ all shortest lines are drawn, in which the initial elements of variation of the y^i are $\alpha dy^1 + \beta\delta y^1 : \alpha dy^2 + \beta\delta y^2 : \ldots : \alpha dy^n + \beta\delta y^n$, α and β denoting arbitrary quantities, these lines make up a surface which can be developed in the space of our common intuition. In this way the expression (III) will measure the curvature of this surface at the point $(y^1,\ldots,y^n)$.

If we now return to the case $n = 3$, the expression(II) is a form of the second degree in

$$dy^2\delta y^3 - dy^3\delta y^2 , \; dy^3\delta y^1 - dy^1\delta y^3 , \; dy^1\delta y^2 - dy^2\delta y^1 ,$$

and in this case we obtain six equations, which the functions g_{ij} are required to satisfy if $\sum_{i,j} g_{ij}dy^i dy^j$ can be transformed into a form with constant coefficients. Given an acquaintance with the traditional methods, it is demonstrated without difficulty that these six conditions, when they are satisfied, suffice. It is to be observed nevertheless that only three of them are independent.

Chapter 4D. The Birth of the Riemann Curvature Tensor

All the developments in this part of the Chapter have their origin in the following question, which Riemann considers in the paper of Part C: When is a Riemann manifold $(M, \langle\ ,\ \rangle)$ flat (locally isometric to $\mathbb{R}^n$ with its usual Riemannian metric)? In other words, when is there a coordinate system $x^1,\ldots,x^n$ on M for which

$$(1) \qquad \langle\ ,\ \rangle = \sum_{i=1}^{n} dx^i \otimes dx^i \quad ?$$

We are going to seek an answer to this question in as straightforward a manner as possible; the quadratic function Q will not be used at all, but at the end it will make a surprise appearance.

We begin by choosing an arbitrary coordinate system y, in terms of which the metric $\langle\ ,\ \rangle$ can be written

$$(2) \qquad \langle\ ,\ \rangle = \sum_{i,j=1}^{n} g_{ij}\, dy^i \otimes dy^j \quad ;$$

and we then seek conditions on the g_{ij} in order for (1) to hold for some coordinate system x. Since this is a purely local question, we can assume that $y^1,\ldots,y^n$ is just the standard coordinate system on $\mathbb{R}^n$.

If we express the dx^i in terms of the dy^j, and equate the coefficients of $dy^i \otimes dy^j$ in (2) with the resulting coefficients in (1), we find that the coordinate system $x^1,\ldots,x^n$ has the desired property if and only if

$$(3) \qquad \sum_{\alpha} \frac{\partial x^\alpha}{\partial y^i} \frac{\partial x^\alpha}{\partial y^j} = g_{ij} \quad .$$

From equation (3) we can immediately derive another, for we obtain

$$\sum_{j,\beta} g_{ij} \frac{\partial y^j}{\partial x^\alpha} \frac{\partial y^k}{\partial x^\beta} = \sum_{j,\beta,\alpha} \frac{\partial x^\alpha}{\partial y^i} \frac{\partial x^\alpha}{\partial y^j} \frac{\partial y^j}{\partial x^\beta} \frac{\partial y^k}{\partial x^\beta}$$

$$= \sum_{\alpha,\beta} \frac{\partial x^\alpha}{\partial y^i} \delta^\alpha_\beta \frac{\partial y^k}{\partial x^\beta}$$

$$= \sum_\alpha \frac{\partial x^\alpha}{\partial y^i} \frac{\partial y^k}{\partial x^\alpha} = \delta^k_i ,$$

which shows that if the coordinate system $x^1,\ldots,x^n$ has the desired property, then

(4) $$\sum_\beta \frac{\partial y^i}{\partial x^\beta} \frac{\partial y^j}{\partial x^\beta} = g^{ij} .$$

Conversely, (4) implies (3). These results are just the equations (3) and (4) which Riemann obtains. Notice that Riemann begins with the square of the norm $\| \|^2 = \Sigma g_{ij} dy^i \cdot dy^j$, and then uses polarization to obtain the inner product, which he writes as $\Sigma g_{ij} dy^i \delta y^j$. Riemann also treats the two coordinate systems x and y on an equal footing throughout, so that his derivations of (3) and (4) are somewhat different. From (4) we obtain

$$\sum_{i,j} g^{ij} \frac{\partial x^\mu}{\partial y^i} \frac{\partial x^\nu}{\partial y^j} = \sum_{\beta,i,j} \frac{\partial y^i}{\partial x^\beta} \frac{\partial y^j}{\partial x^\beta} \frac{\partial x^\mu}{\partial y^i} \frac{\partial x^\nu}{\partial y^j}$$

$$= \sum_\beta \delta^\mu_\beta \delta^\nu_\beta ,$$

and thus the coordinate system $x^1,\ldots,x^n$ has the desired property if and only if

(4') $$\sum_{i,j} g^{ij} \frac{\partial x^\mu}{\partial y^i} \frac{\partial x^\nu}{\partial y^j} = \delta_{\mu\nu} .$$

This equation, which we will find more useful than (4), can be derived directly from (3) in the following way. If $A = (a_{ij}) = (\partial x^i/\partial y^j)$, and $G = (g_{ij})$, then (3) says that

$$A^T \cdot A = G \quad ;$$

this is equivalent to

$$G^{-1} = A^{-1} \cdot (A^T)^{-1} \quad ,$$

and hence to

$$AG^{-1}A^T = I \quad ,$$

which is just (4'). In particular, this shows immediately that (4') is equivalent to (3).

Now equation (3) is a partial differential equation for the functions x^α. In Chapter I.6 we developed a general theory for partial differential equations, but we notice at once that (3) is not an equation of the type to which our theory applies. Our first task will thus be to obtain from (3) an equation which we do know how to handle. The situation is very much like, and may be profitably compared to, that which occurs in Problem I.7-19, where the analysis of a certain set of partial differential equations is reduced to Theorem I.6-1, together with the Poincaré Lemma. To treat equation (3), we begin by differentiating (about all we can do), to obtain

$$\sum_\alpha \frac{\partial^2 x^\alpha}{\partial y^i \partial y^k} \frac{\partial x^\alpha}{\partial y^j} + \sum_\alpha \frac{\partial^2 x^\alpha}{\partial y^j \partial y^k} \frac{\partial x^\alpha}{\partial y^i} = \frac{\partial g_{ij}}{\partial y^k} \quad .$$

By writing down this equation for

$$\frac{\partial g_{ij}}{\partial y^k}, \frac{\partial g_{ik}}{\partial y^j}, \frac{\partial g_{jk}}{\partial y^i},$$

and combining, we obtain an equation equivalent to Riemann's,

$$(5) \quad \sum_\alpha \frac{\partial^2 x^\alpha}{\partial y^j \partial y^k} \frac{\partial x^\alpha}{\partial y^i} = \frac{1}{2}\left(\frac{\partial g_{ij}}{\partial y^k} + \frac{\partial g_{ik}}{\partial y^j} - \frac{\partial g_{jk}}{\partial y^i}\right) = [jk,i] .$$

Thus, the symbols $[jk,i]$, which came up naturally in the calculus of variations, also come up naturally in this different context. After Riemann's Habilitations lecture was published, in 1866, several mathematicians independently derived his results or considered related questions. Christoffel, in particular, introduced these combinations of the partial derivatives of the g_{ij}'s, and the symbols $[ij,k]$ and Γ^k_{ij} are called the Christoffel symbols of the first and second kinds, respectively (Christoffel actually used $\begin{bmatrix} ij \\ k \end{bmatrix}$ and $\begin{Bmatrix} ij \\ k \end{Bmatrix}$, which do not accomodate themselves to the summation convention). In the next Chapter we will see one important use which Christoffel made of these symbols.

At this point we will depart slightly from Riemann's treatment, in order to obtain equations to which Theorem I.6-1 directly applies. From (5) we obtain

$$\sum_{i,\gamma} g^{i\gamma} \frac{\partial x^\lambda}{\partial y^\gamma} [jk,i] = \sum_{\alpha,i,\gamma} \frac{\partial^2 x^\alpha}{\partial y^j \partial y^k} \frac{\partial x^\alpha}{\partial y^i} \frac{\partial x^\lambda}{\partial y^\gamma} g^{i\gamma}$$

$$= \sum_\alpha \frac{\partial^2 x^\alpha}{\partial y^j \partial y^k} \left(\sum_{i,\gamma} \frac{\partial x^\alpha}{\partial y^i} \frac{\partial x^\lambda}{\partial y^\gamma} g^{i\gamma} \right)$$

$$= \sum_\alpha \frac{\partial^2 x^\alpha}{\partial y^j \partial y^k} \delta_{\alpha\lambda} \quad \text{by } (4') ,$$

so we obtain, finally,

$$\text{(6)} \qquad \frac{\partial^2 x^\lambda}{\partial y^j \partial y^k} = \sum_{\gamma=1}^{n} \Gamma^\gamma_{jk} \frac{\partial x^\lambda}{\partial y^\gamma}$$

(which is easily seen to be equivalent to Riemann's equation (6)); we will also write this equation as

$$\frac{\partial\left(\dfrac{\partial x^\lambda}{\partial y^j}\right)}{\partial y^k} = \sum_{\gamma=1}^{n} \Gamma^\gamma_{jk} \frac{\partial x^\lambda}{\partial y^\gamma} \quad .$$

Notice that the index λ plays no special role here; all functions x^λ satisfy the <u>same</u> equation. Thus, for each λ the n-tuple of functions

$$\alpha = \left(\frac{\partial x^\lambda}{\partial y^1}, \ldots, \frac{\partial x^\lambda}{\partial y^n}\right) \qquad \alpha\colon \mathbb{R}^n \longrightarrow \mathbb{R}^n$$

satisfies the set of partial differential equations

$$(*) \qquad \frac{\partial \alpha}{\partial y^k}(y) = f_k(y,\alpha(y)) \quad ,$$

where $f_k\colon \mathbb{R}^n \times \mathbb{R}^n \longrightarrow \mathbb{R}^n$ is given by

$$f^j_k(y,z) = \sum_{\gamma=1}^{n} \Gamma^\gamma_{jk}(y)\cdot z^\gamma \quad .$$

Since this is true for every λ, the equation (*) has n solutions whose initial values at some point, 0 say, are linearly independent. Since constant linear combinations of solutions of (*) are also solutions, it follows that (*) has solutions with arbitrary initial conditions at 0. From Theorem I.6-1 we thus obtain necessary integrability conditions,

$$\frac{\partial f_k}{\partial y^\ell} - \frac{\partial f_\ell}{\partial y^k} + \sum_{\mu=1}^{n} \frac{\partial f_k}{\partial z^\mu} f^\mu_\ell - \sum_{\mu=1}^{n} \frac{\partial f_\ell}{\partial z^\mu} f^\mu_k = 0 \quad .$$

In our case, looking at the $j\underline{th}$ components of these equations, we obtain

$$\sum_{\gamma=1}^{n} \frac{\partial \Gamma^\gamma_{jk}}{\partial y^\ell} z^\gamma - \sum_{\gamma=1}^{n} \frac{\partial \Gamma^\gamma_{j\ell}}{\partial y^k} z^\gamma + \sum_{\mu=1}^{n} \Gamma^\mu_{jk} \sum_{\gamma=1}^{n} \Gamma^\gamma_{\mu\ell} z^\gamma - \sum_{\mu=1}^{n} \Gamma^\mu_{j\ell} \sum_{\gamma=1}^{n} \Gamma^\gamma_{\mu k} z^\gamma = 0 \quad .$$

Since these relations must hold for all $z = (z^1,\ldots,z^n)$, we obtain

$$R^\gamma{}_{j\ell k} = 0 \quad ,$$

(**) where

$$R^\gamma{}_{j\ell k} = \frac{\partial \Gamma^\gamma_{kj}}{\partial y^\ell} - \frac{\partial \Gamma^\gamma_{\ell j}}{\partial y^k} + \sum_{\mu=1}^{n} (\Gamma^\mu_{kj}\Gamma^\gamma_{\ell\mu} - \Gamma^\mu_{\ell j}\Gamma^\gamma_{k\mu})$$

as necessary conditions that $\Sigma g_{ij} dy^i \otimes dy^j = \Sigma dx^i \otimes dx^i$ for some coordinate system $x = (x^1,\ldots,x^n)$. Notice that the set of equations $R^\gamma{}_{j\ell k} = 0$ is equivalent to the set of equations

$$R_{ij\ell k} = \sum_{\gamma=1}^{n} g_{i\gamma} R^\gamma{}_{j\ell k} = 0 \quad .$$

The quantities $R_{ij\ell k}$ can be expressed in another way, after a little calculation. Note first that

$$\sum_{\gamma=1}^{n} g_{i\gamma} \frac{\partial \Gamma^{\gamma}_{jk}}{\partial y^{\ell}} = \frac{\partial}{\partial y^{\ell}} \left(\sum_{\gamma=1}^{n} g_{i\gamma} \Gamma^{\gamma}_{jk} \right) - \sum_{\gamma=1}^{n} \Gamma^{\gamma}_{jk} \frac{\partial g_{i\gamma}}{\partial y^{\ell}}$$

$$= \frac{\partial [jk,i]}{\partial y^{\ell}} - \sum_{\gamma=1}^{n} \Gamma^{\gamma}_{jk}([i\ell,\gamma] + [\gamma\ell,i]) .$$

Substituting into (**), and remembering the definition of $[ij,k]$, we obtain

$$(***) \qquad R_{ij\ell k} = \frac{1}{2}\left(\frac{\partial^2 g_{ik}}{\partial y^j \partial y^{\ell}} + \frac{\partial^2 g_{j\ell}}{\partial y^i \partial y^k} - \frac{\partial^2 g_{i\ell}}{\partial y^j \partial y^k} - \frac{\partial^2 g_{jk}}{\partial y^i \partial y^{\ell}} \right)$$

$$+ \sum_{\alpha,\beta=1}^{n} g^{\alpha\beta}([j\ell,\alpha]\cdot[ik,\beta] - [i\ell,\alpha]\cdot[jk,\beta]) .$$

The condition $R_{ij\ell k} = 0$ is just the condition (I) which Riemann obtains (note that Riemann's p_{ijk} equals $2[jk,i]$) — the quantity which we have denoted by $R_{ij\ell k}$ is what Riemann denotes by $2(ij,k\ell)$; the factor of 2 is not particularly significant, nor is the interchange of ℓ and k, for it is easily seen that $R_{ij\ell k} = - R_{ijk\ell}$.

The notation $R^{i}{}_{jk\ell}$ has been picked in anticipation of the following result.

1. PROPOSITION. On a Riemannian manifold $(M, \langle\ ,\ \rangle)$ there is a tensor of type $\binom{1}{3}$ whose components in any coordinate system y are

$$R^{i}{}_{jk\ell} = \frac{\partial \Gamma^{i}_{j\ell}}{\partial y^{k}} - \frac{\partial \Gamma^{i}_{jk}}{\partial y^{\ell}} + \sum_{\mu=1}^{n} (\Gamma^{\mu}_{j\ell}\Gamma^{i}_{\mu k} - \Gamma^{\mu}_{jk}\Gamma^{i}_{\mu\ell})$$

(where $\langle\ ,\ \rangle = \Sigma g_{ij} dy^i \otimes dy^j$, and the Christoffel symbols Γ are defined as usual).

<u>Proof</u>. We just compute that the components transform "correctly"!! In other words, if $R'^{i}{}_{j\ell k}$ are defined by the same formula, with respect to the coordinate system y', we show that

$$R'^{\alpha}{}_{\beta\gamma\delta} = \sum_{i,j,k,\ell} R^{i}{}_{jk\ell} \frac{\partial y^j}{\partial y'^{\beta}} \frac{\partial y^k}{\partial y'^{\gamma}} \frac{\partial y^\ell}{\partial y'^{\delta}} \frac{\partial y'^{\alpha}}{\partial y^i} \quad .$$

To do this, all one needs is the result from Problem I.9-22,

$$\Gamma'^{\gamma}_{\alpha\beta} = \sum_{i,j,k} \Gamma^{k}_{ij} \frac{\partial y^i}{\partial y'^{\alpha}} \frac{\partial y^j}{\partial y'^{\beta}} \frac{\partial y'^{\gamma}}{\partial y^k} + \sum_{\mu=1}^{n} \frac{\partial^2 y^\mu}{\partial y'^{\alpha} \partial y'^{\beta}} \frac{\partial y'^{\gamma}}{\partial y^\mu} \quad ,$$

and plenty of perserverance.

<u>Slightly more motivated proof</u>. Begin with the equation

$$\sum_{i,j} g_{ij} dy^i \otimes dy^j = \langle\ ,\ \rangle = \sum_{i,j} g'_{ij} dy'^i \otimes dy'^j \quad ,$$

and repeat the whole sequence of computations which we performed in the special case that $g'_{ij} = \delta_{ij}$. The result will be the desired transformation law. (The integrability conditions (**) then follow as a necessary condition for the existence of a coordinate system y' with $g'_{ij} = \delta_{ij}$, for in such a coordinate system we clearly have $R'^{i}{}_{jk\ell} = 0$, which in turn implies that all $R^{i}{}_{jk\ell} = 0$.) ∎

We have thus stumbled onto a new tensor, the <u>Riemannn curvature tensor</u>, which in the coordinate system y equals

$$\sum_{i,j,k,\ell} R^{i}{}_{jk\ell}\, dy^j \otimes dy^k \otimes dy^\ell \otimes \frac{\partial}{\partial y^i} \quad .$$

Eventually we hope to have a useful invariant definition of this tensor; this will involve an enormous amount of exploration. For the time being, we simply accept the classical definition, which arises naturally as an integrability condition, and explain how it is connected with curvature. In the process we will obtain an invariant, but extraordinarily clumsy, definition of the curvature tensor.

It will be convenient to introduce a bit of modern terminology, and denote by R the tensor with components $R^i{}_{jk\ell}$. Since this tensor is of type $\binom{3}{1}$ it may be regarded as a function taking three vectors to another vector. The value of R on $X,Y,Z \in M_p$ will be denoted by

$$R(Y,Z)X \in M_p ,$$

and hence we have

$$R\left(\frac{\partial}{\partial y^k}\bigg|_p , \frac{\partial}{\partial y^\ell}\bigg|_p\right)\frac{\partial}{\partial y^j}\bigg|_p = \sum_{i=1}^{n} R^i{}_{jk\ell}(p)\cdot\frac{\partial}{\partial y^i}\bigg|_p$$

(the reason for choosing the notation $R(Y,Z)X$ comes out in Proposition 6).

The numbers $R_{ijk\ell} = \sum_\gamma g_{i\gamma} R^\gamma{}_{jk\ell}$ are also the components of a tensor, of type $\binom{4}{0}$, but it is unnecessary to perform any calculations to verify this. Clearly

$$R_{ijk\ell}(p) = \left\langle R\left(\frac{\partial}{\partial y^k}\bigg|_p , \frac{\partial}{\partial y^\ell}\bigg|_p\right)\frac{\partial}{\partial y^j}\bigg|_p , \frac{\partial}{\partial y^i}\bigg|_p \right\rangle ,$$

so the tensor in question is just the multilinear map

$$(X,Y,Z,W) \longmapsto \langle R(Z,W)Y,X \rangle .$$

This function of four tangent vectors is closely connected with the quadratic function introduced in Part B of this Chapter:

2. PROPOSITION. Let x be a Riemannian normal coordinate system at p, and Q the quadratic function on $M_p \times M_p$ defined by

$$Q(X,Y) = \sum_{i,j;k,\ell} c_{ij,k\ell}\, dx^i(X)dx^j(X)dx^k(Y)dx^\ell(Y) \quad ,$$

where

$$c_{ij,k\ell} = \frac{1}{2}\frac{\partial^2 g_{ij}}{\partial x^k \partial x^\ell} \quad .$$

Then

$$Q(X,Y) = -\frac{1}{3} < R(X,Y)Y,X > \quad .$$

Proof. We have seen that

$$\begin{aligned}
3Q(X,Y) &= \sum_{i,j,k,\ell} c_{ij,k\ell}(dx^i \wedge dx^k)\cdot(dx^j \wedge dx^\ell)(X,Y) \\
&= \sum_{i,j,k,\ell} c_{ij,k\ell}\, dx^i(X)dx^j(X)dx^k(Y)dx^\ell(Y) \\
&+ \sum_{i,j,k,\ell} c_{ij,k\ell}\, dx^k(X)dx^\ell(X)dx^i(Y)dx^j(Y) \\
&- \sum_{i,j,k,\ell} c_{ij,k\ell}\, dx^j(X)dx^k(X)dx^i(Y)dx^\ell(Y) \\
&- \sum_{i,j,k,\ell} c_{ij,k\ell}\, dx^i(X)dx^\ell(X)dx^j(Y)dx^k(Y) \quad .
\end{aligned}$$

By switching indices we can rewrite this as

$$3Q(X,Y) = \sum_{i,j,k,\ell} c_{ik,j\ell}\, dx^i(X)dx^j(Y)dx^k(X)dx^\ell(Y) \qquad \text{[interchange } j \text{ and } k]$$

$$+ \sum_{i,j,k,\ell} c_{j\ell,ik} \quad \text{"} \qquad \text{[interchange } i \text{ and } \ell]$$

$$- \sum_{i,j,k,\ell} c_{i\ell,jk} \quad \text{"} \qquad [\text{change } i \text{ to } \ell;\ j \text{ to } i;\ \ell \text{ to } j]$$

$$- \sum_{i,j,k,\ell} c_{jk,i\ell} \quad \text{"} \qquad [\text{change } i \text{ to } k;\ k \text{ to } \ell;\ \ell \text{ to } i]$$

$$= \sum_{i,j,k,\ell} (c_{ik,j\ell} + c_{j\ell,ik} - c_{i\ell,jk} - c_{jk,i\ell})dx^i \otimes dx^j \otimes dx^k \otimes dx^\ell(X,Y,X,Y)$$

Now in Riemannian normal coordinates, the Christoffel symbols $[ij,k]$ are all 0 at p, since all $\partial g_{ij}/\partial x^k$ are 0 at p. Referring to equation (***) we thus have

$$3Q(X,Y) = \sum_{i,j,k,\ell} R_{ij\ell k}(p)dx^i \otimes dx^j \otimes dx^k \otimes dx^\ell(X,Y,X,Y)$$

$$= -\sum_{i,j,k,\ell} R_{ijk\ell}(p)dx^i \otimes dx^j \otimes dx^k \otimes dx^\ell(X,Y,X,Y)$$

$$= -\langle R(X,Y)Y,X \rangle \quad . \blacksquare$$

We are now ready to verify some of Riemann's claims.

3. PROPOSITION. Let $(M, \langle\ \rangle)$ be a 2-dimensional Riemannian manifold, and let $X,Y \in M_p$ be linearly independent. Let $\|X,Y\|$ denote the area of the parallelogram spanned by X and Y. Then

$$k(p) = \frac{< R(X,Y)Y,X >}{\|X,Y\|^2} \qquad (= < R(X,Y)Y,X > \text{ if } X \text{ and } Y \text{ are orthonormal})$$

is the same as the Gaussian curvature at p defined by the formula in Theorem 3B-7 (in particular, this proves that the formula in Theorem 3B-7 is indeed independent of the coordinate system).

<u>First Proof</u>. Let (x,y) be a coordinate system on a neighborhood of $p \in M$. It obviously suffices to verify the theorem when $X = \partial/\partial x|_p$ and $Y = \partial/\partial y|_p$, since by Proposition 2, and the results of 4B, the numerator is multiplied by the same factor as the denominator when we change to any other pair of vectors. In this case,

$$< R(X,Y)Y,X > = \left\langle R\left(\frac{\partial}{\partial x}\Big|_p, \frac{\partial}{\partial y}\Big|_p\right)\frac{\partial}{\partial y}\Big|_p, \frac{\partial}{\partial x}\Big|_p \right\rangle$$

$$= R_{1212}(p) \quad .$$

If we write

$$< , > = E\, dx \otimes dx + F\, dx \otimes dy + F\, dy \otimes dx + G\, dy \otimes dy \quad ,$$

so that

$$g_{11} = E$$

$$g_{12} = g_{21} = F$$

$$g_{22} = G \quad ,$$

then (by the formula on page I.9-12)

$$\|X,Y\|^2 = EG - F^2 \quad ,$$

so we must prove that

$$4R_{1212}(EG - F^2) = 4(EG - F^2)^2 k \quad ,$$

where the right side is given by the formula in Theorem 3B-7. This is a fairly straightforward calculation from (***). The first term in (***) corresponds to the last in the formula for $4(EG - F^2)^2 k$, and the second corresponds to the first three in the latter formula. In carrying out the calculation, note that

$$g^{11} = \frac{G}{EG - F^2}$$

$$g^{12} = g^{21} = \frac{-F}{EG - F^2}$$

$$g^{22} = \frac{E}{EG - F^2} \quad ;$$

the denominators cancel out the unwanted factor in $4R_{1212}(EG - F^2)$.

Second Proof (outline). Let (r,φ) be the coordinate system around p which is introduced on page 3B-32. We know that in this coordinate system

$$\langle\ ,\ \rangle = dr \otimes dr + G d\varphi \otimes d\varphi$$

for some function G, and (see page 3B-43) that

$$k(p) = -\frac{\partial^3 \sqrt{G}}{\partial r^3}(p) \quad .$$

Introduce a Riemannian normal coordinate system x^1, x^2 by the equations

$$x^1 = r\cos\varphi \quad , \quad x^2 = r\sin\varphi \ .$$

We can then calculate the g_{ij} in terms of G, and use these results to show that the quantity

$$Q\left(\frac{\partial}{\partial x^1}\bigg|_p , \frac{\partial}{\partial x^2}\bigg|_p\right) = 2c_{11,22}(p)$$

is equal to $-k(p)/3$. The result then follows from Proposition 2. ∎

4. PROPOSITION. Let $(M, < , >)$ be a Riemannian manifold, and W a 2-dimensional subspace of M_p, spanned by $X, Y \in M_p$. Let $\mathcal{O} \subset W$ be a neighborhood of $0 \in M_p$ on which exp is a diffeomorphism, let $i\colon \exp(\mathcal{O}) \longrightarrow M$ be the inclusion, and let $\bar{R}$ be the Riemann curvature tensor for $\exp(\mathcal{O})$ with the induced Riemannian metric $i^* < , >$. Then

$$< \bar{R}(X,Y)Y,X > = < R(X,Y)Y,X > \ .$$

Consequently,

$$\frac{< R(X,Y)Y,X >}{\|X,Y\|^2}$$

is the Gaussian curvature at p of the surface $\exp(\mathcal{O})$.

First Proof. It obviously suffices to prove the theorem when X and Y are orthonormal. Choose a Riemannian normal coordinate system at p with $X = \partial/\partial x^1|_p$, $Y = \partial/\partial x^2|_p$; then x^1, x^2 are a coordinate system on $\exp(\mathcal{O})$. Now we are trying to prove that $\bar{R}_{1212}(p) = R_{1212}(p)$. But in (***), the terms involving Christoffel symbols vanish at p. The theorem is now obvious, since the functions $\bar{g}_{ij}$ $(i,j = 1,2)$ defining the metric $i^* < , >$ are just the corresponding g_{ij} restricted to $\exp(\mathcal{O})$, and they have the same mixed partial derivatives with respect to x^1 and x^2.

Second Proof. It is even more obvious that the quadratic form $\bar{Q}$ associated with $(\exp(\mathcal{O}), i^* < , >)$ is the restriction to W of the quadratic form Q on M_p, for they are the second non-zero terms in the Taylor expansion of the same metric. ∎

5. COROLLARY. Let $(M, < , >)$ be a Riemannian manifold, let $X, Y \in M_p$ span a 2-dimensional subspace W of M_p, and let $\mathcal{O} \subset W$ be a neighborhood of 0 on which exp is a diffeomorphism. If Q is the quadratic form on M defined previously, then

$$\frac{-3Q(X,Y)}{\|X,Y\|^2} = \frac{< R(X,Y)Y,X >}{\|X,Y\|^2} = k \quad ,$$

where k is the Gaussian curvature at p of the surface $\exp(\mathcal{O})$.

The quantity $< R(X,Y)Y,X >/\|X,Y\|^2$ appearing in Corollary 5 is called the sectional curvature $k(W)$ of W. It would seem that the function $(X,Y) \longmapsto < R(X,Y)Y,X >$ contains only a small portion of the total information contained in the curvature tensor, but Propositions 6 and 8, which follow, show that R satisfies certain identities which allow it to be determined in terms of the metric $< , >$ and the quadratic function Q which it determines.

6. PROPOSITION. The curvature tensor satisfies the following identities:

(1) $R(X,Y)Z = -R(Y,X)Z$

(2) $< R(X,Y)Z,W > = - < R(X,Y)W,Z >$

(3) $R(X,Y)Z + R(Y,Z)X + R(Z,X)Y = 0$

(4) $< R(X,Y)Z,W > = < R(Z,W)X,Y >$.

Proof. In a coordinate system x, these relations are equivalent to

(1) $R^i{}_{jk\ell} = -R^i{}_{j\ell k}$

(2) $R_{ijk\ell} = -R_{jik\ell}$

(3) $R^i{}_{jk\ell} + R^i{}_{k\ell j} + R^i{}_{\ell jk} = 0$

(4) $R_{k\ell ij} = R_{ijk\ell}$.

These are immediate from (**) and (***). ∎

Notice that (1) can also be written

(1') $< R(X,Y)Z,W > = - < R(Y,X)Z,W >$ $\qquad R_{ijk\ell} = -R_{jik\ell}$.

Thus $< R(X,Y)Z,W >$ is skew-symmetric in both (X,Y) and (Z,W), which again shows that $< R(X,Y)Y,X >$ changes by $\det(a_{ij})^2$ when X and Y are replaced by $a_{11}X + a_{21}Y, a_{21}X + a_{22}Y$. Property (3) can be written similarly, by taking the inner product with an arbitrary vector W. For later use, we insert a result which shows that the fourth property of R is a formal consequence of the others.

7. PROPOSITION. Let V be a vector space and $\mathcal{R}: V \times V \times V \times V \longrightarrow \mathbb{R}$ a

multilinear map satisfying

$$(1)\quad \mathcal{R}(X,Y,Z,W) = -\mathcal{R}(Y,X,Z,W)$$

$$(2)\quad \mathcal{R}(X,Y,Z,W) = -\mathcal{R}(X,Y,W,Z)$$

$$(3)\quad \mathcal{R}(X,Y,Z,W) + \mathcal{R}(Y,Z,X,W) + \mathcal{R}(Z,X,Y,W) = 0.$$

Then $\mathcal{R}$ also satisfies

$$(4)\quad \mathcal{R}(X,Y,Z,W) = \mathcal{R}(Z,W,X,Y) \quad .$$

Proof. The proof is a tricky manipulation, cleverly systematized by the following diagram from Milnor's Morse Theory.

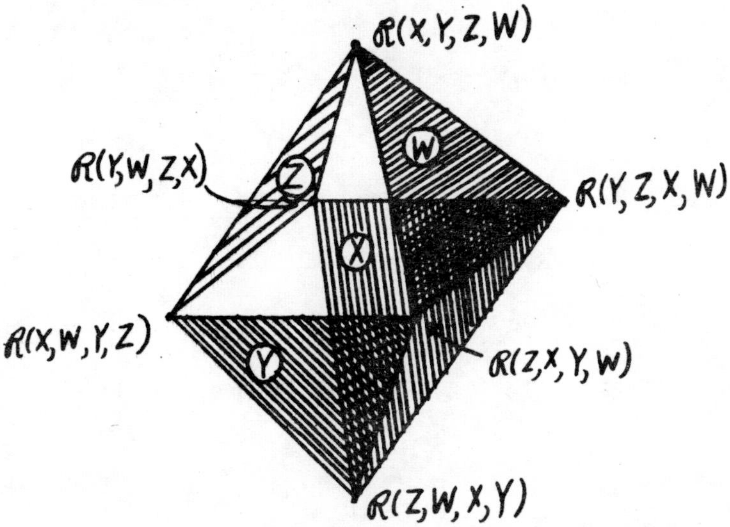

Equation (3) shows that the sum of the numbers at the vertices of triangle W is zero. The sum of the vertices of triangles X, Y, Z is also seen to be zero, using (1) and (2). Adding these identities for the top two triangles, and subtracting the identities for the bottom ones, we see that twice the top vertex minus twice the bottom vertex is zero. ∎

8. PROPOSITION. Let V be a vector space and $\mathcal{R}_i : V \times V \times V \times V \longrightarrow \mathbb{R}$ two multilinear maps satisfying (1) - (4) of Proposition 7. Suppose that

$\mathcal{R}_1(X,Y,X,Y) = \mathcal{R}_2(X,Y,X,Y)$ for all $X,Y \in V$. Then $\mathcal{R}_1 = \mathcal{R}_2$.

Proof. It clearly suffices to prove that a multilinear $\mathcal{R}$ satisfying (1) - (4) is 0 if $\mathcal{R}(X,Y,X,Y) = 0$ for all $X,Y \in V$. Now we have

$$\begin{aligned} 0 &= \mathcal{R}(X,Y+W,X,Y+W) \\ &= \mathcal{R}(X,Y,X,Y) + \mathcal{R}(X,Y,X,W) + \mathcal{R}(X,W,X,Y) + \mathcal{R}(X,W,X,W) \\ &= \mathcal{R}(X,Y,X,W) + \mathcal{R}(X,W,X,Y) \\ &= 2\mathcal{R}(X,Y,X,W) \quad . \end{aligned}$$

Using (1) and (2), we easily see that $\mathcal{R}$ is alternating, and hence skew-symmetric. Consequently, (3) gives

$$3\mathcal{R}(X,Y,Z,W) = 0. \blacksquare$$

Propositions 6 and 8 tell us that the curvature tensor R is completely determined by the values of $\langle R(X,Y)Y,X \rangle$, and hence by the quadratic function Q. [This means that in a sense we can frame a coordinate free definition of the curvature tensor, but it would certainly be an awkward one. Moreover, given a multilinear map $\mathcal{R}: V \times V \times V \times V \longrightarrow \mathbb{R}$, satisfying (1) - (4), it is a fairly difficult exercise to work out a formula for $\mathcal{R}$ in terms of the quantities $\mathcal{R}(X,Y,X,Y)$.] In terms of a coordinate system, we see that R is determined by the components R_{ijij}, of which there are $n\frac{n-1}{2}$ with $i < j$. According to Riemann, these $n\frac{n-1}{2}$ functions must determine the metric completely; in other words, the curvature tensor must determine the metric. Recall that we have selected one special case of this assertion as our Test Case, which can now be restated as follows: If $R = 0$, then the manifold is flat. We are ready to present the first, and longest, of our proofs of the Test Case. It is separated into three Steps, and all our subsequent proofs, no matter how elegant and brief, essentially contain these same three Steps.

9. THEOREM (THE TEST CASE; first version). Let $(M, \langle\ ,\ \rangle)$ be an n-dimensional Riemannian manifold for which the curvature tensor R is 0. Then M is locally isometric to $\mathbb{R}^n$ with its usual Riemannian metric.

Proof. This a purely local question, so we assume that M is $\mathbb{R}^n$, with the standard coordinate system $y^1,\ldots,y^n$, and the Riemannian metric

$$\langle\ ,\ \rangle = \sum_{i,j=1}^{n} g_{ij}\,dy^i \otimes dy^j .$$

Step 1. We claim that there are functions $(h_1,\ldots,h_n)$, with any desired initial conditions $(h_1(0),\ldots,h_n(0))$, satisfying the equations

$$(*) \qquad \frac{\partial h_j}{\partial y^k} = \sum_{\gamma=1}^{n} \Gamma^{\gamma}_{jk} h_\gamma .$$

The reason for this is, of course, that the relations $R^{\gamma}{}_{j\ell k} = 0$, which express the vanishing of R, are just the integrability conditions for (*), as we have already seen.

In particular, for $\alpha = 1,\ldots,n$ we can choose such a set $(h^{(\alpha)}{}_1,\ldots,h^{(\alpha)}{}_n)$ satisfying the initial condition

$$(h^{(\alpha)}{}_1(0),\ldots,h^{(\alpha)}{}_n(0))_0 = X_\alpha ,$$

where $X_1,\ldots,X_n \in \mathbb{R}^n{}_0$ is orthonormal with respect to $\langle\ ,\ \rangle_0$.

Step 2. We claim that if $(h_1,\ldots,h_n)$ satisfies (*), then $h = dx$ for some function x, i.e. $h_j = \partial x/\partial y^j$. In terms of the form

$$\eta = h_1 dy^1 + \ldots + h_n dy^n \quad ,$$

we are just saying that η is exact. We know (Corollary I.7-15) that this is true if and only if

$$\frac{\partial h_j}{\partial y^k} = \frac{\partial h_k}{\partial y^j} \quad .$$

Glancing at (*), we see that this is indeed true, since $\Gamma^\gamma_{jk} = \Gamma^\gamma_{kj}$.

Now choose functions x^α with $h^{(\alpha)}{}_j = \partial x^\alpha/\partial y^j$. Then the functions x^α satisfy

$$(\dagger) \qquad \frac{\partial^2 x^\alpha}{\partial y^j \partial y^k} = \sum_{\gamma=1}^{n} \Gamma^\gamma_{jk} \frac{\partial x^\alpha}{\partial y^\gamma} \quad ,$$

[these are the equations (6), obtained earlier]

and

$$\left(\frac{\partial x^\alpha}{\partial y^j}(0)\right) = \begin{pmatrix} X_1 \\ \vdots \\ X_n \end{pmatrix} \quad ;$$

this matrix is non-singular, so $x^1,\ldots,x^n$ is a coordinate system in a neighborhood of 0.

Step 3. We claim that x is the desired coordinate system, i.e., that

$$(\dagger\dagger) \qquad g_{\mu\nu} = \sum_{i,j=1}^{n} g^{ij} \frac{\partial x^\mu}{\partial y^i} \frac{\partial x^\nu}{\partial y^j}$$

[equation (4')] .

We know that this equation holds at 0, by the choice of the initial conditions $\partial x^{\alpha}/\partial y^{j}(0)$. So it suffices to show that the right side of (††) has all partial derivatives $\partial/\partial y^{k}$ equal to 0. To do this, we will first need a formula for $\partial g^{ij}/\partial y^{k}$. Recall that by the equation on page I.9-41 (which is equivalent to the definition of the Christoffel symbols), we have

$$(\ddagger) \qquad \frac{\partial g_{ij}}{\partial y^{k}} = [ik,j] + [jk,i] \quad .$$

Now, differentiating

$$\sum_{m=1}^{n} g_{\ell m} g^{mj} = \delta_{m}^{j}$$

yields

$$\sum_{m=1}^{n} g_{\ell m} \frac{\partial g^{mj}}{\partial y^{k}} = - \sum_{m=1}^{n} \frac{\partial g_{\ell m}}{\partial y^{k}} g^{mj} \quad ,$$

which gives

$$\begin{aligned} \frac{\partial g^{ij}}{\partial y^{k}} &= \sum_{\ell,m} g^{i\ell} g_{\ell m} \frac{\partial g^{mj}}{\partial y^{k}} \\ &= - \sum_{\ell,m} g^{i\ell} g^{mj} \frac{\partial g_{\ell m}}{\partial y^{k}} \\ &= - \sum_{\ell,m} g^{i\ell} g^{mj} ([\ell k,m] + [mk,\ell]) \quad \text{by } (\ddagger) \\ &= - \sum_{\ell} g^{i\ell} \Gamma_{\ell k}^{j} - \sum_{m} g^{mj} \Gamma_{mk}^{i} \quad , \end{aligned}$$

or

$$(\ddagger\ddagger) \qquad \frac{\partial g^{ij}}{\partial y^{k}} = - \sum_{\ell=1}^{n} (g^{i\ell} \Gamma_{\ell k}^{j} + g^{\ell j} \Gamma_{\ell k}^{i}) \quad .$$

So

$$\frac{\partial}{\partial y^k}\left(\sum_{i,j} g^{ij}\frac{\partial x^\mu}{\partial y^i}\frac{\partial x^\nu}{\partial y^j}\right) = \sum_{i,j}\frac{\partial g^{ij}}{\partial y^k}\frac{\partial x^\mu}{\partial y^i}\frac{\partial x^\nu}{\partial y^j}$$

$$+ \sum_{i,j} g^{ij}\frac{\partial^2 x^\mu}{\partial y^i \partial y^k}\frac{\partial x^\nu}{\partial y^j} + \sum_{i,j} g^{ij}\frac{\partial x^\mu}{\partial y^i}\frac{\partial^2 x^\nu}{\partial y^j \partial y^k}$$

$$= \sum_{i,j}\frac{\partial g^{ij}}{\partial y^k}\frac{\partial x^\mu}{\partial y^i}\frac{\partial x^\nu}{\partial y^j}$$

$$+ \sum_{i,j} g^{ij}\sum_\gamma \Gamma^\gamma_{ik}\frac{\partial x^\mu}{\partial y^\gamma}\frac{\partial x^\nu}{\partial y^j}$$

$$+ \sum_{i,j} g^{ij}\sum_\gamma \Gamma^\gamma_{jk}\frac{\partial x^\mu}{\partial y^i}\frac{\partial x^\nu}{\partial y^\gamma} \qquad \text{by } (\dagger) .$$

Switching some indices, we thus have

$$\frac{\partial}{\partial y^k}\left(\sum_{i,j} g^{ij}\frac{\partial x^\mu}{\partial y^i}\frac{\partial x^\nu}{\partial y^j}\right) = \sum_{i,j}\frac{\partial x^\mu}{\partial y^i}\frac{\partial x^\nu}{\partial y^j}\left(\frac{\partial g^{ij}}{\partial y^k} + \sum_\gamma g^{\gamma j}\Gamma^i_{\gamma k} + \sum_\gamma g^{i\gamma}\Gamma^j_{\gamma k}\right)$$

$$= 0 \qquad \text{by } (\ddagger) .$$

This completes the proof of the theorem. ∎

As a brief review of the proof, we note that

<u>Step 1</u> uses the <u>integrability conditions, $R = 0$,</u> to obtain certain forms $\sum h_i^{(\alpha)} dy^i$, with any desired initial conditions;

Step 2 uses symmetry of the Christoffel symbols, $\Gamma_{ij}^{k} = \Gamma_{ji}^{k}$, to prove that $\Sigma h_i^{(\alpha)} dy^i = dx^{(\alpha)}$ for some x^α;

Step 3 uses the definition of the Christoffel symbols $[ij,k]$ to prove that the vectors $\partial/\partial x^\alpha$ are orthonormal.

Despite its length, the proof is essentially a straightforward application of the integrability conditions for partial differential equations. As Riemann says, at the very end of the section in Part C, "we obtain ... equations, which the functions g_{ij} are required to satisfy if $\sum_{i,j} g_{ij} dy^i dy^j$ can be transformed into a form with constant coefficients. Given an acquaintance with the traditional methods, it is demonstrated without difficulty that these ... conditions, when they are satisfied, suffice."

We have thus proved one special case of Riemann's assertion that the curvature determines the metric. We will not return to the more general assertion until Chapter 7, for our immediate task will be to begin systematizing all the results which have been uncovered so far.

Addendum. Riemann's Invariant Definition of the Curvature Tensor.

Between the derivation of the necessary conditions for a flat manifold, $R_{ijk\ell} = 0$, and the final remark about their sufficiency, Riemann inserts one more remarkable result, which will play no further role in our development of Riemannian geometry, but which is nevertheless extremely interesting, for it amounts to another invariant definition of the curvature tensor. Riemann's description is particularly difficult to decipher because he uses classical notation from the calculus of variations. To state it in modern terms, consider a function $s\colon \mathbb{R}^2 \longrightarrow M$, with $s(0) = p \in M$. This function s can be thought of as a "2-parameter variation of the point p", and on M it gives rise to 2 vector fields,

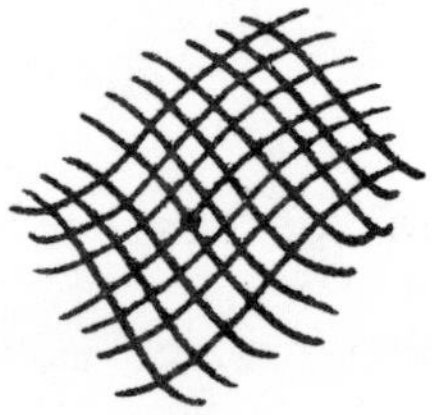

$$\frac{\partial s}{\partial x} = s_*\left(\frac{\partial}{\partial x}\right)$$

$$\frac{\partial s}{\partial y} = s_*\left(\frac{\partial}{\partial y}\right) .$$

This notation involves the usual ambiguities, and in conformity with this, the function

$$(x,y) \longmapsto \left< \frac{\partial s}{\partial x}(x,y),\ \frac{\partial s}{\partial x}(x,y) \right>$$

on $\mathbb{R}^2$ will be denoted simply by

$$< \frac{\partial s}{\partial x} , \frac{\partial s}{\partial x} > \quad .$$

At the top of page 4C-4, Riemann directs us to consider the expression

$$\text{(A)} \quad \frac{\partial^2}{\partial x^2} < \frac{\partial s}{\partial y} , \frac{\partial s}{\partial y} > -2 \frac{\partial^2}{\partial x \partial y} < \frac{\partial s}{\partial x} , \frac{\partial s}{\partial y} > + \frac{\partial}{\partial y^2} < \frac{\partial s}{\partial x} , \frac{\partial s}{\partial x} > \quad .$$

If we let

$$X = \frac{\partial s}{\partial x} , \ Y = \frac{\partial s}{\partial y} \quad ,$$

then the value of this expression at $(x,y) = (0,0)$ can be written

$$X_p(X(< Y,Y >)) - 2 X_p(Y(< X,Y >)) + Y_p(Y(< X,X >)) \quad ,$$

or simply

$$\text{(B)} \qquad [XX < Y,Y > - 2 XY < X,Y > + YY < X,X >](p) \quad .$$

On the other hand, we can be sure that the expression (B) will equal (A) for some $s: R^2 \longrightarrow M$ only when $[X,Y] = 0$. Riemann does not consider all $s: R^2 \longrightarrow M$, but only those with the following property: If $\sigma: R^3 \longrightarrow M$ and $\sigma(x,y,0) = s(x,y)$, then

$$\left.\begin{aligned}
&\frac{\partial}{\partial z}<\frac{\partial\sigma}{\sigma x},\frac{\partial\sigma}{\partial y}> - \frac{\partial}{\partial y}<\frac{\partial\sigma}{\partial x},\frac{\partial\sigma}{\partial z}> - \frac{\partial}{\partial x}<\frac{\partial\sigma}{\partial y},\frac{\partial\sigma}{\partial z}> = 0\\
&\frac{\partial}{\partial z}<\frac{\partial\sigma}{\partial x},\frac{\partial\sigma}{\partial x}> -2\frac{\partial}{\partial x}<\frac{\partial\sigma}{\partial x},\frac{\partial\sigma}{\partial z}> = 0\\
&\frac{\partial}{\partial z}<\frac{\partial\sigma}{\partial y},\frac{\partial\sigma}{\partial y}> -2\frac{\partial}{\partial y}<\frac{\partial\sigma}{\partial y},\frac{\partial\sigma}{\partial z}> = 0 \quad .
\end{aligned}\right\}\quad \begin{aligned}&\text{at } (x,y,z)\\ &= (0,0,0).\end{aligned}$$

Riemann claims that if $s: \mathbb{R}^2 \longrightarrow M$ has this property, then the value of (A) at $(x,y) = (0,0)$ is $-2 < R(X_p,Y_p)Y_p,X_p >$. (To account for the factor of -2 recall that Riemann's $(ij,k\ell)$ equals $Rij\ell k/2 = -Rijk\ell/2$, and note that Riemann has $(dy^i \wedge dy^j) \otimes (dy^k \wedge dy^\ell)$ instead of $dy^i \otimes dy^j \otimes dy^k \otimes dy^\ell$.) This assertion can be rephrased as follows.

> Let X and Y be vector fields on M with $[X,Y] = 0$. Suppose also that for every vector field Z with $[X,Z](p) = [Y,Z](p) = 0$, we have
>
> $$\left.\begin{aligned}
> &Z < X,Y > - Y < X,Z > - X < Y,Z > = 0\\
> &Z < X,X > - 2X < X,Z > = 0\\
> &Z < Y,Y > - 2Y < Y,Z > = 0
> \end{aligned}\right\}\quad \text{at } p\ .$$
>
> Then
>
> $$-2 < R(X_p,Y_p)Y_p,X_p > = [XX < Y,Y > - 2XY < X,Y > + < X,X >](p).$$

Admittedly, it is not even <u>a priori</u> clear that we can find vector fields X and Y with the desired property and arbitrary values X_p,Y_p. However, the existence of such vector fields, and the conclusion itself, can be established by a computation. Rather than performing such a computation, we will save the proof of this result until Addendum 2 of Chapter 6, where it will follow from a more significant expression for the curvature tensor.

Chapter 5. The Absolute Differential Calculus (The Ricci Calculus); or, The Debauch of Indices

Although Riemann gave an invariant description of his curvature tensor, the (classical) notion of a tensor did not even exist in his time. The development of the "Calculus of Tensors" is due mainly to Ricci, and was carried out in the years 1887-1896; in 1901 he and his student Levi-Civita gave a detailed description in a memoir "Methods de calcul differential absolu et leurs applications." In addition to a comprehensive use of tensors, what distinguished the Absolute Differential Calculus, and gave it its name, was an important construction which greatly simplified all the concepts of Riemannian geometry, especially the curvature tensor. Instead of checking that the horrible formula used to define $R^i{}_{jk\ell}$ transforms correctly, we are going to check that an equally mysterious — but not quite so horrible formula — transforms correctly, and thus defines a tensor; then we will define the curvature tensor in terms of this one.

In 1869, in one of the earliest papers which took up Riemann's ideas, Christoffel had already made the observation which Ricci made the basis for his calculus. Suppose that Y is a vector field, and

$$Y = \sum_{j=1}^{n} \lambda^j \frac{\partial}{\partial x^j}$$

$$= \sum_{j=1}^{n} \lambda'^j \frac{\partial}{\partial x'^j} .$$

Consider the following symbols, defined in terms of the Christoffel symbols of the second kind for the coordinate systems x and x':

$$\lambda^j{}_{;h} = \frac{\partial \lambda^j}{\partial x^h} + \sum_{\nu=1}^{n} \lambda^\nu \Gamma^j_{h\nu}$$

$$\lambda'^j{}_{;h} = \frac{\partial \lambda'^j}{\partial x'^h} + \sum_{\nu=1}^{n} \lambda'^\nu \Gamma'^j_{h\nu} .$$

It is easy to check that for these symbols — the sums of partial derivatives of the λ^j and certain linear combinations of them — we have

$$\lambda'^{\alpha}{}_{;\beta} = \sum_{h,j=1}^{n} \lambda^j{}_{;h} \frac{\partial x^h}{\partial x'^{\beta}} \frac{\partial x'^{\alpha}}{\partial x^j} .$$

For the calculation we use the formula on page 4D-8 together with the formula

$$\lambda'^{\alpha} = \sum_{j=1}^{n} \lambda^j \frac{\partial x'^{\alpha}}{\partial x^j} ;$$

when we compute $\partial\lambda'^{\alpha}/\partial x'^{\beta}$, the extra terms involving second partial derivatives of the x'^{α} just cancel out with the extra terms in the transformation formula for the Christoffel symbols. This calculation shows that there is a certain tensor field of type $\binom{1}{1}$ which equals

$$\sum_{h,j=1}^{n} \lambda^j{}_{;h}\, dx^h \otimes \frac{\partial}{\partial x^j}$$

in the x coordinate system. In classical terms, "if the λ^j transform like a tensor of type $\binom{0}{1}$, then the $\lambda^j{}_{;h}$ transform like a tensor of type $\binom{1}{1}$." Similarly, if the λ_i transform like a tensor of type $\binom{1}{0}$, then the quantities

$$\lambda_{i;h} = \frac{\partial \lambda_i}{\partial x^h} - \sum_{\nu=1}^{n} \lambda_{\nu} \Gamma^{\nu}_{hi}$$

are easily seen to transform like a tensor of type $\binom{2}{0}$, so that there is a tensor of type $\binom{2}{0}$ which equals

$$\sum_{i,h=1}^{n} \lambda_{i;h}\, dx^i \otimes dx^h$$

in the x coordinate system. Generally, given a tensor of type $\binom{k}{\ell}$, with components

$$A_{i_1 \dots i_k}^{j_1 \dots j_\ell} ,$$

there is a new tensor, of type $\binom{k+1}{\ell}$, with components

$$A_{i_1 \dots i_k ; h}^{j_1 \dots j_\ell} = \frac{\partial A_{i_1 \dots i_k}^{j_1 \dots j_\ell}}{\partial x^h} + \sum_{s=1}^{\ell} \sum_{\nu=1}^{n} A_{i_1 \dots i_k}^{j_1 \dots j_{s-1} \nu j_{s+1} \dots j_\ell} \Gamma_{h\nu}^{j_r}$$

$$- \sum_{r=1}^{n} \sum_{\nu=1}^{n} A_{i_1 \dots i_{r-1} \nu i_{r+1} \dots i_k}^{j_1 \dots j_\ell} \Gamma_{h i_r}^{\nu} .$$

The proof is again just an enormous calculation. At the other extreme from this most general case is the special case of a tensor field of type $\binom{0}{0}$ on M, i.e., a function $f\colon M \longrightarrow \mathbb{R}$. Here we simply define

$$f_{;h} = \frac{\partial f}{\partial x^h} ,$$

so that in this case at least we know what the tensor field

$$\sum_{h=1}^{n} f_{;h} dx^h = \sum_{h=1}^{n} \frac{\partial f}{\partial x^h} dx^h$$

actually is — it is just df.

The tensor of type $\binom{k+1}{\ell}$ which we thus obtain from a tensor A of type $\binom{k}{\ell}$ is called the <u>covariant derivative</u> of A, since it is covariant of one order greater than A. It is also called the "absolute derivative" of A; here the word "absolute" means that it doesn't depend on a particular coordinate system. Various notations for the covariant derivative are encountered. Sometimes one sees $\lambda^j_{i|h}$, while the simpler λ^j_{ih} is often used when no misunderstanding is possible. The most common convention is not to use commas between the indices of the original components $A^{j_1\ldots j_\ell}_{i_1\ldots i_k}$, and then to use the comma to indicate the covariant derivative, with components $A^{j_1\ldots j_\ell}_{i_1\ldots i_k,h}$. Although we have now stopped using commas between the original indices, we will still use semi-colons, to avoid confusion.

A partial answer to the question "What does the covariant derivative really mean?" is given by the following observation.

1. PROPOSITION. Let x be a Riemannian normal coordinate system at $p \in M$, and A a tensor of type $\binom{k}{\ell}$ on M, with

$$A = \sum A^{j_1\ldots j_\ell}_{i_1\ldots i_k}\, dx^{i_1} \otimes \ldots \otimes dx^{i_k} \otimes \frac{\partial}{\partial x^{j_1}} \otimes \ldots \otimes \frac{\partial}{\partial x^{j_\ell}} .$$

Then the components at p of the covariant derivative of A are just ordinary partial derivatives

$$(*) \qquad A^{j_1\ldots j_\ell}_{i_1\ldots i_k;h}(p) = \frac{\partial A^{j_1\ldots j_\ell}_{i_1\ldots i_k}}{\partial x^h}(p) .$$

Proof. In a Riemannian normal coordinate system at p, the Christoffel symbols $[ij,k]$, and hence also the Γ^k_{ij}, are all 0 at p. ∎

Proposition 1 can even be used to define covariant derivatives. The Riemannian normal coordinate systems at p are a natural set of coordinate systems, determined by the metric < , >; any two such systems at p differ only by an element of O(n), and it is easy to see from this that a definition of the covariant derivative by means of (*) would not depend on which one was picked. With a little work, one could then deduce the general expression for the components of the covariant derivative of A in any coordinate system. In succeeding chapters we will be giving still other interpretations of the covariant derivative.

The operation of covariant differentiation obeys many rules analagous to those for ordinary differentiation.

2. PROPOSITION. Let A and B be tensors of type $\binom{k}{\ell}$, and C a tensor of type $\binom{k'}{\ell'}$. Then

$$(1)\quad \left(A^{j_1\ldots j_\ell}_{i_1\ldots i_k} + B^{j_1\ldots j_\ell}_{i_1\ldots i_k}\right)_{;h} = A^{j_1\ldots j_\ell}_{i_1\ldots i_k;h} + B^{j_1\ldots j_\ell}_{i_1\ldots i_k;h}$$

$$(2)\quad \left(A^{j_1\ldots j_\ell}_{i_1\ldots i_k}\, C^{v_1\ldots v_{\ell'}}_{u_1\ldots u_{k'}}\right)_{;h} = A^{j_1\ldots j_\ell}_{i_1\ldots i_k;h}\, C^{v_1\ldots v_{\ell'}}_{u_1\ldots u_{k'}} + A^{j_1\ldots j_\ell}_{i_1\ldots i_k}\, C^{v_1\ldots v_{\ell'}}_{u_1\ldots u_{k'};h} .$$

$$(3)\quad \left(\sum_{\nu=1}^{n} A^{j_1\ldots j_{s-1}\nu j_{s+1}\ldots j_\ell}_{i_1\ldots i_{r-1}\nu i_{r+1}\ldots i_k}\right)_{;h} = \sum_{\nu=1}^{n} A^{j_1\ldots j_{s-1}\nu j_s\ldots j_\ell}_{i_1\ldots i_{r-1}\nu i_r\ldots i_k;h}$$

(Covariant differentiation is a derivation and commutes with sums and contractions.)

Consequently, we have, for example,

$$(4)\quad \Big(\sum_{\substack{\alpha_1,\ldots,\alpha_r \\ \beta_1,\ldots,\beta_s}} A^{\beta_1\cdots\beta_s j_{s+1}\cdots j_\ell}_{\alpha_1\cdots\alpha_r i_{r+1}\cdots i_k} \; C^{\alpha_1\cdots\alpha_r v_{r+1}\cdots v_{\ell'}}_{\beta_1\cdots\beta_s u_{s+1}\cdots u_{k'}} \Big)_{;h}$$

$$= \sum_{\substack{\alpha_1,\ldots,\alpha_r \\ \beta_1,\ldots,\beta_s}} \Bigg[\Big(A^{\beta_1\cdots\beta_s j_{s+1}\cdots j_\ell}_{\alpha_1\cdots\alpha_r i_{r+1}\cdots i_k} \Big)_{;h} \; C^{\alpha_1\cdots\alpha_r v_{r+1}\cdots v_{\ell'}}_{\beta_1\cdots\beta_s u_{s+1}\cdots u_{k'}}$$

$$+ A^{\beta_1\cdots\beta_s j_{s+1}\cdots j_\ell}_{\alpha_1\cdots\alpha_r i_{r+1}\cdots i_k} \Big(C^{\alpha_1\cdots\alpha_r v_{r+1}\cdots v_{\ell'}}_{\beta_1\cdots\beta_s u_{s+1}\cdots u_{k'}} \Big)_{;h} \Bigg] .$$

Proof. Compute Because of Proposition 1, the computations become trivial if one uses Riemannian normal coordinates. ∎

Aside from these general formulas, there are two of crucial importance. The first of these is about the covariant derivative of the tensors $\langle\, ,\, \rangle$ and $\langle\, ,\, \rangle^*$.

3. PROPOSITION (RICCI'S LEMMA). The g_{ij} and g^{ij} behave like constants in covariant differentiation; that is,

$$g_{ij;k} = 0$$

$$g^{ij}{}_{;k} = 0 \quad .$$

Proof. The proof is, of course, a calculation. For the g_{ij} we have

$$g_{ij;k} = \frac{\partial g_{ij}}{\partial x^k} - \sum_{\nu=1}^{n} g_{\nu j}\Gamma^{\nu}_{ki} - \sum_{\nu=1}^{n} g_{i\nu}\Gamma^{\nu}_{kj}$$

$$= \frac{\partial g_{ij}}{\partial x^k} - [ik,j] - [jk,i]$$

$$= 0 \quad ,$$

by equation (‡) on page 4-D21. Similarly, the second equation is equivalent to (‡‡).

It can also be obtained from the first equation, the relation $\sum_j g^{ij} g_{j\ell} = \delta^i_\ell$, Proposition 2, and the easily verified relation $\delta^i_{\ell;k} = 0$. ∎

If A is a tensor of type $\binom{k}{\ell}$, with components $A^{j_1 \cdots j_\ell}_{i_1 \cdots i_k}$, then the operation of covariant differentiation can be applied to the tensor B with components

$$B^{j_1 \cdots j_\ell}_{i_1 \cdots i_k h} = A^{j_1 \cdots j_\ell}_{i_1 \cdots i_k ;h} \quad ;$$

there results a tensor C with components

$$C^{j_1 \cdots j_\ell}_{i_1 \cdots i_j h\eta} = B^{j_1 \cdots j_\ell}_{i_1 \cdots i_k h;\eta} = \left(A^{j_1 \cdots j_\ell}_{i_1 \cdots i_k ;h}\right)_{;\eta} .$$

These components are denoted by

$$A^{j_1 \cdots j_\ell}_{i_1 \cdots i_k ;h;\eta} \text{ or simply } A^{j_1 \cdots j_\ell}_{i_1 \cdots i_k ;h\eta} .$$

For example, if we start with a function f [a tensor of type $\binom{0}{0}$], then

$$f_{;i} = \frac{\partial f}{\partial x^i}$$

and

$$f_{;ij} = (f_{;i})_{;j} = \frac{\partial\left(\frac{\partial f}{\partial x^i}\right)}{\partial x^j} - \sum_{\nu=1}^{n} \frac{\partial f}{\partial x^\nu} \Gamma^\nu_{ji}$$

$$= \frac{\partial^2 f}{\partial x^i \partial x^j} - \sum_{\nu=1}^{n} \frac{\partial f}{\partial x^\nu} \Gamma^\nu_{ji} .$$

Notice that

$$f_{;ij} = f_{;ji} \quad ,$$

by symmetry of the Γ^{ν}_{ij}. The same result definitely does not hold for other tensors; instead we have the following basic result.

4. PROPOSITION (RICCI'S IDENTITIES). If the λ^i and λ_i are components of tensors of type $\binom{0}{1}$ and $\binom{1}{0}$, respectively, then

$$\lambda^i{}_{;jk} - \lambda^i{}_{;kj} = -\sum_{\ell=1}^{n} \lambda^{\ell} R^i{}_{\ell jk}$$

$$\lambda_{i;jk} - \lambda_{i;kj} = \sum_{\ell=1}^{n} \lambda_{\ell} R^{\ell}{}_{ijk} \quad ,$$

where

$$R^i{}_{jk\ell} = \frac{\partial \Gamma^i_{\ell j}}{\partial x^k} - \frac{\partial \Gamma^i_{kj}}{\partial x^{\ell}} + \sum_{\mu=1}^{n} \Gamma^{\mu}_{\ell j}\Gamma^i_{k\mu} - \Gamma^{\mu}_{kj}\Gamma^i_{\ell\mu} \quad .$$

(There are similar identities for tensors of type $\binom{k}{\ell}$, but we will ignore them.*)

---

* For those who cannot bear to be left in ignorance, the general Ricci identity is

$$A^{j_1\ldots j_\ell}_{i_1\ldots i_k;h\eta} - A^{j_1\ldots j_\ell}_{i_1\ldots i_k;\eta h} = \sum_{r=1}^{k}\sum_{\nu=1}^{n} A^{j_1\ldots j_\ell}_{i_1\ldots i_{r-1}\nu i_{r+1}\ldots i_k} R^{\nu}{}_{i_r h\eta}$$

$$- \sum_{s=1}^{\ell}\sum_{\nu=1}^{n} A^{j_1\ldots j_{s-1}\nu j_{s+1}\ldots j_\ell}_{i_1\ldots i_k} R^{j_s}{}_{\nu h\eta}$$

Proof. Compute. [The second identity is a consequence of the first (and *vice versa*), for if we are given λ_i and define

$$\lambda^i = \sum_{\alpha=1}^{n} g^{i\alpha}\lambda_\alpha \quad , \qquad \text{so that } \lambda_i = \sum_{\alpha=1}^{n} g_{i\alpha}\lambda^\alpha \quad ,$$

then

$$\lambda_{i;jk} - \lambda_{i;kj} = \Big(\sum_{\alpha=1}^{n} g_{i\alpha}\lambda^\alpha\Big)_{;jk} - \Big(\sum_{\alpha=1}^{n} g_{i\alpha}\lambda^\alpha\Big)_{;kj}$$

$$= \sum_{\alpha=1}^{n} g_{i\alpha}(\lambda^\alpha{}_{;jk} - \lambda^\alpha{}_{;kj}) \qquad \text{using Proposition 2}$$

$$= -\sum_{\alpha=1}^{n}\sum_{\ell=1}^{n} g_{i\alpha}\lambda^\ell R^\alpha{}_{\ell jk}$$

$$= -\sum_{\ell=1}^{n} \lambda^\ell R_{i\ell jk} \qquad \text{(by definition of } R_{i\ell jk})$$

$$= \sum_{\ell=1}^{n} \lambda^\ell R_{\ell ijk} \qquad \text{by Proposition 4D-6}$$

$$= \sum_{\ell=1}^{n}\sum_{\alpha=1}^{n} g_{\ell\alpha}\lambda^\ell R^\alpha{}_{ijk}$$

$$= \sum_{\alpha=1}^{n} \lambda_\alpha R^\alpha{}_{ijk} . \,]$$ ∎

5. COROLLARY. The $R^i{}_{jk\ell}$ are the components of a tensor of type $\binom{3}{1}$.

Proof. Let Z be a vector field, with

$$Z = \sum_{i=1}^{n} \lambda^i \frac{\partial}{\partial x^i} \quad ,$$

and let A be the tensor field with components $\lambda^i{}_{;jk}$. The first equation in Proposition 4 shows that

$$(*)\quad B(\frac{\partial}{\partial x^j}, \frac{\partial}{\partial x^k}) = A(\frac{\partial}{\partial x^j}, \frac{\partial}{\partial x^k}) - A(\frac{\partial}{\partial x^k}, \frac{\partial}{\partial x^j}) = \sum_{i=1}^{n} (\lambda^i{}_{;jk} - \lambda^i{}_{;kj}) \frac{\partial}{\partial x^i}$$

$$= - \sum_{i=1}^{n} (\sum_{\ell=1}^{n} \lambda^{\ell} R^i{}_{\ell jk}) \frac{\partial}{\partial x^i}$$

$$= - \sum_{\ell=1}^{n} \lambda^{\ell} (\sum_{i=1}^{n} R^i{}_{\ell jk} \frac{\partial}{\partial x^i}) .$$

This shows, in particular, that $B(\partial/\partial x^j|_p, \partial/\partial x^k|_p)$ does not depend on the vector field Z, but on the vector $Z(p)$ alone. So if we define

$$R(X,Y)Z = - B(X,Y) ,$$

then R is a tensor of type $\binom{3}{1}$, and

$$B(\frac{\partial}{\partial x^j}, \frac{\partial}{\partial x^k}) = - R(\frac{\partial}{\partial x^j}, \frac{\partial}{\partial x^k})Z$$

$$= - \sum_{\ell=1}^{n} \lambda^{\ell} R(\frac{\partial}{\partial x^j}, \frac{\partial}{\partial x^k})\frac{\partial}{\partial x^{\ell}} .$$

Since this is true for all n-tuples $\{\lambda^{\ell}\}$, comparison with equation (*) shows that the components of R are indeed $R^i{}_{\ell jk}$. ∎

Classically, this corollary would be deduced from the following general principle (compare Problem I.4-5(i)):

> Suppose we are given a set of numbers $T^i{}_{\ell jk}$ for the coordinate system x, a set $T'^i{}_{\ell jk}$ for the coordinate system x', etc. Suppose also that

$$C^i_{jk} = \sum_{\ell=1}^{n} \lambda^\ell T^i_{\ell jk}\ , \quad C'^i_{jk} = \sum_{\ell=1}^{n} \lambda'^\ell T'^i_{\ell jk}\ , \text{ etc.,}$$

are the components of a tensor for all tensors of type $\binom{0}{1}$ with components λ^ℓ in the coordinate system x, and components λ'^ℓ in the coordinate system x', etc. Then $T^i_{\ell jk}$ are the components of a tensor.

The classical proof is by a calculation. We have

$$\sum_{\delta=1}^{n} \lambda'^\delta T'^\alpha_{\delta\beta\gamma} = C'^\alpha_{\beta\gamma} = \sum_{i,j,k} C^i_{jk} \frac{\partial x'^\alpha}{\partial x^i} \frac{\partial x^j}{\partial x'^\beta} \frac{\partial x^k}{\partial x'^\gamma}$$

$$= \sum_{i,j,k,\ell} \lambda^\ell T^i_{\ell jk} \frac{\partial x'^\alpha}{\partial x^i} \frac{\partial x^j}{\partial x'^\beta} \frac{\partial x^k}{\partial x'^\gamma}$$

$$= \sum_{i,j,k,\ell,\delta} \lambda'^\delta T^i_{\ell jk} \frac{\partial x^\ell}{\partial x'^\delta} \frac{\partial x'^\alpha}{\partial x^i} \frac{\partial x^j}{\partial x'^\beta} \frac{\partial x^k}{\partial x'^\gamma} .$$

Since this is true for arbitrary λ, we can choose all λ'^δ but one equal to zero; this gives the desired transformation formulas.]

Corollary 5 represents only one minor application of the Ricci identities. A more significant application is obtained when we consider a manifold with vanishing curvature tensor. In this case, we have

$$\lambda^i{}_{;jk} = \lambda^i{}_{;kj}$$

$$\lambda_{i;jk} = \lambda_{i;kj} .$$

Thus, in manifolds with a vanishing curvature tensor (and only in such manifolds) the order of covariant differentiation is immaterial, so that in this respect covariant differentiation behaves like ordinary partial differentiation.* This enables us to give a simpler proof of

6. THEOREM (THE TEST CASE; second version) Let $(M, \langle\ ,\ \rangle)$ be an n-dimensional Riemannian manifold for which the curvature tensor R is 0. Then M is locally isometric to $\mathbb{R}^n$ with its usual Riemannian metric.

Proof. As before, we assume we are in $\mathbb{R}^n$, and choose the standard coordinate system $y^1,\dots,y^n$ around 0.

Step 1. We claim that we can find 1-forms $\eta = \sum_i \lambda_i dy^i$, with any desired value $\eta(0)$, satisfying

$$(*) \qquad \lambda_{i;j} = 0 \qquad i,j = 1,\dots,n .$$

To prove this, we begin by finding $\lambda_i(y,0,\dots,0)$, with the prescribed value for $y = 0$, and such that

$$(*) \quad 0 = \lambda_{i;1}(y,0,\dots 0) = \frac{\partial\lambda_i(y,0,\dots,0)}{\partial y^1} - \sum_{\nu=1}^{n} \lambda_\nu(y,0,\dots,0)\Gamma^\nu_{1i}(y,0,\dots,0) .$$

This just involves solving a set of ordinary differential equations, which is even linear, so that solutions exist for all y where the Γ's are defined.

*At first sight, it might appear that this should always be the case, since covariant differentiation is the same as ordinary partial differentiation in a Riemannian normal coordinate system around p. However, the relation $\lambda^i{}_{;j} = \partial\lambda^i/\partial x^j$ holds only at p, so generally $\lambda^i{}_{;jk}(p) \neq \partial^2\lambda^i/\partial x^j\partial x^k(p)$.

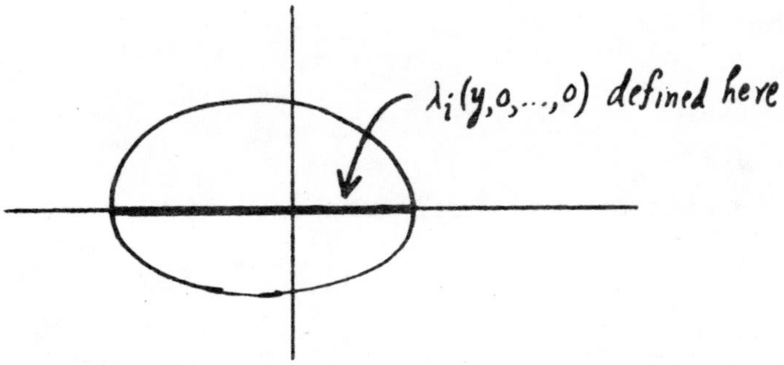

Next we find $\lambda_i(y_1,y,0,\ldots,0)$, with the initial values $\lambda_i(y_1,0,\ldots,0)$ just obtained, satisfying

$$(*_2) \qquad 0 = \lambda_{i;2}(y_1,y,\ldots,0) = \frac{\partial\lambda_i(y_1,y,0,\ldots,0)}{\partial y^2} - \sum_{\nu=1}^{n} \lambda_\nu(y_1,y,0,\ldots,0)\Gamma^\nu_{2i}(y_1,y,0,\ldots,0)$$

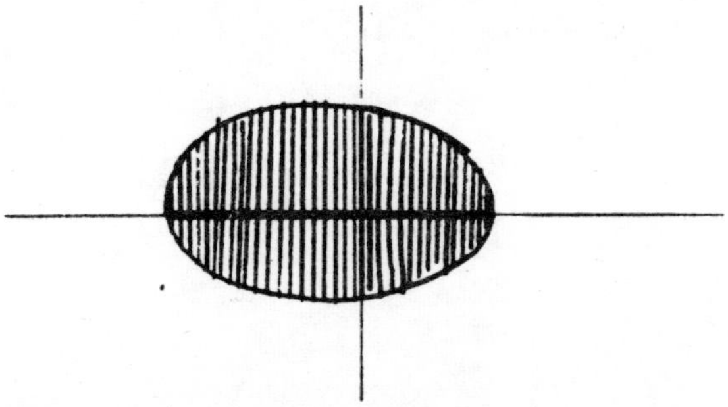

We continue in this way until we eventually obtain $\lambda_i(y_1,\ldots,y_{n-1},y)$ satisfying

$$(*_n) \qquad 0 = \lambda_{i;n}(y_1,\ldots,y_{n-1},y) \quad .$$

We now claim that, in addition to the relation $\lambda_{i;1}(y,0,\ldots,0) = 0$ given by $(*_1)$, we actually have

$$\lambda_{i;1}(y_1,y,0,\ldots,0) = 0 \quad .$$

To see this, we first note that we have

$$\lambda_{i;21} = \lambda_{i;2;1} = \frac{\partial\lambda_{i;2}}{\partial y^1} - \sum_{\nu=1}^{n} \lambda_{\nu;2}\Gamma^\nu_{1i} - \sum_{\nu=1}^{n} \lambda_{i;\nu}\Gamma^\nu_{12} \quad .$$

Since $0 = \lambda_{i;2}(y_1,y,0,\ldots,0)$ by $(*_2)$, we obtain

$$\lambda_{i;21} = - \sum_{\nu=1}^{n} \lambda_{i;\nu} \Gamma^{\nu}_{12} \qquad \text{at} \quad (y_1, y, 0, \ldots, 0) \quad .$$

Since $R = 0$, the Ricci identities then imply that

$$\lambda_{i;12} = \lambda_{i;21} = - \sum_{\nu=1}^{n} \lambda_{i;\nu} \Gamma^{\nu}_{12} \qquad \text{at} \quad (y_1, y, 0, \ldots, 0) \quad ,$$

i.e., that

$$\frac{\partial \lambda_{i;1}}{\partial y^2} - \sum_{\nu=1}^{n} \lambda_{\nu;1} \Gamma^{\nu}_{2i} - \sum_{\nu=1}^{n} \lambda_{i;\nu} \Gamma^{\nu}_{21}$$

$$= - \sum_{\nu=1}^{n} \lambda_{i;\nu} \Gamma^{\nu}_{12} \qquad \text{at} \quad (y_1, y, 0, \ldots, 0) \quad ,$$

so that

$$\frac{\partial \lambda_{i;1}}{\partial y^2} - \sum_{\nu=1}^{n} \lambda_{\nu;1} \Gamma^{\nu}_{2i} = 0 \qquad \text{at} \quad (y_1, y, 0, \ldots, 0).$$

Since we have the initial conditions $\lambda_{i;1}(y_1, 0, \ldots, 0) = 0$, it is clear that the solution of this equation is just the desired one,

$$\lambda_{i;1}(y_1, y, 0, \ldots, 0) = 0 \quad .$$

Proceeding in the same way, we next obtain

$$\lambda_{i;1}(y_1, y_2, y, 0, \ldots, 0) = \lambda_{i;2}(y_1, y_2, y, 0, \ldots, 0) = \lambda_{i;3}(y_1, y_2, y, 0, \ldots, 0) = 0 \quad ,$$

and, eventually,

$$0 = \lambda_{i;1} = \lambda_{i;2} = \dots = \lambda_{i;n} \qquad \text{at } (y_1,\dots,y_{n-1},y) \ .$$

This completes the proof of the claim.

For $\alpha = 1,\dots,n$ we now choose $\eta^{(\alpha)} = \Sigma\lambda^{(\alpha)}{}_i dy^i$, so that

$$X_\alpha = (\lambda^{(\alpha)}{}_1(0),\dots,\lambda^{(\alpha)}{}_n(0))_0$$

are orthonormal with respect to $\langle\ ,\ \rangle_0$.

<u>Step 2</u> We claim that if $\lambda_{i;j} = 0$, then $\Sigma\lambda_i dy^i = dx$ for some function x. This is because we have

$$0 = \lambda_{i;j} = \frac{\partial\lambda_i}{\partial y^j} - \sum_{\nu=1}^{n} \lambda_\nu \Gamma^\nu_{ji} \ ,$$

which shows that

$$\frac{\partial\lambda_i}{\partial y^j} = \frac{\partial\lambda_j}{\partial y^i} \ .$$

Now choose functions x^α with $\lambda^{(\alpha)}{}_i = \partial x^\alpha/\partial y^i$. As before, the x^α are a coordinate system in a neighborhood of 0.

<u>Step 3</u> We claim that x is the desired coordinate system, i.e., that

$$\delta_{\mu\nu} = \sum_{i,j=1}^{n} g^{ij} \frac{\partial x^\mu}{\partial y^i} \frac{\partial x^\nu}{\partial y^j} = \sum_{i,j=1}^{n} g^{ij} \lambda^{(\mu)}{}_i \lambda^{(\nu)}{}_j \ .$$

As before, we just have to prove that the right side of this equation has all partial derivatives $\partial/\partial y^k$ equal to 0. But

$$\frac{\partial}{\partial y^k}\left(\sum_{i,j} g^{ij}\lambda^{(\mu)}{}_i\lambda^{(\nu)}{}_j\right) = \left(\sum_{i,j} g^{ij}\lambda^{(\mu)}{}_i\lambda^{(\nu)}{}_j\right)_{;k}$$

$$= \sum_{i,j} g^{ij}{}_{;k}\lambda^{(\mu)}{}_i\lambda^{(\nu)}{}_j + g^{ij}\lambda^{(\mu)}{}_{i;k}\lambda^{(\nu)}{}_j + g^{ij}\lambda^{(\mu)}{}_i\lambda^{(\nu)}{}_{j;k}$$

by Proposition 2

$= 0$ by Ricci's Lemma and equations (*). ∎

Comparing this proof of the Test Case with the first proof, we see that

<u>Step 1</u> uses the conditions <u>$R = 0$</u> to obtain the forms $\Sigma\lambda^{(\alpha)}{}_i dy^i$ satisfying (*). Instead of appealing to Theorem I.6-1, we essentially reprove this theorem; the Ricci identities make the proof almost as easy as the proof of Theorem I.6-0, the only complication being that the "mixed covariant derivatives" $\lambda_{i;12}$ depend on all $\lambda_{i;j}$, not just on $\lambda_{i;1}$.

<u>Step 2</u>, precisely the same as before, uses <u>symmetry of the Christoffel symbols</u>, $\Gamma^k_{ij} = \Gamma^k_{ji}$.

<u>Step 3</u> uses the <u>definition of the Christoffel symbols</u> $[ij,k]$ to prove that the vectors $\partial/\partial x^\alpha$ are orthonormal. The proof is much simpler because some of the calculations have been absorbed into the proof of Proposition 2, while the calculations for equations (‡‡) on page 4D-21 have been incorporated into Ricci's Lemma.

The absolute differential calculus turned out to be so useful (for many other applications besides the one just given) that it was soon exploited in the way all successful mathematical theories are — it was generalized. Notice that the possibility of defining covariant derivatives depends only on the equation

$$(*) \qquad \Gamma'^{\gamma}_{\alpha\beta} = \sum_{i,j,k} \Gamma^{k}_{ij} \frac{\partial x^i}{\partial x'^{\alpha}} \frac{\partial x^j}{\partial x'^{\beta}} \frac{\partial x'^{\gamma}}{\partial x^k} + \sum_{\mu=1}^{n} \frac{\partial^2 x^{\mu}}{\partial x'^{\alpha} \partial x'^{\beta}} \frac{\partial x'^{\gamma}}{\partial x^{\mu}} \quad ;$$

it does not depend on the particular way that the Γ^k_{ij} are defined in terms of the g_{ij}. This observation suggests that we focus our attention on the transformation law (*) itself. Quantities which transform in this way are classically called connections. More precisely:

> A **(classical) connection** on a manifold M is an assignment of n^3 numbers to each coordinate system, such that equation (*) holds between the n^3 numbers Γ^k_{ij} assigned to the coordinate system x and the n^3 numbers Γ'^k_{ij} assigned to the coordinate system x'.

Although this definition is exceedingly unappealing, classically it was motivated in the following way. If f is a function, then the quantities $\partial f/\partial x^i$ are the components of a tensor, but if λ^i are the components of a tensor (of type $\binom{0}{1}$), then the quantities $\partial\lambda^i/\partial x^j$ are not. If we attempt to construct a tensor by adding linear combinations of the λ^i, thus obtaining

$$\frac{\partial\lambda^i}{\partial x^j} + \sum_{\mu=1}^{n} \lambda^{\mu}\Gamma^i_{j\mu} \ ,$$

we find that these quantities are the components of a tensor provided that the Γ^k_{ij} transform according to (*).

Once we are given a connection (whatever in the world this connection may mean), we can imitate most of the work already done in this chapter for the special case where the Γ^k_{ij} are the Christoffel symbols, and in addition we can generalize other considerations from Riemannian geometry. What follows is a brief outline of this program.

We note first that, in contrast to the special case where the Γ^k_{ij} are the Christoffel symbols, a general connection need not satisfy $\Gamma^k_{ij} = \Gamma^k_{ji}$. However, it is easy to see that the quantities

$$T^k_{ij} = \Gamma^k_{ij} - \Gamma^k_{ji}$$

are the components of a tensor T, the torsion tensor of the connection. (In the next Chapter, we will see the reason for the term "connection", but no one seems to have a good explanation for the term "torsion" in this case.) A connection Γ^k_{ij} is called symmetric if the torsion tensor is zero. In this case, $T^k_{ij} = 0$ for every coordinate system, so $\Gamma^k_{ij} = \Gamma^k_{ji}$ for every coordinate system. Conversely, if $\Gamma^k_{ij} = \Gamma^k_{ji}$ in a set of coordinate systems which cover M, then the connection is symmetric.

The following result gives at least a little geometric significance to symmetry of a connection.

7. PROPOSITION. The torsion tensor T of a connection satisfies $T(p) = 0$ if and only if there is a coordinate system around p with

$$\Gamma^k_{ij}(p) = 0 \qquad \text{for all } i,j,k .$$

Proof. If $\Gamma^k_{ij}(p) = 0$, then $T(p) = 0$, so $T'^k_{ij} = 0$ in any coordinate system x', which means that $\Gamma'^{\gamma}_{\alpha\beta} = \Gamma'^{\gamma}_{\beta\alpha}$.

Conversely, suppose that $\Gamma^k_{ij}(p) = \Gamma^k_{ji}(p)$ for all i,j,k in a coordinate system x. Define x'^k by

$$x'^k(q) = [x^k(q) - x^k(p)] + \frac{1}{2}\sum_{i,j=1}^{n} \Gamma^k_{ij}(p)[x^i(q) - x^i(p)]\cdot[x^j(q) - x^j(p)].$$

Using $\Gamma^k_{ij}(p) = \Gamma^k_{ji}(p)$, we compute that

$$\frac{\partial x'^k}{\partial x^\ell} = \delta^k_\ell + \sum_{i=1}^{n} \Gamma^k_{i\ell}(p)[x^i - x^i(p)]$$

$$\frac{\partial x'^k}{\partial x^\ell}(p) = \delta^k_\ell \quad .$$

This shows that x' is a coordinate system in a neighborhood of p. Moreover,

$$\frac{\partial^2 x'^k}{\partial x^i \partial x^\ell} = \Gamma^k_{i\ell}(p) \quad .$$

Substituting into (*), we see that $\Gamma'^\gamma_{\alpha\beta}(p) = 0$. ∎

Proposition 7 becomes more significant when we introduce the concept of geodesics. In our treatment of Riemannian metrics, we defined geodesics as critical points for the energy function. For a general connection Γ^k_{ij}, we can simply define a <u>geodesic</u> as a path γ satisfying

$$\frac{d^2\gamma^k}{dt^2} + \sum_{i,j} \Gamma^k_{ij} \frac{d\gamma^i}{dt}\frac{dx^j}{dt} = 0 \quad ;$$

a calculation shows that this condition does not depend on the coordinate system. The basic theorems on differential equations show that geodesics through p are uniquely determined by their tangent vectors $\gamma'(0) \in M_p$; we can thus define

$\exp\colon M_p \longrightarrow M$ as before.* We can also introduce "Riemannian normal coordinates" at p; we choose any basis $X_1,\dots,X_n$ for M_p, define $\chi\colon M_p \longrightarrow \mathbb{R}^n$ by $\chi(\sum_{i=1}^n a^i X_i) = (a^1,\dots,a^n)$, and let $x = \chi\circ\exp^{-1}$. As in Proposition 4B-1, we note that the geodesic $\gamma^k(t) = \xi^k t$ satisfies

$$\sum_{i,j=1}^n \Gamma_{ij}^k(\gamma(t))\xi^i\xi^j = 0 \quad ,$$

so that

$$\sum_{i,j=1}^n \Gamma_{ij}^k(p)\xi^i\xi^j = 0 \qquad \text{for all n-tuples } (\xi^1,\dots,\xi^n) \quad .$$

This implies that

$$\Gamma_{ij}^k(p) + \Gamma_{ji}^k(p) = 0 \quad ;$$

if we also have $T(p) = 0$, then we deduce that $\Gamma_{ij}^k(p) = 0$.

We define covariant derivatives of tensors and the curvature tensor R for a connection by the formulas on page 5-3 and 5-8. Notice that Proposition 1 holds for a symmetric connection, while Proposition 2 holds for any connection. Naturally, Proposition 3 has no analogue for general connections. For non-symmetric connections, the comparison of mixed covariant derivatives becomes a little more complicated. Recall that for a function $f\colon M \longrightarrow \mathbb{R}$ we have

$$f_{;ij} = \frac{\partial^2 f}{\partial x^i \partial x^j} - \sum_{\nu=1}^n \frac{\partial f}{\partial x^\nu}\Gamma_{ji}^\nu \quad .$$

When the connection is not symmetric, we can express $f_{;ij} - f_{;ji}$ in terms of the torsion tensor, which is also involved in the new Ricci identities.

---

*Many global results about geodesics for a Riemannian metric do not hold for more general connections; see Chapter 8, Addendum 2.

8. PROPOSITION. For a connection Γ^k_{ij}, with curvature tensor R and torsion tensor T, we have

$$f_{;ij} - f_{;ji} = \sum_{\nu=1}^{n} \frac{\partial f}{\partial x^\nu} T^\nu_{ij}$$

$$\lambda^i_{;jk} - \lambda^i_{;kj} = -\sum_{\ell=1}^{n} \lambda^\ell R^i_{\ell jk} + \sum_{\ell=1}^{n} \lambda^i_{;\ell} T^\ell_{jk}$$

$$\lambda_{i;jk} - \lambda_{i;kj} = \sum_{\ell=1}^{n} \lambda_\ell R^\ell_{ijk} + \sum_{\ell=1}^{n} \lambda_{i;\ell} T^\ell_{jk} .$$

Proof. Compute, compute. ∎

(For all these relations it is, of course, extremely important that, in the definitions, care is given to the order of the subscripts in the Γ's.)

Finally, we wish to consider the properties of the curvature tensor which are given in Proposition 4D-6. Two of these have no analogue for a general connection, since they involve a metric, but there is an additional relation, involving the covariant derivative, which holds for any connection.

9. PROPOSITION. The curvature tensor for any connection satisfies the following identities:

(1) $R^i_{jk\ell} = -R^i_{j\ell k}$

(2) (Bianchi's 1st identity)

$$R^i_{jk\ell} + R^i_{k\ell j} + R^i_{\ell jk}$$

$$= (T^i_{k\ell;j} + T^i_{\ell j;k} + T^i_{jk;\ell}) + \sum_{\mu=1}^{n} (T^\mu_{jk} T^i_{\mu\ell} + T^\mu_{k\ell} T^i_{\mu j} + T^\mu_{\ell j} T^i_{\mu k}) .$$

(3) (Bianchi's 2nd identity)

$$(R^h{}_{ijk;\ell} + R^h{}_{ik\ell;j} + R^h{}_{i\ell j;k})$$

$$+ \sum_{\mu=1}^{n} (T^\mu_{jk} R^h{}_{i\mu\ell} + T^\mu_{k\ell} R^h{}_{i\mu j} + T^\mu_{\ell j} R^h{}_{i\mu k}) = 0.$$

In particular, if the connection is symmetric, then we have the much simpler relations

(2') $R^i{}_{jk\ell} + R^i{}_{k\ell j} + R^i{}_{\ell jk} = 0$

(3') $R^h{}_{ijk;\ell} + R^h{}_{ik\ell;j} + R^h{}_{i\ell j;k} = 0;$

classically, (3') alone is known as "Bianchi's identity".

<u>Proof.</u> Equation (1) follows immediately from the definition. In the case of a symmetric connection, equation (2') is also easy to verify (we have already done it in Proposition 4D-6). It is even simpler to verify if we use Riemannian normal coordinates at p; in this case the definition (pg.5-8) gives

$$R^i{}_{jk\ell}(p) = \frac{\partial\Gamma^i_{\ell j}}{\partial x^k}(p) - \frac{\partial\Gamma^i_{kj}}{\partial x^\ell}(p) ,$$

which yields (2') at once. Using Proposition 1, for a symmetric connection, we also obtain

$$R^h{}_{ijk;\ell}(p) = \frac{\partial^2\Gamma^h_{ki}}{\partial x^\ell \partial x^j}(p) - \frac{\partial^2\Gamma^h_{ji}}{\partial x^\ell \partial x^k}(p) ,$$

which gives (3').

The proof of (2) and (3) in general is considerably more complicated. To begin with, notice that $T^i_{\mu\ell} = -T^i_{\ell\mu}$, so

$$\sum_\mu T^\mu_{jk} T^i_{\mu\ell} = \sum_\mu \Gamma^\mu_{jk} T^i_{\mu\ell} + \sum_\mu \Gamma^\mu_{kj} T^i_{\ell\mu} .$$

We also have

$$T^i_{jk;\ell} = \frac{\partial T^i_{jk}}{\partial x^\ell} + \sum_\mu \Gamma^i_{\ell\mu} T^\mu_{jk} - \sum_\mu \Gamma^\mu_{\ell j} T^i_{\mu k} - \sum_\mu \Gamma^\mu_{\ell k} T^i_{j\mu} .$$

From these equations we see that the right side of (2) equals

$$\left(\frac{\partial T^i_{jk}}{\partial x^\ell} + \sum_\mu T^\mu_{jk} \Gamma^i_{\ell\mu} \right) + \text{the two terms obtained by cyclically permuting } j,k,\ell.$$

Using the definition of T^i_{jk}, this is easily seen to equal the left side of (2).

To prove (3), we note first that (1) gives

$$\sum_\mu T^\mu_{jk} R^h{}_{i\mu\ell} = \sum_\mu \Gamma^\mu_{jk} R^h{}_{i\mu\ell} + \Gamma^\mu_{kj} R^h{}_{i\ell\mu} .$$

We also have

$$R^h{}_{ijk;\ell} = \frac{\partial R^h{}_{ijk}}{\partial x^\ell} + \sum_\mu R^\mu{}_{ijk} \Gamma^h_{\ell\mu} - \sum_\mu R^h{}_{\mu jk} \Gamma^\mu_{\ell i}$$

$$- \sum_\mu R^h{}_{i\mu k} \Gamma^\mu_{\ell j} - \sum_\mu R^h{}_{ij\mu} \Gamma^\mu_{\ell k} .$$

From these equations we see that the left side of (3) equals

$$\left(\frac{\partial R^{h}{}_{ijk}}{\partial x^{\ell}} + \sum_{\mu} R^{\mu}{}_{ijk}\, \Gamma^{h}_{\ell\mu} - \sum_{\mu} R^{h}{}_{\mu jk}\, \Gamma^{\mu}_{\ell i} \right) + \text{the two terms obtained by cyclically permuting } j,k,\ell.$$

Plugging back into the definition, we find that this is zero. $\blacksquare$

Believe it or not, the Bianchi identity will be useful at some points later on. Even more surprising, in Chapter 7 we will present a derivation of the Bianchi identity which will make it seem like a natural result. With the present proof of the Bianchi identity we end our summary of the classical theory of connections. The presentation was made mainly as background for the succeeding chapters, in which the same results will begin to take on a more modern appearance.

Chapter 6. The ∇ Operator

The contents of this chapter really differ very little from those of the previous one, but everything will look quite different. The clean modern symbolism which gets introduced here is abandoned only in those parts of the chapter which compare the present treatment with that given previously. This refurbishment of the classical theory, due to Koszul, is effected by singling out for invariant treatment just one of the concepts introduced previously, and then defining the other concepts in terms of it. We will begin with a definition, and then compare it to the classical one.

> A (Koszul) connection on a C^∞ manifold M is a function ∇ (read "dell") which associates a C^∞ vector field $\nabla_X Y$ to any two C^∞ vector fields X and Y, and which satisfies
>
> (1) $\nabla_{X_1+X_2} Y = \nabla_{X_1} Y + \nabla_{X_2} Y$
>
> (2) $\nabla_X(Y_1 + Y_2) = \nabla_X Y_1 + \nabla_X Y_2$
>
> (3) $\nabla_{fX} Y = f \cdot \nabla_X Y$
>
> (4) $\nabla_X(fY) = f \cdot \nabla_X Y + X(f) \cdot Y$.

Notice that ∇ is assumed linear over the C^∞ functions in the argument X. By our standard theorem (I.4-2) this means that for any given vector field Y we can define

$$\nabla_{X_p} Y \in M_p$$

for every $X_p \in M_p$ so that

$$\nabla_X Y = p \mapsto \nabla_{X_p} Y \ .$$

Alternatively, one can define a Koszul connection to be a map ∇ which assigns a vector $\nabla_{X_p} Y \in M_p$ to every $X_p \in M_p$ and every C^∞ vector field Y, and which satisfies

(1') $\nabla_{X_p + X'_p} Y = \nabla_{X_p} Y + \nabla_{X'_p} Y$

(2') $\nabla_{X_p}(Y_1 + Y_2) = \nabla_{X_p} Y_1 + \nabla_{X_p} Y_2$

(3') $\nabla_{aX_p} Y = a\, \nabla_{X_p} Y$ for all $a \in \mathbb{R}$

(4') $\nabla_{X_p}(fY) = f(p)\cdot\nabla_{X_p} Y + X_p(f)\cdot Y_p$

(5') if X and Y are C^∞ vector fields, then so is

$$p \longmapsto \nabla_{X_p} Y .$$

We then define $\nabla_X Y$ by $\nabla_X Y(p) = \nabla_{X_p} Y$.

As one example of a Koszul connection, we take M to be $\mathbb{R}^n$ and let $\nabla_{X_p} Y$ be the directional derivative of Y in the direction X_p (computed by taking the directional derivative of the component functions of Y). It is clear that properties (1') - (5') hold for this ∇. However, the most important justification for the particular conditions required of a Koszul connection comes from a comparison with classical connections. If $x^1,\dots,x^n$ is a coordinate system on M, and we define Γ^k_{ij} by

$$(*) \qquad \nabla_{\frac{\partial}{\partial x^i}} \frac{\partial}{\partial x^j} = \sum_{k=1}^{n} \Gamma^k_{ij} \frac{\partial}{\partial x^k} ,$$

then from (1) - (4) it is an easy exercise to deduce that Γ^k_{ij} are the components of a (classical) connection; conversely, given a classical connection, we can use (*) and (1) - (4) to determine a well-defined ∇. In the coordinate system x, if

$$X = \sum_{i=1}^{n} a^i \frac{\partial}{\partial x^i} \quad , \quad Y = \sum_{j=1}^{n} \lambda^j \frac{\partial}{\partial x^j} \quad ,$$

then

$$\nabla_X Y = \sum_{k=1}^{n} \left(\sum_{i=1}^{n} a^i \lambda^k{}_{;i} \right) \frac{\partial}{\partial x^k} \quad .$$

For a given vector field Y we have a tensor ∇Y of type $\binom{1}{1}$, that is, a collection of linear transformations $\nabla Y(p)\colon M_p \longrightarrow M_p$, given by

$$\nabla Y(p)(X_p) = \nabla_{X_p} Y \quad (= \nabla_X Y(p)) \quad .$$

Clearly

$$\nabla Y\left(\frac{\partial}{\partial x^i}\right) = \sum_{k=1}^{n} \lambda^k{}_{;i} \frac{\partial}{\partial x^k} \quad ;$$

we can also write this as

$$\nabla Y = \sum_{i,k} \lambda^k{}_{;i} dx^i \otimes \frac{\partial}{\partial x^k} \quad ,$$

which shows that ∇Y is the tensor classically described in terms of its components $\lambda^k{}_{;i}$.

The covariant derivative of all other tensors are now going to be defined in terms of this covariant derivative, which we have made the cornerstone of our new definition of a connection (and <u>not</u> merely in terms of the Γ^k_{ij} which it determines). There are two completely different ways of doing this, each of which has its advantages. The first way is purely formal:

1. PROPOSITION. Let X be a C^{∞} vector field on a C^{∞} manifold M with a Koszul connection ∇. Then there is a unique operator

$$A \mapsto \nabla_X A$$

from C^{∞} tensor fields to C^{∞} tensor fields, preserving the type $\binom{k}{\ell}$, such that

(1) $\nabla_X f = X(f)$

(2) $\nabla_X Y$ is the vector field given by the connection ∇

(3) $A \mapsto \nabla_X A$ is linear over $\mathbb{R}$

(4) $\nabla_X(A \otimes B) = \nabla_X A \otimes B + A \otimes \nabla_X B$

(5) for any contraction C, we have $\nabla_X \circ C = C \circ \nabla_X$.

Moreover, each $\nabla_X A$ is linear over the C^{∞} functions in the argument X, so for every tensor field A of type $\binom{k}{\ell}$ and every $X_p \in M_p$ we can define

$$\nabla_{X_p} A \in \mathcal{T}^k_\ell(M_p) \qquad (\text{and } \nabla_{X_p+X'_p} A = \nabla_{X_p} A + \nabla_{X'_p} A \,;\ \nabla_{aX_p} A = a\nabla_{X_p} A$$

Proof. Essentially, this is Problem I.5-15. If we define

$$Df = X(f)$$

$$DY = \nabla_X Y \qquad ,$$

then D does satisfy the condition

$$D(fY) = fDY + Df \cdot Y \qquad \text{i.e.,}\ D(f \otimes Y) = f \otimes DY + Df \otimes Y$$

which is assumed for this Problem. Briefly, the proof that $D = \nabla$ can be extended uniquely is as follows. For a 1-form ω we want

$$\begin{aligned} X(\omega(Y)) = D(\omega(Y)) &= D\ (\text{contraction of}\ \omega \otimes Y) \\ &= \text{contraction of}\ [D\omega \otimes Y + \omega \otimes DY]\ , \\ &= D\omega(Y) + \omega(DY) \\ &= D\omega(Y) + \omega(\nabla_X Y)\ , \end{aligned}$$

so we want

$$(\nabla_X \omega)(Y) = D\omega(Y) = X(\omega(Y)) - \omega(\nabla_X Y) \quad \text{for all}\ Y.$$

Since any A is a sum of functions times tensor products of vector fields and 1-forms, conditions (3) and (4) determine $\nabla_X A$.

Following through the proof in detail, it is easily checked that $\nabla_X A$ is linear over the C^∞ functions in the argument X. ∎

In view of Proposition 1, for any tensor A of type $\binom{k}{\ell}$ we can define a new tensor ∇A of type $\binom{k+1}{\ell}$ by

$$\nabla A(p)(X_{1p},\ldots,X_{kp},X_p) = \nabla_{X_p} A(X_{1p},\ldots,X_{kp}) \ .$$

If $\omega = \Sigma\, \lambda_j dx^j$ is a tensor of type $\binom{1}{0}$, and we set

$$\nabla_{\frac{\partial}{\partial x^i}} \omega = \sum_{\ell=1}^{n} a_\ell dx^\ell \ ,$$

then we have

$$\frac{\partial}{\partial x^i}\lambda_k = \nabla_{\frac{\partial}{\partial x^i}}\left(\text{contraction of } \omega \otimes \frac{\partial}{\partial x^k}\right)$$

$$= \text{contraction of}\left(\left[\nabla_{\frac{\partial}{\partial x^i}}\omega\right] \otimes \frac{\partial}{\partial x^k} + \omega \otimes \nabla_{\frac{\partial}{\partial x^i}}\frac{\partial}{\partial x^k}\right)$$

$$= a_k + \sum_{\mu=1}^{n} \Gamma^{\mu}_{ik}\lambda_{\mu} \quad ,$$

so we obtain

$$\nabla_{\frac{\partial}{\partial x^i}}\omega = \sum_{k=1}^{n}\left(\frac{\partial \lambda_k}{\partial x^i} - \sum_{\mu=1}^{n}\Gamma^{\mu}_{ik}\lambda_{\mu}\right)dx^k$$

$$= \sum_{k=1}^{n}\lambda_{k;i}dx^k \quad ;$$

we can also write this as

$$\nabla\omega = \sum_{i,k}\lambda_{k;i}dx^k \otimes dx^i \quad ,$$

which shows that $\nabla\omega$ is the tensor classically described in terms of its components $\lambda_{k;i}$. Similarly, we easily see that if

$$A = \sum A^{j_1 \ldots j_\ell}_{i_1 \ldots i_k}\, dx^{i_1} \otimes \ldots \otimes dx^{i_k} \otimes \frac{\partial}{\partial x^{j_1}} \otimes \ldots \otimes \frac{\partial}{\partial x^{j_\ell}} \quad ,$$

then

$$\nabla A = \sum A^{j_1 \ldots j_\ell}_{i_1 \ldots i_k;h}\, dx^{i_1} \otimes \ldots \otimes dx^{i_k} \otimes dx^h \otimes \frac{\partial}{\partial x^{j_1}} \otimes \ldots \otimes \frac{\partial}{\partial x^{j_\ell}} \quad .$$

The uniqueness clause in Proposition 1 is precisely what accounts for its usefulness: all properties of ∇A should be derivable from properties (1) - (5), since these properties characterize ∇A. On the other hand, the proposition gives no idea what is going on geometrically. To obtain such a picture we introduce another extremely important concept.

Let $c\colon [a,b] \longrightarrow M$ be a curve. By a vector field V along c we mean a function V on $[a,b]$ with $V_t = V(t) \in M_{c(t)}$. In a coordinate system

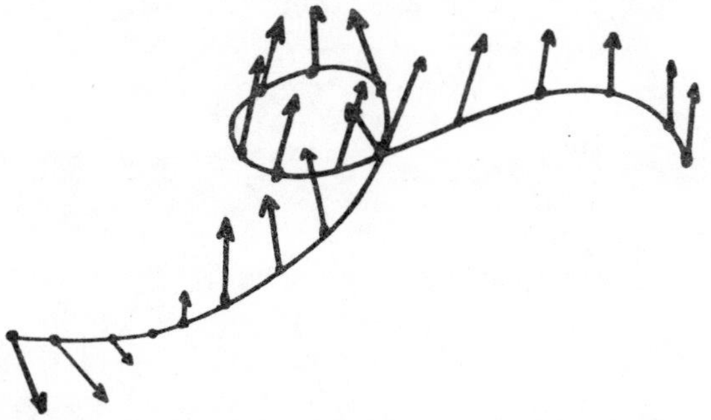

(x,U) we can write

$$V_t = \sum_{i=1}^{n} v_i(t)\cdot \frac{\partial}{\partial x^i}\bigg|_{c(t)} .$$

We call V a C^∞ vector field along c if the functions v^i are C^∞ on $[a,b]$; this is equivalent to saying that $t \mapsto V_t(f)$ is C^∞ for every C^∞ function f on M.

Now suppose that V is a C^∞ vector field on a neighborhood of $c([0,1])$. Then

$$t \mapsto \nabla_{\frac{dc}{dt}} V$$

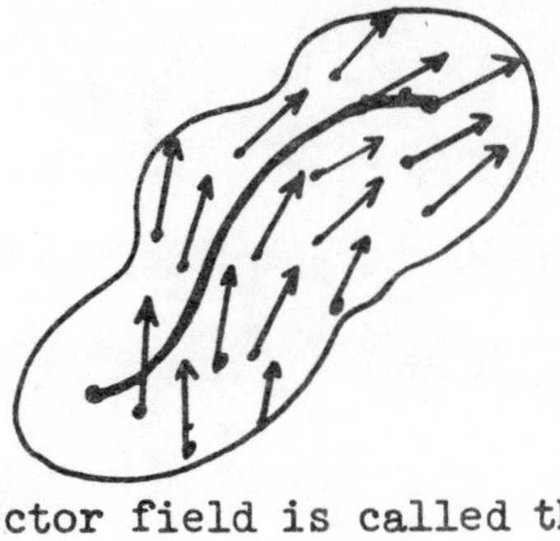

is a C^∞ vector field along c. This vector field is called the covariant derivative of V along c; we will denote it by the convenient symbolism

$$\frac{DV}{dt} ,$$

which involves all the classical ambiguities. We would like to generalize this covariant derivative along c to vector fields V which are themselves defined only along c.

2. PROPOSITION. There is precisely one operation $V \mapsto DV/dt$, from C^∞ vector fields V along c to C^∞ vector fields along c, with the property that

a) If $V_s = Y_{c(s)}$ for some C^∞ vector field Y defined in a neighborhood of $c(t)$, then

$$\frac{DV}{dt} = \nabla_{\frac{dc}{dt}} Y \ ,$$

b) If $dc/dt = 0$, then

$$\frac{DV}{dt}(t) = 0 \ .$$

For this operation we have

$$\frac{D(V+W)}{dt} = \frac{DV}{dt} + \frac{DW}{dt}$$

$$\frac{D(fV)}{dt} = \frac{df}{dt} V + f \frac{DV}{dt} \qquad \text{for } f\colon [a,b] \longrightarrow \mathbb{R} \ .$$

Proof. If $dc/dt \neq 0$, then c is an imbedding in a neighborhood of t, and it follows easily that we have $V_s = Y_{c(s)}$ for some Y defined in a neighborhood of $c(t)$. Choose a coordinate system x on (a sufficiently small part of)

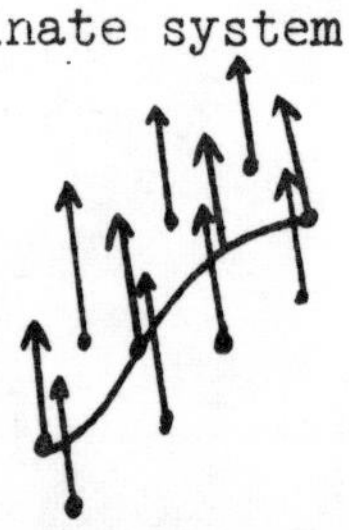

this neighborhood, and let

$$Y = \sum_{i=1}^{n} Y^i \frac{\partial}{\partial x^i} ,$$

$$c^i(t) = x^i(c(t)), \quad \text{so that} \quad \frac{dc}{dt} = \sum_{i=1}^{n} \frac{dc^i}{dt} \left. \frac{\partial}{\partial x^i} \right|_{c(t)} .$$

For a) to hold we must have

$$\frac{DV}{dt} = \nabla_{\frac{dc}{dt}} Y = \sum_{i=1}^{n} \frac{dc^i}{dt} \nabla_{\frac{\partial}{\partial x^i}} \Big(\sum_{j=1}^{n} Y^j \frac{\partial}{\partial x^j} \Big) (c(t))$$

$$= \sum_{i=1}^{n} \frac{dc^i}{dt} \left\{ \sum_{j=1}^{n} \frac{\partial Y^j}{\partial x^i}(c(t)) \cdot \left. \frac{\partial}{\partial x^j} \right|_{c(t)} + \sum_{j,k=1}^{n} Y^j(c(t)) \Gamma_{ij}^k(c(t)) \cdot \left. \frac{\partial}{\partial x^k} \right|_{c(t)} \right\}$$

$$= \sum_{k=1}^{n} \left(\frac{dc}{dt}(Y^k) + \sum_{i,j} \Gamma_{ij}^k(c(t)) \frac{dc^i}{dt} Y^j(c(t)) \right) \left. \frac{\partial}{\partial x^k} \right|_{c(t)}$$

$$= \sum_{k=1}^{n} \left(\frac{dY^k(c(t))}{dt} + \sum_{i,j} \Gamma_{ij}^k(c(t)) \frac{dc^i}{dt} Y^j(c(t)) \right) \left. \frac{\partial}{\partial x^k} \right|_{c(t)} .$$

This formula shows that the value of $\nabla_{dc/dt} Y$ at $c(t)$ does not depend on Y, but only on the values of Y along c. It also shows that the vector field defined by the formula does not depend on the coordinate system picked. This shows that DV/dt is a C^∞ vector field along c (even where $dc/dt = 0$, since the formula does give 0 in this case). The last two properties of DV/dt are clear. ∎

Now we say that a vector field V along c is <u>parallel along</u> c (with respect to ∇) if $DV/dt = 0$ along c. When $M = \mathbb{R}^n$ and ∇ is just the directional derivative, we obtain the standard picture of a parallel vector field.

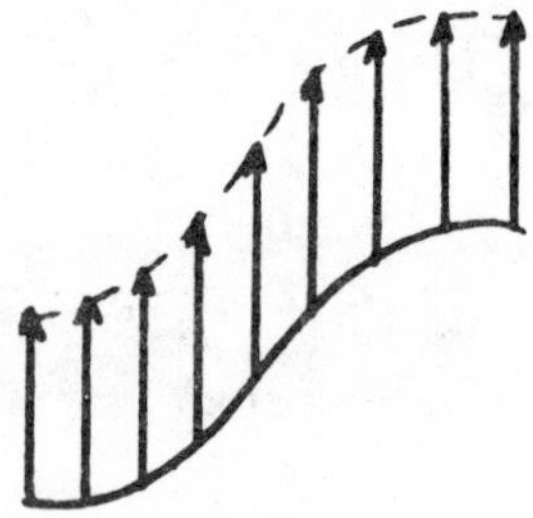

In general, given a curve $c\colon [a,b] \longrightarrow M$, and a vector $V_a \in M_{c(a)}$, there is a unique vector field V along c which is parallel along c. This is because the equations

$$(*) \quad \frac{dv^k(t)}{dt} + \sum_{i,j=1}^{n} \frac{dc^i(t)}{dt}\,\Gamma_{ij}^k(c(t))v^j(t) = 0$$

are linear differential equations with unique solutions v^j, defined on all of $[a,b]$, for given initial conditions; the desired vector field V is then

$$V_t = \sum_{j=1}^{n} v^j(t)\cdot\left.\frac{\partial}{\partial x^j}\right|_{c(t)} .$$

The vector $V_t \in M_{c(t)}$ is said to be obtained from V_a by <u>parallel translation along</u> c. It is clear from equations (*) that $(V + W)_t = V_t + W_t$ and $(\lambda\cdot V)_t = \lambda\cdot V_t$ for $\lambda \in \mathbb{R}$. We therefore obtain a linear transformation

$$\tau_t\colon M_{c(a)} \longrightarrow M_{c(t)} \qquad\qquad V_a \longmapsto V_t \quad .$$

Clearly, τ_t is one-one, for its inverse is just parallel translation along the reversed portion of c from t to a. Thus, along any curve c we obtain an isomorphism between any two tangent spaces $M_{c(t_1)}$ and $M_{c(t_2)}$; this possibility of comparing, or "connecting", tangent spaces at different points gives rise to the term "connection". It was invented by Levi-Civita, who used the equations (*) as the definition.

The parallel translation τ_t is defined in terms of ∇, but we can also reverse the process:

3. PROPOSITION. Let c be a curve with $c(0) = p$ and $c'(0) = X_p$. Then

$$\nabla_{X_p} Y = \lim_{h \to 0} \frac{1}{h}({\tau_h}^{-1} Y_{c(h)} - Y_p) .$$

Proof. For a fixed h, let Z be the parallel vector field along c with

$$(1) \quad Z_0 = {\tau_h}^{-1} Y_{c(h)} ,$$

so that we also have

$$(2) \quad Z_h = Y_{c(h)} .$$

Writing

$$Z_t = \sum_{i=1}^{n} z^i(t) \left.\frac{\partial}{\partial x^i}\right|_{c(t)} \qquad Y = \sum_{i=1}^{n} Y^i \frac{\partial}{\partial x^i} ,$$

we have

$$(3) \quad \frac{dz^k(t)}{dt} + \sum_{i,j=1}^{n} \frac{dc^i}{dt} z^j(t)\Gamma_{ij}^k(c(t)) = 0$$

and

$$(4) \quad z^k(h) = y^k(c(h)), \text{ by (2).}$$

By the mean value theorem,

(5) $\qquad z^k(h) = z^k(0) + hz^{k}{}'(\xi) \qquad$ for some $\xi \in (0,h)$ (depending on k).

So the $k^{\underline{th}}$ component of $(1/h)(\tau_h{}^{-1}Y_{c(h)} - Y_p)$ is

$$\frac{1}{h}\Big(z^k(0) - Y^k(c(0))\Big) = \frac{1}{h}\Big(z^k(h) - hz^{k}{}'(\xi) - Y^k(c(0))\Big) \quad \text{by (5)}$$

$$= -z^{k}{}'(\xi) + \frac{1}{h}\Big(Y^k(c(h)) - Y^k(c(0))\Big) \quad \text{by (4)}$$

$$= \sum_{i,j=1}^{n} \frac{dc^i}{dt}(\xi) z^j(\xi)\Gamma^k_{ij}(c(\xi)) + \frac{1}{h}\Big[Y^k(c(h)) - Y^k(c(0))\Big]$$

by (3) .

As $h \to 0$, this expression (which involves a different Z for each h) clearly approaches

$$\frac{dY^k(c(t))}{dt}(0) + \sum_{i,j=1}^{n} \frac{dc^i}{dt}(0)Y^j(c(0))\Gamma^k_{ij}(c(0)) \quad ;$$

in the proof of Proposition 2 we saw that this is indeed the kth component of $\nabla_{X_p} Y$. ∎

Motivated by Proposition 3, we now define $\nabla_{X_p} A$ where A is any tensor field A of type $\binom{k}{\ell}$. We have

$$A(q) \in \mathcal{J}^k_\ell(M_q) \quad \text{for all } q \quad ,$$

and the isomorphism

$$\tau_t \colon M_{c(0)} \to M_{c(t)}$$

gives rise to an isomorphism

$$\mathcal{J}^k_\ell(\tau_t): \mathcal{J}^k_\ell(M_{c(0)}) \to \mathcal{J}^k_\ell(M_{c(t)}) \quad ;$$

therefore we can define

$$\nabla_{X_p} A = \lim_{h \to 0} \frac{1}{h}(\mathcal{J}^k_\ell(\tau_h)^{-1} A(c(h)) - A(p)) \quad .$$

If we regard $A(q)$ as a function of k tangent vectors in M_q and ℓ vectors in M_q^*, this means that for $v_1,\ldots,v_k \in M_p$ and $\lambda_1,\ldots,\lambda_\ell \in M_p^*$ we have

$$(\nabla_{X_p} A)(v_1,\ldots,v_k, \lambda_1,\ldots,\lambda_\ell)$$

$$= \lim_{h \to 0} \frac{1}{h} \Big[A(c(h))(\tau_h v_1,\ldots,\tau_h v_k, \tau_h{}^* \lambda_1,\ldots,\tau_h{}^* \lambda_k)$$

$$- A(p)(v_1,\ldots,v_k,\lambda_1,\ldots,\lambda_k) \Big] \quad .$$

4. PROPOSITION. This $\nabla_{X_p} A$ coincides with that given by Proposition 1.

Proof. It suffices to prove properties (1) -(5) for the new $\nabla_{X_p} A$. This is left to the reader [the proof of (4) and (5) involves the usual trick which one uses in the proof of the product rule for derivatives] . ∎

(Note that without this result it would not be at all obvious that $\nabla_{X_p} A$ is linear in X_p.)

5. COROLLARY. Let A be a tensor field of type $\binom{k}{1}$. If $Y_1,\ldots,Y_k$ are vector fields, then

$$(\nabla_{X_p} A)(Y_{1p},\ldots,Y_{kp}) = \nabla_{X_p}(A(Y_1,\ldots,Y_k)) - \sum_{i=1}^{k} A(Y_{1p},\ldots,\nabla_{X_p} Y_i,\ldots,Y_{kp}) \quad .$$

Proof. Use Proposition 1, noting that $A(Y_1,\ldots,Y_k)$ can be obtained by applying k contractions to

$$A \otimes Y_1 \otimes \ldots \otimes Y_k \quad .$$

(It is also instructive to obtain a proof from the definition in terms of parallel translations).∎

Our definition of $\nabla_X A$ may be compared to the definition, in Problem I.5-14, of $L_X A$. In the latter case the maps φ^*_h given by the vector field X play the roles of the parallel translations τ_h in the present instance. Notice that Problem I.5-15 can be used in both cases to formally extend the operation from vector fields to arbitrary tensors.

After these preliminaries, the further study of Koszul connections proceeds quite rapidly. We first define, for vector fields X and Y,

$$T(X,Y) = \nabla_X Y - \nabla_Y X - [X,Y] \; .$$

A simple calculation shows that T is linear over the C^∞ functions, so that it determines a tensor. Clearly this is just the classical torsion tensor; as usual, the invariant definition involves a bracket, which disappears in the expression in coordinates.

We now want to distinguish the connection determined by the Christoffel symbols for a metric. Suppose we are in a Riemannian manifold $(M, \langle\ ,\ \rangle)$. We will call a connection compatible with $\langle\ ,\ \rangle$ if the parallel translations $\tau_t\colon M_{c(a)} \longrightarrow M_{c(t)}$ along any curve $c\colon [a,b] \longrightarrow M$ are isometries (with respect to $\langle\ ,\ \rangle_{c(a)}$ and $\langle\ ,\ \rangle_{c(t)}$).

6. LEMMA. A connection ∇ is compatible with a metric $\langle\ ,\ \rangle$ if and only if it satisfies the following condition: If V and W are vector fields along any curve c, then

$$\frac{d}{dt}\langle V,W\rangle = \langle\frac{DV}{dt},W\rangle + \langle V,\frac{DW}{dt}\rangle .$$

Proof. Suppose ∇ satisfies this condition. Then if V is parallel along c we have

$$\frac{d}{dt}\langle V,V\rangle = 2\langle\frac{DV}{dt},V\rangle = 0 ,$$

so $\langle V,V\rangle$ is constant along c. Thus each τ_t is norm-preserving, and hence an isometry.

Conversely, suppose ∇ is compatible with the metric. Choose parallel vector fields $P_1,\ldots,P_n$ along c which are orthonormal at one point of c, and hence at every point of c. Let

$$V_t = \sum_{i=1}^{n} v^i(t)P_{i_t} \qquad W_t = \sum_{j=1}^{n} w^j(t)P_{j_t} .$$

Then

$$\langle V,W\rangle = \sum_{i=1}^{n} v^i\cdot w^i ,$$

and by Proposition 2 we have, remembering that $DP_i/dt = 0$,

$$\frac{DV}{dt} = \sum_{i=1}^{n}\frac{dv^i}{dt}P_i \qquad \frac{DW}{dt} = \sum_{j=1}^{n}\frac{dw^j}{dt}P_j .$$

So

$$< \frac{DV}{dt}, W > + < V, \frac{DW}{dt} > = \sum_{i=1}^{n} \left(\frac{dv^i}{dt} w^i + v^i \frac{dw^i}{dt}\right) = \frac{d}{dt} < V, W > .$$ ∎

7. COROLLARY. The connection ∇ is compatible with $< , >$ if and only if

$$X_p < Y, Z > = < \nabla_{X_p} Y, Z_p > + < Y_p, \nabla_{X_p} Z >$$

for all vector fields Y, Z and vectors $X_p \in M_p$.

Proof. Apply the Lemma to a curve c with $c'(0) = X_p$. ∎

8. LEMMA (Fundamental Lemma of Riemannian Geometry). On a Riemannian manifold $(M, < , >)$ there is a unique symmetric connection compatible with $< , >$.

Proof. Suppose ∇ is compatible with $< , >$. Choose a coordinate system (x, U). By Corollary 7

$$(*) \quad \frac{\partial g_{jk}}{\partial x^i} = \frac{\partial}{\partial x^i} < \frac{\partial}{\partial x^j}, \frac{\partial}{\partial x^k} > = < \nabla_{\frac{\partial}{\partial x^i}} \frac{\partial}{\partial x^j}, \frac{\partial}{\partial x^k} > + < \frac{\partial}{\partial x^j}, \nabla_{\frac{\partial}{\partial x^i}} \frac{\partial}{\partial x^k} > .$$

Cyclically permuting i, j, k, and using

$$\nabla_{\frac{\partial}{\partial x^i}} \frac{\partial}{\partial x^j} = \nabla_{\frac{\partial}{\partial x^j}} \frac{\partial}{\partial x^i} \qquad \text{(from symmetry)} ,$$

we obtain

$$\sum_{\ell=1}^{n} \Gamma_{ij}^{\ell} g_{\ell k} = \left\langle \nabla_{\frac{\partial}{\partial x^i}} \frac{\partial}{\partial x^j}, \frac{\partial}{\partial x^k} \right\rangle = [ij,k] \quad ,$$

which implies that

$$\Gamma_{ij}^{\ell} = \sum_{k=1}^{n} g^{k\ell}[ij,k] \quad .$$

Thus the Γ's for ∇ must be the Christoffel symbols.

We know that the Christoffel symbols do indeed satisfy (*), which shows that the equation in Corollary 7 does hold. (In fact, this equation is equivalent to Ricci's Lemma, 5-3.) ▮

The unique connection of Lemma 8 is sometimes called the <u>Levi-Civita connection</u> for the metric $\langle\ ,\ \rangle$. Naturally, Lemma 8 is more impressive before reading the proof than after. Nevertheless, it is still a very nice result. Perhaps its only defect is the restriction to <u>symmetric</u> connections; in Addendum 1 to this chapter we present some justification for this restriction. We have already given one interpretation of symmetry; but the following will be more useful for present purposes. For a C^∞ function $s\colon \mathbb{R}^2 \to M$ (a "parameterized surface" in M), we define a **vector field** V **along** s to be a function V with

$$V_{(x,y)} \in M_{s(x,y)} \quad .$$

In particular, we have the vector fields

$$\frac{\partial s}{\partial x} = s_*\left(\frac{\partial}{\partial x}\right) \qquad \frac{\partial s}{\partial y} = s_*\left(\frac{\partial}{\partial y}\right) \quad .$$

For any C^∞ vector field V along s, we define

$$\left(\frac{DV}{\partial x}\right)_{(x,y)} = \text{covariant derivative along } c(t) = s(t,y) \text{ of } t \mapsto V_{(t,y)} ,$$

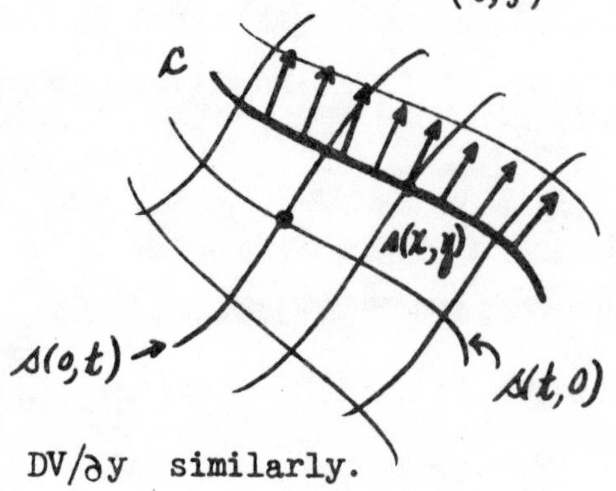

and we define $DV/\partial y$ similarly.

9. PROPOSITION. If ∇ is symmetric, then

$$\frac{D}{\partial x}\frac{\partial s}{\partial y} = \frac{D}{\partial y}\frac{\partial s}{\partial x} .$$

Proof. Express both sides in terms of a coordinate system and compute. ∎

We are now ready to define the curvature tensor for a Koszul connection ∇. If X,Y,Z are vector fields, we consider the vector field

$$R(X,Y)Z = \nabla_X(\nabla_Y Z) - \nabla_Y(\nabla_X Z) - \nabla_{[X,Y]}Z .$$

A straightforward computation shows that (because of the bracket term) R is linear over the C^∞ functions in all three variables, so it defines a tensor R, the curvature tensor of ∇. Obviously, this definition is somehow related to the Ricci identity (Proposition 5-8)

$$(*) \qquad \lambda^i{}_{;jk} - \lambda^i{}_{;kj} = -\sum_{\ell=1}^n \lambda^\ell R^i{}_{\ell jk} + \sum_{\ell=1}^n \lambda^i{}_{;\ell} T^\ell{}_{jk} ,$$

but at first sight we seem to be missing a term for the torsion. The reason is easily discovered. Recall that if

$$Z = \sum_{i=1}^{n} \lambda^i \frac{\partial}{\partial x^i} ,$$

then $\lambda^i{}_{;j}$ are the components of ∇Z, i.e.,

$$\nabla_{\frac{\partial}{\partial x^j}} Z = \sum_{i=1}^{n} \lambda^i{}_{;j} \frac{\partial}{\partial x^i} .$$

Now $\lambda^i{}_{;jk} = \lambda^i{}_{;j;k}$ are not the components of $\nabla_{\partial/\partial x^k}(\nabla_{\partial/\partial x^j} Z)$; rather they are the components of $\nabla(\nabla Z)$. So

$$\sum_{i=1}^{n} \lambda^i{}_{;jk} \frac{\partial}{\partial x^i} = \nabla\nabla Z\left(\frac{\partial}{\partial x^j}, \frac{\partial}{\partial x^k}\right)$$

$$= \left[\nabla_{\frac{\partial}{\partial x^k}} (\nabla Z)\right]\left(\frac{\partial}{\partial x^j}\right)$$

$$= \nabla_{\frac{\partial}{\partial x^k}} \left(\nabla Z\left(\frac{\partial}{\partial x^j}\right)\right) - \nabla Z\left(\nabla_{\frac{\partial}{\partial x^k}} \frac{\partial}{\partial x^j}\right) \quad \text{by Corollary 5}$$

$$= \nabla_{\frac{\partial}{\partial x^k}} \left(\nabla_{\frac{\partial}{\partial x^j}} Z\right) - \nabla_{\left(\nabla_{\frac{\partial}{\partial x^k}} \frac{\partial}{\partial x^j}\right)} Z .$$

Consequently,

$$\nabla_{\frac{\partial}{\partial x^j}}\left(\nabla_{\frac{\partial}{\partial x^k}} Z\right) - \nabla_{\frac{\partial}{\partial x^k}}\left(\nabla_{\frac{\partial}{\partial x^j}} Z\right) = \sum_{i=1}^{n} (\lambda^i{}_{;kj} - \lambda^i{}_{;jk}) \frac{\partial}{\partial x^i} + \nabla_{\left(\nabla_{\frac{\partial}{\partial x^j}} \frac{\partial}{\partial x^k} - \nabla_{\frac{\partial}{\partial x^k}} \frac{\partial}{\partial x^j}\right)} Z$$

$$= \sum_{i=1}^{n}\left(\sum_{\ell=1}^{n} \lambda^\ell R^i{}_{\ell jk} - \sum_{\ell=1}^{n} \lambda^i{}_{;\ell} T^\ell{}_{jk}\right)\frac{\partial}{\partial x^i} + \nabla_{\left(\sum_{\ell=1}^{n} T^\ell{}_{jk} \frac{\partial}{\partial x^\ell}\right)} Z \quad \text{by } (*) .$$

Since

$$\nabla_{\sum_{\ell=1}^{n} T^{\ell}_{jk} \frac{\partial}{\partial x^{\ell}}} Z = \sum_{\ell=1}^{n} T^{\ell}_{jk} \nabla_{\frac{\partial}{\partial x^{\ell}}} Z = \sum_{i,\ell} T^{\ell}_{jk} \lambda^{i}{}_{;\ell} \frac{\partial}{\partial x^{i}} ,$$

we obtain simply

$$\nabla_{\frac{\partial}{\partial x^j}} \left(\nabla_{\frac{\partial}{\partial x^k}} Z \right) - \nabla_{\frac{\partial}{\partial x^k}} \left(\nabla_{\frac{\partial}{\partial x^j}} Z \right) = \sum_{i=1}^{n} \left(\sum_{\ell=1}^{n} \lambda^{\ell} R^{i}{}_{\ell jk} \right) \frac{\partial}{\partial x^i}$$

$$= R\left(\frac{\partial}{\partial x^j}, \frac{\partial}{\partial x^k} \right) Z ,$$

so the definition does agree with the classical one. At the same time, it is clearly preferable, in that it does not involve the torsion. [Classically, there is practically no way to even name the quantity

$$\nabla_{\frac{\partial}{\partial x^j}} \left(\nabla_{\frac{\partial}{\partial x^k}} Z \right) ,$$

since $\lambda^{i}{}_{;kj}$ is automatically interpreted as the components of the tensor $\nabla\nabla Z$. For a vector field $Y = \Sigma b^i \, \partial/\partial x^i$, we can write the components of

$$\nabla_{\frac{\partial}{\partial x^j}} (\nabla_Y Z) \qquad \text{as} \qquad \left(\sum_{\ell=1}^{n} \lambda^{i}{}_{;\ell} b^{\ell} \right)_{;j} ,$$

the summation over ℓ making it clear that we are taking covariant derivatives of the tensor field with components $\mu^i = \Sigma \lambda^{i}{}_{;\ell} b^{\ell}$. However, for the special case

$$\nabla_{\frac{\partial}{\partial x^j}}\left(\nabla_{\frac{\partial}{\partial x^k}} Z\right),$$

about all we can write is

$$(\sum_{\ell=1}^{n} \lambda^i{}_{;\ell}\,\delta^{\ell}_k)_{;j} \,.\;]$$

The following result will be our analogue of the Ricci identities.

10. PROPOSITION. Let $s\colon \mathbb{R}^2 \longrightarrow M$ be a parameterized surface, and V a C^∞ vector field along s. Then

$$\frac{D}{\partial y}\frac{D}{\partial x}V - \frac{D}{\partial x}\frac{D}{\partial y}V = R(\frac{\partial s}{\partial x}, \frac{\partial s}{\partial y})V \;.$$

Proof. Compute in a coordinate system. ∎

It should come as no surprise to learn that we can now prove the Test Case; it may be surpising, however, to see how simple the proof becomes.

11. THEOREM (THE TEST CASE; third version). Let $(M, \langle\;,\;\rangle)$ be an n-dimensional Riemannian manifold for which the curvature tensor R (for the Levi-Civita connection) is 0. Then M is locally isometric to $\mathbb{R}^n$ with its usual Riemannian metric.

Proof. We assume we are in $\mathbb{R}^n$, with the standard coordinate system $y^1,\ldots,y^n$.

Step 1. We claim that we can find vector fields X, with arbitrary values

$X(0) \in R^n{}_0$, satisfying

$$\nabla_{\frac{\partial}{\partial y^i}} X = 0 \qquad \text{for all } i ,$$

and hence $\nabla_Z X = 0$ for any Z. To do this we first choose $X(y,0,\ldots,0)$, with the desired initial value, so that it is parallel along the y^1-axis.

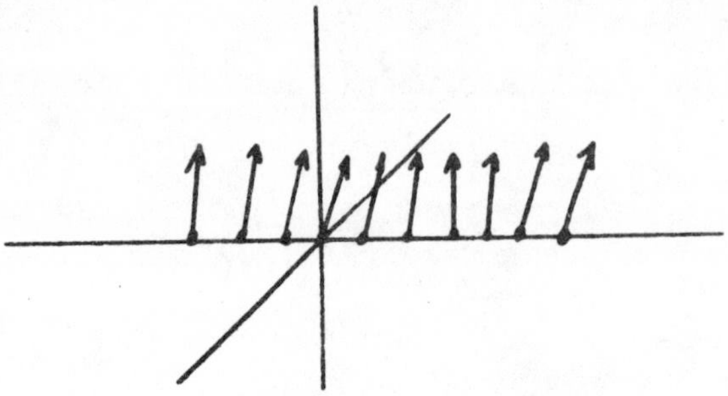

For each fixed y^1, we then choose $X(y^1,y,0,\ldots,0)$, with the values $X(y^1,0,\ldots,0)$ just obtained, so that X is parallel along the curves $y \longmapsto (y^1,y,0,\ldots,0)$. The vector field X is now defined on the surface

$$s(y^1,y^2) = (y^1,y^2,0,\ldots,0) \quad .$$

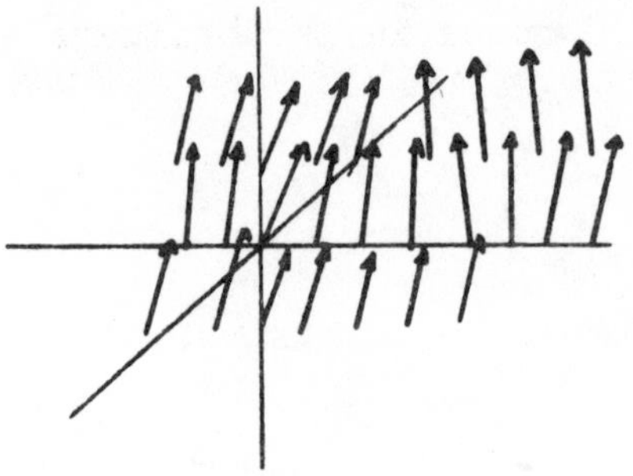

Clearly $DX/\partial y^2$ is 0 along s, while $DX/\partial y^1$ is 0 along $\{s(y,0)\}$. Now we have

$$\frac{D}{\partial y^2}\frac{D}{\partial y^1} X - \frac{D}{\partial y^1}\frac{D}{\partial y^2} X = R\left(\frac{\partial s}{\partial y^1}, \frac{\partial s}{\partial y^2}\right) X = 0 \quad ,$$

so

$$\frac{D}{\partial y^2}\frac{D}{\partial y^1} X = 0 \quad .$$

This means that $DX/\partial y^1$ is parallel along the curves $y \mapsto s(y^1,y)$. Since $DX/\partial y^1$ is 0 at $s(y^1,0)$, we have $DX/\partial y^1 = 0$ along s.

We can clearly continue in this way to obtain the desired X. Now choose $X_1,\ldots,X_n$ with this property so that $X_1(0),\ldots,X_n(0)$ are orthonormal with respect to $\langle\ ,\ \rangle_0$. Clearly $X_1,\ldots,X_n$ are linearly independent in a neighborhood of 0.

Step 2. Since the connection ∇ associated with $\langle\ ,\ \rangle$ is symmetric, we have

$$0 = \nabla_{X_i} X_j - \nabla_{X_j} X_i - [X_i,X_j] \ .$$

But $\nabla_Z X_i = 0$ for all Z. So $[X_i,X_j] = 0$. This means that there is a coordinate system $x^1,\ldots,x^n$ with $X_i = \partial/\partial x^i$.

Step 3. We claim that x is the desired coordinate system, i.e., that the X_i are everywhere orthonormal. This is obvious, for they are orthonormal at 0 and parallel along any curve, and parallel translation preserves the inner product $\langle\ ,\ \rangle$. ∎

This proof is completely analogous to the first two proofs of the Test Case, except that it is "dual" to them in the sense that we find the vector fields $\partial/\partial x^i$ instead of the 1-forms dx^i.

Step 1 uses the conditions $R = 0$ to obtain the parallel vector fields X_i, in a way completely analagous to, but simpler than, Step 1 in the second version.

Step 2 now uses symmetry of the connection to prove that $[X_i,X_j] = 0$, a condition which is dual to exactness of the 1-forms obtained in the previous versions.

<u>Step 3</u> uses the <u>definition of the Levi-Civita connection</u> to prove that the vectors X_i are everywhere orthonormal. As we have already pointed out, the fact that parallel translation preserves the inner product $\langle\ ,\ \rangle$ is equivalent to Ricci's Lemma, which we used in Step 3 of the second version.

Notice that the proof shows that if we can find n everywhere linearly independent vector fields $X_1,\ldots,X_n$, which are <u>parallel</u> (i.e., which satisfy $\nabla_Z X_i$ for all Z), then the manifold is flat. Consequently, such vector fields generally <u>cannot</u> be found. This means that parallel translation of a vector along two different curves with the same end-points generally gives <u>different</u> results; for if X_p were the parallel translate of X_0 along both the curves shown below, we would clearly have both

$$\nabla_{\frac{\partial}{\partial y^1}} X(p) = 0 \quad \text{and} \quad \nabla_{\frac{\partial}{\partial y^2}} X(p) = 0 \ .$$

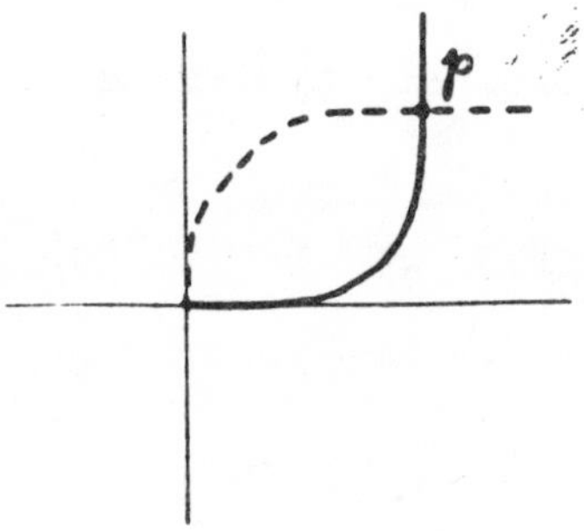

This phenomenon can also be described by saying that parallel translation of a vector along a closed curve generally brings it back to a different vector. Later we will see some more quantitative statements of this fact.

Our next two results are simply Proposition 5-9 and 4D-6. They are reproved here in a spirit more in keeping with the present treatment of connections, but it is not hard to see that they are essentially the proofs given before. As a matter of notation, if we are given an expression $A(X,Y,Z)$, we will let $\mathfrak{S}A(X,Y,Z)$ denote the cyclic sum

$$\mathfrak{S}A(X,Y,Z) = A(X,Y,Z) + A(Y,Z,X) + A(Z,X,Y) \ .$$

12. PROPOSITION. Let ∇ be a connection with torsion T and curvature R. Then for all vector fields X,Y,Z,W we have

(1) $R(X,Y)Z = -R(Y,X)Z$

(2) (Bianchi's 1st identity)

$$\mathfrak{S}\{R(X,Y)Z\} = \mathfrak{S}\{(\nabla_X T)(Y,Z)\} + \mathfrak{S}\{T(T(X,Y),Z)\}$$

(3) (Bianchi's 2nd identity)

$$\mathfrak{S}\{(\nabla_Z R)(X,Y,W)\} + \mathfrak{S}\{R(T(X,Y),Z)W\} = 0 \quad .$$

In particular, if $T = 0$, then

(2') $\mathfrak{S}\{(X,Y)Z\} = 0$

(3') $\mathfrak{S}\{(\nabla_Z R)(X,Y,W)\} = 0$.

Proof. (1) is clear from the definition.

To prove (2), we first note that

$$T(T(X,Y),Z) = T(\nabla_X Y,Z) - T(\nabla_Y X,Z) - T([X,Y],Z)$$

$$= T(\nabla_X Y,Z) + T(Z,\nabla_Y X) - T([X,Y],Z).$$

We also have, by Corollary 5,

$$(\nabla_Z T)(X,Y) = \nabla_Z(T(X,Y)) - T(\nabla_Z X,Y) - T(X,\nabla_Z Y) \quad .$$

From these equations we obtain

$$\mathfrak{S}\{T(T(X,Y),Z)\} = -\mathfrak{S}\{(\nabla_Z T)(X,Y)\} + \mathfrak{S}\{\nabla_Z(T(X,Y)) - T([X,Y],Z)\} .$$

The second term on the right side equals

$$\mathfrak{S}\{\nabla_X(\nabla_Y Z) - \nabla_Y(\nabla_X Z) - \nabla_{[X,Y]}Z\} + \mathfrak{S}\{[[X,Y],Z]\}$$

$$= \mathfrak{S}\{R(X,Y)Z\} + 0 ,$$

since the Jacobi identity states that $\mathfrak{S}\{[[X,Y],Z]\} = 0$.

To prove (3), we first use (1) to obtain

$$\mathfrak{S}\{R(T(X,Y),Z)W\} = \mathfrak{S}\{R(\nabla_X Y,Z)W + R(Z,\nabla_Y X)W - R([X,Y],Z)W\}$$

$$= \mathfrak{S}\{R(\nabla_Z X,Y)W + R(X,\nabla_Z Y)W\} - \mathfrak{S}\{R([X,Y],Z)W\} .$$

By Corollary 5 we also have

$$(\nabla_Z R)(X,Y,W) = \nabla_Z(R(X,Y)W) - R(X,Y)\nabla_Z W - R(\nabla_Z X,Y)W - R(X,\nabla_Z Y)W .$$

From these equations we have

$$\mathfrak{S}\{R(T(X,Y),Z)W\} = -\mathfrak{S}\{(\nabla_Z R)(X,Y,W)\} + \mathfrak{S}\{\nabla_Z(R(X,Y)W) - R(X,Y)\nabla_Z W - R([X,Y],Z)W\} .$$

Now

$$\nabla_Z(R(X,Y)W) - R(X,Y)\nabla_Z W - R([X,Y],Z)W$$

$$= \nabla_Z\nabla_X\nabla_Y W - \nabla_Z\nabla_Y\nabla_X W \boxed{- \nabla_Z\nabla_{[X,Y]}W}$$

$$- \nabla_X\nabla_Y\nabla_Z W + \nabla_Y\nabla_X\nabla_Z W + \nabla_{[X,Y]}\nabla_Z W$$

$$- \nabla_{[X,Y]}\nabla_Z W + \boxed{\nabla_Z\nabla_{[X,Y]}W} + \nabla_{[[X,Y],Z]}W .$$

Writing $[\nabla_X, \nabla_Y]$ for the operation $W \longmapsto \nabla_X \nabla_Y W - \nabla_Y \nabla_X W$ (as on page I.5-32), we can write this as

$$\nabla_Z([\nabla_X, \nabla_Y]W) - [\nabla_X, \nabla_Y](\nabla_Z W) + \nabla_{[[X,Y],Z]} W \quad .$$

The cyclic sum of these quantities is 0, because of the Jacobi identity $\mathfrak{S}\{[[X,Y],Z]\} = 0$, and the "Jacobi identity"

$$\mathfrak{S}\{[\nabla_Z, [\nabla_X, \nabla_Y]]W\} = 0$$

(recall that in any ring, if we define $[a,b] = ab - ba$, then $[\ ,\ ]$ satisfies the Jacobi identity). ∎

13. PROPOSITION. For the curvature tensor R of the Levi-Civita connection ∇ associated to a Riemannian metric $\langle\ ,\ \rangle$ we also have

$$(1) \qquad \langle R(X,Y)Z, W \rangle = - \langle R(X,Y)W, Z \rangle$$

$$(2) \qquad \langle R(X,Y)Z, W \rangle = \langle R(Z,W)X, Y \rangle \quad .$$

Proof. Equation (1) is equivalent to

$$\langle R(X,Y)Z, Z \rangle = 0 \qquad \text{for all } X, Y, Z \quad .$$

It suffices to prove this when $[X,Y] = 0$. In this case

$$\langle R(X,Y)Z, Z \rangle = \langle \nabla_X(\nabla_Y Z) - \nabla_Y(\nabla_X Z), Z \rangle \quad ,$$

so we must show that $\langle \nabla_X(\nabla_Y Z), Z \rangle$ is symmetric in X and Y.

Now $YX < Z,Z >$ is symmetric in X and Y, since $[X,Y] = 0$. But

$$X < Z,Z > = 2 < \nabla_X Z, Z > ,$$

so

$$YX < Z,Z > = 2 < \nabla_Y \nabla_X Z, Z > + 2 < \nabla_X Z, \nabla_Y Z > .$$

Since the right-most term is symmetric in X and Y, so is $< \nabla_Y \nabla_X Z, Z >$.

Equation (2) follows from (1), Proposition 12, and Proposition 4D-7. ∎

To complete our treatment of Koszul connections, we define a <u>geodesic</u> for ∇ to be a path $\gamma\colon [a,b] \longrightarrow M$ with

$$\frac{D}{dt}\frac{d\gamma}{dt} = 0 ;$$

thus, the tangent vector $d\gamma/dt$ must be parallel along γ. In the coordinate system x we immediately obtain the equations for a geodesic,

$$\frac{d^2\gamma^k}{dt^2} + \sum_{i,j=1}^{n} \Gamma^k_{ij}(\gamma(t)) \frac{d\gamma^i}{dt}\frac{d\gamma^j}{dt} = 0 .$$

The existence of a unique geodesic, with given initial vector $\gamma'(0) \varepsilon M_p$ follows immediately. Since parallel translation is an isomorphism, the tangent vector $d\gamma/dt$ of a geodesic γ is nowhere zero (except when γ is a constant path). If V is any vector field along γ, then we can write

$$V = f\frac{d\gamma}{dt} \qquad f\colon [a,b] \longrightarrow \mathbb{R} ,$$

and

$$\frac{DV}{dt} = \frac{df}{dt}\cdot\frac{d\gamma}{dt} + f\,\frac{D}{dt}\,\frac{d\gamma}{dt}$$
$$= \frac{df}{dt}\,\frac{d\gamma}{dt} \quad .$$

This is 0 only if f is linear. Consequently, a reparameterization $\bar{\gamma} = \gamma\circ p$ of γ is also a geodesic only if p is linear*.

We know that this result can be made more precise for the Levi-Civita connection ∇ associated to a metric $\langle\ ,\ \rangle$, and the proof of the more precise result is now especially easy: We have

$$\frac{d}{dt} \left\langle \frac{d\gamma}{dt}, \frac{d\gamma}{dt} \right\rangle = 2 \left\langle \frac{D}{dt}\frac{d\gamma}{dt}, \frac{d\gamma}{dt} \right\rangle = 0 \quad ,$$

so $\|d\gamma/dt\|$ is constant, i.e., γ is parameterized proportionally to arc length (Theorem I.9-12). The reader may see for himself how the proof of Gauss' Lemma (I.9-15) is simplified in the present set up. We will complete our study of the ∇ operator by providing the invariant description of the First Variation Formula promised so long ago.

<u>14. THEOREM (FIRST VARIATION FORMULA)</u>. Let $\gamma\colon [a,b] \longrightarrow M$ be a piecewise C^∞ path and $\alpha\colon (-\varepsilon,\varepsilon) \times [a,b] \longrightarrow M$ a variation. Let

$W_t = \dfrac{\partial\alpha}{\partial u}(0,t)$ the "variation vector field"

$V_t = \dfrac{D}{dt}\dfrac{d\gamma}{dt}$ the "velocity vector of γ"

$A_t = \dfrac{D}{dt} V_t$ the "acceleration vector of γ".

*Again we point out that certain global theorems about geodesics for Riemannian metrics do not generalize to arbitrary connections; see Chapter 8, Addendum 2.

Also choose $a = t_0 < \ldots < t_n = b$ to include all discontinuity points of V, and set

$$\nabla_{t_i} V = V(t_i+) - V(t_i-) \qquad i = 1,\ldots,N-1$$

$$\nabla_{t_0} V = V(t_0+)$$

$$\nabla_{t_N} V = - V(t_N-) \quad .$$

Then

$$\left.\frac{dE(\bar{\alpha}(u))}{du}\right|_{u=0} = - \int_a^b < W_t, A_t > dt - \sum_{i=0}^{N} < W_{t_i}, \nabla_{t_i} V > \quad .$$

<u>Proof</u>. We will give the proof when V has no discontinuities, leaving to the reader the simple auxiliary argument for the general case.

From

$$\frac{\partial}{\partial u} < \frac{\partial \alpha}{\partial t}, \frac{\partial \alpha}{\partial t} > = 2 < \frac{D}{\partial u}\frac{\partial \alpha}{\partial t}, \frac{\partial \alpha}{\partial t} >$$

we obtain

$$\frac{dE(\bar{\alpha}(u))}{du} = \frac{1}{2}\frac{d}{du}\int_a^b < \frac{\partial \alpha}{\partial t}, \frac{\partial \alpha}{\partial t} > dt = \int_a^b < \frac{D}{\partial u}\frac{\partial \alpha}{\partial t}, \frac{\partial \alpha}{\partial t} > dt$$

$$= \int_a^b < \frac{D}{\partial t}\frac{\partial \alpha}{\partial u}, \frac{\partial \alpha}{\partial t} > dt \quad \text{by Proposition 9.}$$

Now the identity

$$\frac{\partial}{\partial t} < \frac{\partial \alpha}{\partial u}, \frac{\partial \alpha}{\partial t} > = < \frac{D}{\partial t}\frac{\partial \alpha}{\partial u}, \frac{\partial \alpha}{\partial t} > + < \frac{\partial \alpha}{\partial u}, \frac{D}{\partial t}\frac{\partial \alpha}{\partial t} >$$

implies the following analogue of integrating by parts:

$$\int_a^b < \frac{D}{\partial t}\frac{\partial\alpha}{\partial t}, \frac{\partial\alpha}{\partial t} > dt = < \frac{\partial\alpha}{\partial u}, \frac{\partial\alpha}{\partial t} > \Big|_{t=a}^{t=b} - \int_a^b < \frac{\partial\alpha}{\partial u}, \frac{D}{\partial t}\frac{\partial\alpha}{\partial t} > dt \quad .$$

Thus

$$\frac{dE(\bar{\alpha}(u))}{du}\Big|_{u=0} = -\int_a^b < W_t, A_t > dt - < W(a), V(a) > + < W(b), V(b) > \quad ,$$

which is the desired formula. ▮

Addendum 1. Connections with the same Geodesics

Let ∇ and $\bar{\nabla}$ be two connections on M. We define

$$D(X,Y) = \bar{\nabla}_X Y - \nabla_X Y.$$

A simple computation shows that D is linear over the C^∞ functions in both arguments, so it determines a tensor D, the difference tensor of the two connections. In a coordinate system x we have

$$D = \Sigma\, \bar{\Gamma}_{ij}^{k} - \Gamma_{ij}^{k}\, dx^i \otimes dx^j \otimes \frac{\partial}{\partial x^k}\,.$$

As is well known, there is a unique way to write $D = S + A$ with S symmetric and A alternating, namely

$$S(X,Y) = \frac{1}{2}\,[D(X,Y) + D(Y,X)]$$

$$A(X,Y) = \frac{1}{2}\,[D(X,Y) - D(Y,X)]\,.$$

Note that if $\bar{T}$ and T are the torsion tensors of $\bar{\nabla}$ and ∇, then

$$(*) \qquad 2A(X,Y) = \bar{\nabla}_X Y - \nabla_X Y - \bar{\nabla}_Y X + \nabla_Y X$$

$$= \bar{T}(X,Y) + [X,Y] - T(X,Y) - [X,Y]$$

$$= \bar{T}(X,Y) - T(X,Y),$$

so $\bar{\nabla}$ and ∇ have the same torsion if and only if $A = 0$.

<u>15. PROPOSITION.</u> The following are equivalent:

(a) The connections $\bar{\nabla}$ and ∇ have the same geodesics [with the same parametrizations]

(b) $D(X,X) = 0$ for all X

(c) $S = 0$.

<u>Proof.</u> (a) => (b): Given $0 \neq X_p \in M$, let γ be the geodesic (for the connections $\bar{\nabla}$ and ∇) with $\gamma'(0) = X_p$. Let X be a vector field in a neighborhood of p which equals $d\gamma/dt$ along the part of γ in the neighborhood. Then

$$D(X_p,X_p) = \bar{\nabla}_{X_p} X - \nabla_{X_p} X = 0 - 0.$$

(b) => (a): Let γ be a geodesic for ∇, and let X be a vector field which equals $d\gamma/dt$ along a portion of γ. Then

$$\bar{\nabla}_{X_p} X = D(X_p,X_p) + \nabla_{X_p} X = 0 + 0,$$

which shows that γ is a geodesic for $\bar{\nabla}$.

(b) <=> (c): $D(X,X) = 0 \iff S(X,X) = 0$. The latter condition implies that $S = 0$, for we have

$$\begin{aligned} 0 = S(X+Y,\ X+Y) &= S(X,X) + S(Y,Y) + S(X,Y) + S(Y,X) \\ &= 2S(X,Y). \blacksquare \end{aligned}$$

16. COROLLARY. If the connections ∇ and $\bar{\nabla}$ have the same geodesics and the same torsion, then $\bar{\nabla} = \nabla$.

17. COROLLARY. For every connection $\bar{\nabla}$, there is a unique connection ∇ with the same geodesics and with torsion 0.

Proof. Uniqueness follows from Corollary 16. For existence we define $\nabla_X Y = \bar{\nabla}_X Y - \frac{1}{2}\bar{T}(X,Y)$, checking easily that ∇ is a connection. Since $\bar{T}$ is skew-symmetric, $D = \frac{1}{2}\bar{T}$ must simply be A, so $S = 0$, so ∇ and $\bar{\nabla}$ have the same geodesics. Also, $T = \bar{T} - 2A = 0$, so ∇ has torsion 0. ∎

We thus see that if we divide all connections into equivalence classes, putting all connections with the same geodesics in the same class, then each class has exactly one connection with zero torsion. We can also say exactly how large each class is: If ∇ is a connection with torsion 0 and $\bar{T}$ is a skew-symmetric tensor of type $\binom{2}{0}$, then there is a connection $\bar{\nabla}$ in the same class as ∇, but with torsion $\bar{T}$, namely

$$\bar{\nabla}_X Y = \nabla_X Y + \frac{1}{2}\bar{T}(X,Y).$$

It is also of interest to inquire when two connections $\bar{\nabla}$ and ∇ have the same geodesics, with possibly different parametrizations. Such connections are called "projectively equivalent", because in ordinary space a projective map is one that takes straight lines to straight lines.

18. PROPOSITION (H. WEYL; 1921). The following are equivalent:

(a) The connections $\bar{\nabla}$ and ∇ have the same geodesics, with possibly different parametrizations

(b) For every X there is λ_X with $D(X,X) = \lambda_X \cdot X$

(c) There is a (unique) 1-form ω with $S(X,Y) = \omega(X)Y + \omega(Y)X$.

Proof. (a) => (b): Given $X_p \in M$, let γ be the geodesic for the connection ∇ with $\gamma' = X_p$, and let $\bar{\gamma}$ be the reparameterization which makes it a geodesic for $\bar{\nabla}$. Let X be a vector field which equals $d\gamma/dt$ along γ and $\bar{X}$ a vector field which equals $d\bar{\gamma}/dt$ along γ. Then along γ we have $X = f\bar{X}$ for some f. So

$$\begin{aligned} D(X_p,X_p) &= \bar{\nabla}_{X_p} X - \nabla_{X_p} X \\ &= \bar{\nabla}_{f(p)\bar{X}_p} f\bar{X} - 0 \\ &= f(p)[\bar{X}_p(f)\cdot\bar{X}_p + f(p)\, \bar{\nabla}_{\bar{X}_p} \bar{X}] \\ &= f(p)\bar{X}_p(f)\cdot\bar{X}_p = \bar{X}_p(f)\cdot X_p. \end{aligned}$$

(b) => (a): Let γ be a geodesic for ∇, and let X be a vector field which equals $d\gamma/dt$ along γ. Then for $p = \gamma(t)$ we have

$$(1) \quad \bar{\nabla}_{X_p} X = D(X_p,X_p) + \nabla_{X_p} X = \lambda_{X_p} X_p = g(t)X_p, \text{ say.}$$

Let $f(t) = e^{G(t)} \neq 0$, where $G'(t) = g(t)$ so that

(2) $\frac{df}{dt} = g(t)f(t)$,

and let $\bar{X}$ be a vector field such that

(3) $\bar{X}_{\gamma(t)} = \frac{1}{f(t)} X_{\gamma(t)} = \frac{1}{f(t)} \frac{d\gamma}{dt}$.

Then

$$\bar{\nabla}_{\bar{X}_p} \bar{X} = \frac{1}{f(t)} \bar{\nabla}_{X_p} \frac{1}{f} X$$

$$= \frac{1}{f(t)} \left[X_p(\tfrac{1}{f})X_p + \frac{1}{f(t)} \bar{\nabla}_{X_p} X \right]$$

$$= \frac{1}{f(t)} \left[-\frac{df/dt}{f(t)^2} X_p + \frac{1}{f(t)} g(t)X_p \right] \qquad \text{by (1)}$$

$$= \frac{1}{f(t)} \left[-\frac{df/dt}{f(t)^2} X_p + \frac{df/dt}{f(t)^2} X_p \right] \qquad \text{by (2)}$$

$$= 0.$$

Consequently, γ is a geodeisc for $\bar{\nabla}$ if we reparameterize it as $\bar{\gamma} = \gamma \circ p^{-1}$, where p is chosen so that

$$\frac{d\bar{\gamma}}{dt} = \bar{X}_{\bar{\gamma}(t)} ,$$

i.e., so that

$$(p^{-1})'(t) \left.\frac{d\gamma}{dt}\right|_{p^{-1}(t)} = \frac{1}{f(p^{-1}(t))} \left.\frac{d\gamma}{dt}\right|_{p^{-1}(t)} \qquad \text{by (3).}$$

For this we need

$$\frac{1}{p'(p^{-1}(t))} = \frac{1}{f(p^{-1}(t))} ,$$

or simply $p'(s) = f(s) > 0$ [compare Problem I. 9-27].

(c) => (b) is clear.

(b) => (c): We first establish the following algebraic

Lemma. Let V be a vector space, and $S: V \times V \longrightarrow V$ a symmetric bilinear map such that for each $v \in V$ there is $\lambda_v \in \mathbb{R}$ with

$$S(v,v) = \lambda_v v .$$

Then there is a unique $\varphi \in V^*$ such that

$$S(v,w) = \varphi(v)w + \varphi(w)v .$$

Proof. If φ exists, clearly $\varphi(v) = \lambda_v/2$ for $v \neq 0$. Conversely, if this definition makes φ linear we will be done, for then

$$S(v,v) + S(w,w) + 2S(v,w) = S(v+w,v+w) = \lambda_{v+w}(v+w) = (\lambda_v + \lambda_w)(v+w),$$

so

$$2\varphi(v)v + 2\varphi(w)w + 2S(v,w) = 2\varphi(v)v + 2\varphi(w)w + 2\varphi(v)w + 2\varphi(w)v ,$$

which yields

$$S(v,w) = \varphi(v)w + \varphi(w)v .$$

We prove that φ is linear as follows. From

$$\lambda_{v+w}(v+w) = \lambda_v v + \lambda_w w + 2S(v,w)$$

$$\lambda_{v-w}(v-w) = \lambda_v v + \lambda_w w - 2S(v,w)$$

we obtain

$$(\lambda_{v+w} + \lambda_{v-w} - 2\lambda_v)v + (\lambda_{v+w} - \lambda_{v-w} - 2\lambda_w)\, w = 0.$$

For linearly independent v and w we thus have

$$\lambda_{v+w} + \lambda_{v-w} - 2\lambda_v = 0$$

$$\lambda_{v+w} - \lambda_{v-w} - 2\lambda_w = 0 ,$$

and hence $\lambda_{v+w} = \lambda_v + \lambda_w$. This is clearly also true if v and w are linearly dependent. Homogeneity is likewise trivial.

When V is n-dimensional, we have the explicit formula

$$(*) \quad \varphi(v) = \frac{1}{n+1} \text{ trace of } w \longmapsto S(v,w),$$

which can be deduced as follows. Choose a basis $v_1, \ldots, v_n$ with $v = v_1$. Then

$$S(v_1,v_j) = \varphi(v_1)v_j + \varphi(v_j)\varphi_1 \quad ,$$

so

$$\text{trace of } w \mapsto S(v_1,w) = \sum_{j=1}^{n} j^{\text{th}} \text{ component of } \varphi(v_1)v_j + \varphi(v_j)v_1;$$

this component is $\varphi(v_1)$ for $j \neq 1$, and $2\varphi(v_1)$ for $j = 1$.

Formula (*) clearly shows that

$$\omega(X) = \frac{1}{n+1} \text{ trace } Y \mapsto S(X,Y)$$

is the desired C^∞ 1-form on M. ∎

19. COROLLARY The connections $\bar{\nabla}$ and ∇ have the same torsion and the same geodesics (suitably reparameterized) if and only if there is a (unique) 1-form ω with

$$D(X,Y) = \omega(X)Y + \omega(Y)X.$$

Note, by the way, that for any connection ∇ and any 1-form ω, the function

$$\bar{\nabla}_X Y = \nabla_X Y + \omega(X)Y + \omega(Y)X$$

always *is* a connection, with the same torsion as ∇.

Addendum 2. Riemann's Invariant Definition of the Curvature Tensor.

Using the results of this Chapter, we now prove the result announced in the Addendum to Chapter 4-D.

20. THEOREM (RIEMANN). Let X and Y be vector fields with $[X,Y] = 0$. Suppose also that for every vector field with $[X,Z](p) = [Y,Z](p) = 0$, we have

$$\left.\begin{aligned}
&(1)\quad Z < X,Y > - Y < X,Z > - X < Y,Z > = 0\\
&(2)\quad Z < X,X > - 2X < X,Z > = 0\\
&(3)\quad Z < Y,Y > - 2Y < Y,Z > = 0
\end{aligned}\right\}\quad \text{at } p\ .$$

Then

$$-2 < R(X_p,Y_p)Y_p,X_p > = [XX < Y,Y > - 2XY < X,Y > + YY < X,X >](p) \quad .$$

(Moreover, these are vector fields X and Y with these properties and arbitrary values X_p, Y_p.)

Proof. Using Corollary 7, we see that equation (2) is equivalent to

$$2 < \nabla_Z X, X > - 2 < \nabla_X X, Z > - 2 < X, \nabla_X Z > = 0 \qquad \text{at } p$$

and hence to

$$< \nabla_X X, Z > = 0 \qquad \text{at } p ,$$

since $\nabla_Z X(p) - \nabla_X Z(p) = [X,Z](p) = 0$. Thus equation (2) is equivalent to

$$(2') \qquad \nabla_X X(p) = 0 \quad .$$

Similarly, (3) is equivalent to

$$(3') \qquad \nabla_Y Y(p) = 0 \quad .$$

Finally, (1) is equivalent to

$$< \nabla_Z X, Y > + < X, \nabla_Z Y > - < \nabla_Y X, Z > - < X, \nabla_Y Z > - < \nabla_X Y, Z > - < Y, \nabla_X Z > = 0 \quad \text{at } p.$$

and hence to

$$< \nabla_X Y + \nabla_Y X, Z > = 0 \qquad \text{at } p.$$

This is equivalent to

$$0 = \nabla_X Y(p) + \nabla_Y X(p) = 2\, \nabla_X Y(p), \qquad \text{since } \nabla_X Y - \nabla_Y X = [X,Y] = 0 \ ,$$

so (1) is equivalent to

$$(1') \qquad \nabla_X Y(p) = \nabla_Y X(p) = 0 \quad .$$

We can obtain such vector fields X and Y, with given values $X_p, Y_p \in M_p$, by mapping $\mathbb{R}^2$ into M in such a way that the x and y axes go into two geodesics in M with $s_*(\partial/\partial y)$ parallel along the image of the x-axis and $s_*(\partial/\partial x)$ parallel along the image of the y-axis.

Now for such vector fields we have

$$XX < Y,Y > (p) = 2X < \nabla_X Y, Y > (p) = 2 < \nabla_X \nabla_X Y, Y > (p) + 2 < \nabla_X Y, \nabla_X Y > (p)$$

$$= 2 < \nabla_X \nabla_X Y, Y > (p)$$

$$YY < X,X > (p) = 2 < \nabla_Y \nabla_Y X, X > (p)$$

$$-2XY < X,Y > (p) = -2X(< \nabla_Y X, Y > + < X, \nabla_Y Y >)(p)$$

$$= -2 < \nabla_X \nabla_Y X, Y > (p) - 2 < \nabla_Y X, \nabla_X Y > (p) - 2 < \nabla_X X, \nabla_Y Y > (p) - 2 < X, \nabla_X \nabla_Y Y > (p)$$

$$= -2 < \nabla_X \nabla_Y X, Y > (p) - 2 < X, \nabla_X \nabla_Y Y > (p) .$$

So the sum is

$$2 \left[< \nabla_X \nabla_X Y, Y > - < \nabla_X \nabla_Y X, Y > \right] (p) + 2 \left[< \nabla_Y \nabla_Y X, X > - < \nabla_X \nabla_Y Y, X > \right] (p)$$

$$= 0 + 2 \left[< \nabla_Y \nabla_X Y, X > - < \nabla_X \nabla_Y Y, X > \right] (p),$$ since $\nabla_Y X - \nabla_X Y = [X,Y] = 0$ everywhere

$$= 2 < R(Y_p, X_p) Y_p, X_p >$$

$$= -2 < R(X_p, Y_p) Y_p, X_p > . \blacksquare$$

Chapter 7. The Repère Mobile (The Moving Frame)

The previous Chapter betrayed the historical development which we have been following, for the ∇ operator did not appear until very late in the game, around 1954*. In the meantime, Elie Cartan had elaborated a completely different theory, the method of the repère mobile. Despite the fact that this theory was invented soon after the Ricci calculus, some of its features are most easily understood by referring to the ∇ operator which historically came so much later.

Roughly speaking, the relationship between the results of Chapter 5 and those of Chapter 6 can be characterized as follows. When working with ∇ operators we express results in terms of arbitrary vector fields, while in the classical theory we always use the vector fields $X_i = \partial/\partial x^i$ given by a coordinate system (x,U). At each point $p \in U$, these vector fields provide us with an ordered basis

$$(X_1(p),\ldots,X_n(p)) \quad \text{for } M_p \quad .$$

In general, an ordered basis $(v_1,\ldots,v_n)$ for a vector space V will also be called a <u>frame</u> in V. Now the vector fields X_i may be used to determine a function

$$p \longmapsto (X_1(p),\ldots,X_n(p)) \quad ,$$

whose values are frames in the various tangent spaces M_p; such a function is called a <u>moving frame.</u> Conversely, every moving frame $p \longmapsto F_p$

---

*Nomizu, <u>Invariant affine connections on homogeneous spaces</u>, Amer. J. Math. 76 (1954), pp. 33-65.

determines vector fields $X_1,\ldots,X_n$ by $F_p = (X_1(p),\ldots,X_n(p))$. Consequently, as a matter of convenience we will not distinguish between everywhere linearly independent vector fields $X_1,\ldots,X_n$, and the moving frame they determine. In E. Cartan's theory, the basic idea, whose ramifications turn out to be extremely significant, is to express everything in terms of an arbitrary moving frame $X_1,\ldots,X_n$, and not just in terms of the "natural moving frame" $X_i = \partial/\partial x^i$ given by a coordinate system x.

Notice that a moving frame $X_1,\ldots,X_n$ need not be the natural frame for any coordinate system, since we need not have $[X_i,X_j] = 0$. Also, a moving frame may exist on a region which cannot even be included in a coordinate system. For example, there is a moving frame X_1,X_2 on the whole torus. On a Riemannian manifold $(M, \langle\ ,\ \rangle)$ we define an <u>orthonormal</u>

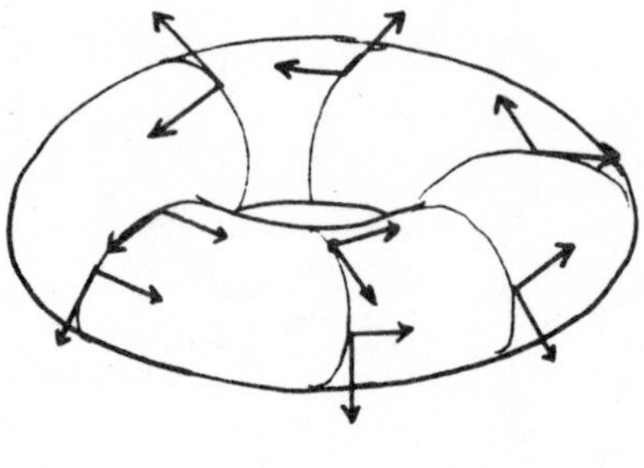

moving frame $X_1,\ldots,X_n$ to be one such that each $(X_1(p),\ldots,X_n(p))$ is an orthonormal frame for M_p. The frame illustrated above, on the torus, is an example. An arbitrary moving frame $X_1,\ldots,X_n$ gives rise to an orthonormal moving frame if we apply the Gram-Schmidt orthonormalization process to it (Problem I.9-11).

Given a moving frame $X_1,\ldots,X_n$ on (an open subset U of) M, we now ask if it is possible to describe quantitiatively just how the frame is moving. As a guide, we first consider a moving frame $X_1,\ldots,X_n$ in $\mathbb{R}^n$. In this case, we can, with a little abuse of notation, consider each X_i as an $\mathbb{R}^n$-valued function $X_i : \mathbb{R}^n \longrightarrow \mathbb{R}^n$. If we consider the identity map as an $\mathbb{R}^n$-valued function $P: \mathbb{R}^n \longrightarrow \mathbb{R}^n$, then the $\mathbb{R}^n$-valued 1-form dP is just

$$dP(X_a) = X \quad .$$

Consequently, if we introduce 1-forms θ^i by

$$\text{(I)} \quad dP = \sum_{i=1}^{n} \theta^i \cdot X_i \qquad \text{i.e.,} \quad dP(X_a) = \sum_{i=1}^{n} \theta^i(X_a) \cdot X_i(a) \in \mathbb{R}^n \quad ,$$

then θ^i are just the "dual forms" to the X_i, that is, $\theta^i(X_j) = \delta^i_j$.
We also introduce 1-forms ω^i_j by

$$\text{(II)} \qquad dX_j = \sum_{i=1}^{n} \omega^i_j \cdot X_i \quad .$$

Clearly $\omega^i_j(X_a)$ is the X_i component of $dX_j(X_a)$. Since $dX_j(X_a)$ is just the directional derivative of X_j in the direction X_a, we can interpret $\omega^i_j(X_a)$ as the <u>rate at which</u> X_j <u>rotates toward</u> $X_i(a)$ <u>as we move along a curve with tangent vector</u> X_a.

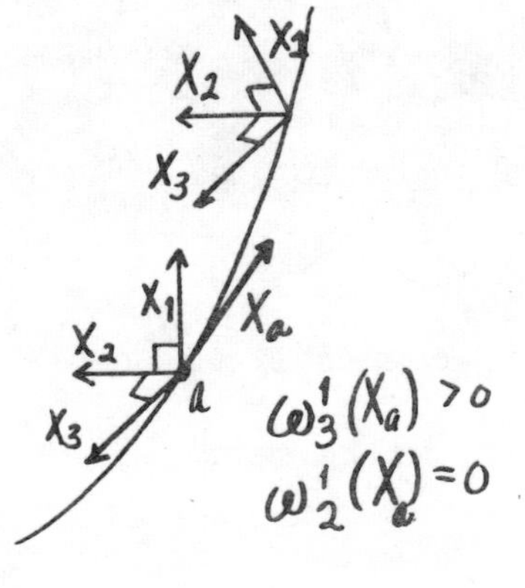

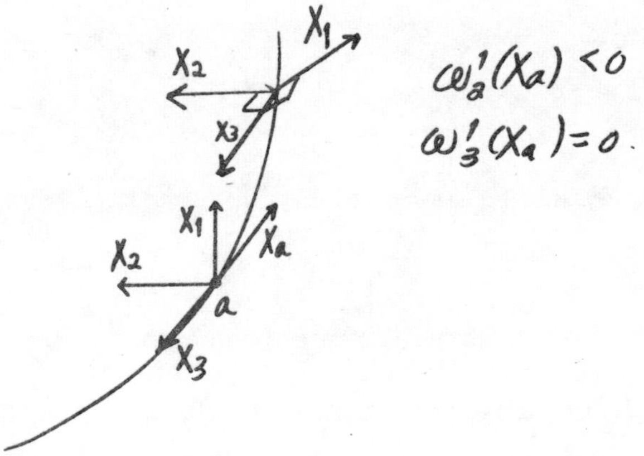

Now it is not possible to find a moving frame with arbitrary θ^i and ω^i_j; certain integrability conditions must be satisfied:

1. PROPOSITION. The 1-forms θ^i and ω^i_j for a moving frame $X_1,\ldots,X_n$ in $\mathbb{R}^n$ satisfy the structural equations of Euclidean space:

$$d\theta^i = \sum_k \theta^k \wedge \omega^i_k = -\sum_k \omega^i_k \wedge \theta^k .$$

$$d\omega^i_j = -\sum_k \omega^i_k \wedge \omega^k_j .$$

Proof. Noting that the equation $dP = \sum \theta^i \cdot X_i$ can also be written $dP = \sum \theta^i \wedge X_i$ (for the $\mathbb{R}^n$-valued 0-form X_i), we have

$$0 = d^2P = d(\sum_i \theta^i \wedge X_i) = \sum_i d\theta^i \wedge X_i - \sum_k \theta^k \wedge dX_k$$

$$= \sum_i d\theta^i X_i - \sum_k \theta^k \wedge \sum_i \omega^i_k X_i \qquad \text{by (II)} .$$

Setting the coefficient of each X_i equal to 0, we obtain the first structural equation.

We also have

$$0 = d^2X_j = \sum_i d\omega^i_j X_i - \sum_k \omega^k_j \wedge dX_k$$

$$= \sum_i d\omega^i_j X_i - \sum_k \omega^k_j \wedge \sum_i \omega^i_k X_i ,$$

from which we immediately deduce the second structural equation. ∎

The structural equations can be written much more compactly if we use slightly modified matrix notation. Henceforth, we will write matrices as $A = (A^i_j)$, and define

$$(A \cdot B)^i_j = \sum_k A^i_k B^k_j ,$$

so that our new A^i_k corresponds to the old A_{ik}. If $\mathbf{v} = (v_1, \ldots, v_n)$ is an ordered n-tuple of vectors, and we define

$$w_j = \sum_i A^i_j v_i \quad ,$$

then the n-tuple $\mathbf{w} = (w_1, \ldots, w_n)$ will be denoted by $\mathbf{v} \cdot A$. We put the A on the right because we have

$$\begin{aligned}
[\mathbf{v} \cdot (A \cdot B)]_j &= \sum_i (A \cdot B)^i_j v_i = \sum_i \sum_k A^i_k B^k_j v_i \\
&= \sum_k B^k_j (\sum_i A^i_k v_i) = \sum_k B^k_j (\mathbf{v} \cdot A)_k \\
&= [(\mathbf{v} \cdot A) \cdot B]_j \quad ,
\end{aligned}$$

which we can write simply as*

$$\mathbf{v} \cdot (AB) = (\mathbf{v} \cdot A) \cdot B \quad .$$

Naturally, if $\mathbf{X} = (X_1, \ldots, X_n)$ is an n-tuple of vector fields on a manifold M, and $A = (A^i_j)$ is a matrix of functions (or a matrix-valued function, whichever way you prefer to look at it), then $\mathbf{X} \cdot A$ denotes the n-tuple of

---

*We express this by saying that the $n \times n$ matrices <u>act on the right</u> on the set of all n-tuples of vectors of V. On the other hand, suppose we choose a <u>fixed</u> basis $v_1, \ldots, v_n$ for V, and then define $A \cdot v$ for $v \in V$ by defining $A \cdot v_j = \sum_i A^i_j v_i$ and extending by linearity. In this case we will have $(A \cdot B) \cdot v = A \cdot (B \cdot v)$; so we say that the $n \times n$ matrices <u>act on the left</u> on V.

vector fields $(\mathbf{X}\cdot A)_j = \sum_i A^i_j X_i$.

We extend this notation to forms in the natural way. If $\omega = (\omega^i_j)$ and $\eta = (\eta^i_j)$ are matrices of forms, of degree k and ℓ, respectively, then $\omega \wedge \eta$ is the matrix of $(k + \ell)$-forms

$$(\omega \wedge \eta)^i_j = \sum_k \omega^i_k \wedge \eta^k_j \quad ;$$

if $(\theta^1,\dots,\theta^n)$ is an n-tuple of ℓ-forms, then $\omega \wedge \theta$ and $\theta \wedge \omega$ are the n-tuples of $(k + \ell)$-forms

$$(\omega \wedge \theta)^i = \sum_j \omega^i_j \wedge \theta^j = (-1)^{k\ell} \sum_j \theta^j \wedge \omega^i_j = (-1)^{k\ell}(\theta \wedge \omega)^i .$$

With these conventions we can write the structural equations of Euclidean space as

$$d\theta = \theta \wedge \omega = -\omega \wedge \theta$$

$$d\omega = -\omega \wedge \omega \quad .$$

Henceforth, we will use this notation whenever convenient; for a while the reader may feel more secure rewriting things in standard form. Our very next result justifies our characterization of the structural equations as "integrability conditions."

2. PROPOSITION. Let $\omega = (\omega^i_j)$ be a matrix of 1-forms on $\mathbb{R}^n$ which satisfy the second structural equations

$$d\omega = -\omega \wedge \omega \qquad \text{i.e.} \quad d\omega^i_j = -\sum_k \omega^i_k \wedge \omega^k_j \quad .$$

1. In a neighborhood of 0 there is a matrix $A = (A^i_j)$ of functions, with arbitrary initial condition $A(0)$, such that

$$dA = A \wedge \omega \qquad \text{i.e.,} \quad dA^i_j = \sum_k A^i_k \cdot \omega^k_j \quad .$$

2. In a neighborhood of 0 there is a moving frame $X_1, \ldots, X_n$ with arbitrary initial conditions $X_1(0), \ldots, X_n(0)$, so that the dual 1-forms θ^i satisfy the first structural equation

$$d\theta = \theta \wedge \omega \qquad \text{i.e.,} \quad d\theta^i = \sum_k \theta^k \wedge \omega^i_k \quad .$$

<u>Proof</u>. 1. Let $y^1, \ldots, y^n$ be the standard coordinate system on $\mathbb{R}^n$, and let $y^1, \ldots, y^n, z^i_j$ be the standard coordinate system on $\mathbb{R}^{n+n^2}$. Let Z be the matrix of functions $Z = (z^i_j)$ and consider the matrix of 1-forms

$$(*) \qquad \Lambda = dZ - (\omega \wedge Z) \; .$$

We have

$$\begin{aligned} d\Lambda &= -(d\omega \wedge Z) + (\omega \wedge dZ) \\ &= -(\omega \wedge \omega) \wedge Z + [\omega \wedge (\Lambda + \omega \wedge Z)] \quad \text{by } (*) \text{ and the hypothesis} \\ &= \omega \wedge \Lambda \quad . \end{aligned}$$

By Proposition I.7-14, the n-dimensional distribution $\Delta_p = \bigcap \ker \Lambda^i_j(p)$ in $\mathbb{R}^{n+n^2}$ is integrable. Since $\Delta_{(0, z_0)} = (y^1, \ldots, y^n)$-plane, the integral manifold through any point $(0, z_0)$ is locally the graph of a function $p \longmapsto A(p) \in \mathbb{R}^{n^2}$ with $A(0) = z_0$. Since $Z = A$ along the graph, and $\Lambda = 0$ on this graph, equation $(*)$ gives the desired equation

(**) $$dA = \omega \wedge A \,.$$

2. If A is non-singular at 0, then in a neighborhood of 0 we can choose 1-forms θ^i with $dy^i = \sum_j A^i_j \theta^j$, which we can write simply as

$$dy = A \wedge \theta = \theta \wedge A \,.$$

Then

$$0 = d\theta \wedge A - (\theta \wedge dA) \quad ,$$

so

$$\begin{aligned} d\theta &= \theta \wedge dA \wedge A^{-1} \\ &= \theta \wedge \omega \qquad \text{by } (**) \,. \end{aligned}$$

So we just define the moving frame $X_1, \ldots, X_n$ by $\theta^i(X_j) = \delta^i_j$. ∎

For orthonormal moving frames we have one more relation:

<u>3. PROPOSITION</u>. The forms ω^i_j for an orthonormal moving frame $X_1, \ldots, X_n$ in $\mathbb{R}^n$ satisfy

$$\omega^i_j = -\,\omega^j_i \,,$$

i.e., the matrix ω is skew-symmetric.

<u>Proof</u>. We have

$$0 = d(\langle X_i, X_j \rangle) = \langle dX_i, X_j \rangle + \langle X_i, dX_j \rangle \quad .$$

Here $\langle dX_i, X_j \rangle (X_a)$ means $\langle dX_i(X_a), X_j(a) \rangle$. Since the X_i are orthogonal, clearly $\langle dX_i, X_j \rangle (X_a)$ is just the X_j component of $dX_i(X_a)$. This means that $\langle dX_i, X_j \rangle = \omega_i^j$. ∎

Now consider a moving frame $X_1, \ldots, X_n$ on a manifold M. We can still define the <u>dual 1-forms</u> θ^i by $\theta^i(X_j) = \delta_j^i$; equation (I) can now be rewritten as $X_q = \Sigma\, \theta^i(X_q) \cdot X_i(q)$ for any $X_q \in M_q$, or simply $dP = \Sigma\, \theta^i \cdot X_i$ where "dP" denotes the identity map of a tangent space into itself. We cannot use equation (II) to define the forms ω_j^i, since "dX_i" makes no sense on a general manifold. However, Propositions 1 and 3 suggest a way of defining these forms on a Riemannian manifold.

<u>4. PROPOSITION</u>. Let $X_1, \ldots, X_n$ be a moving frame on a manifold M, and let θ^i be the dual 1-forms. Then there exist unique 1-forms ω_j^i such that

(a) $$\omega_j^i = -\omega_i^j$$

(b) $$d\theta^i = \sum_k \theta^k \wedge \omega_k^i \ .$$

<u>Proof</u>. Suppose ω_j^i satisfy (a) and (b). There are unique functions a_{jk}^i and b_{jk}^i with

$$\omega_j^i = \sum_k a_{jk}^i \theta^k \ ,$$

$$d\theta^i = \frac{1}{2} \sum_{j,k} b_{jk}^i \theta^j \wedge \theta^k \ , \qquad b_{jk}^i = -b_{kj}^i \quad .$$

Then (a) is equivalent to

$$(a') \qquad a^j_{ik} = - a^i_{jk} \quad ,$$

while (b) gives

$$\frac{1}{2} \sum_{j,k} b^i_{jk} \theta^j \wedge \theta^k = d\theta^i = \sum_j \theta^j \wedge \omega^i_j$$

$$= \sum_{j,k} a^i_{jk} \theta^j \wedge \theta^k \quad ,$$

and hence

$$(b') \qquad a^i_{jk} - a^i_{kj} = b^i_{jk} \quad .$$

Cyclically permuting i,j,k, we obtain from (a') and (b')

$$a^i_{jk} = \frac{1}{2}(b^i_{jk} + b^j_{ki} - b^k_{ij}) \quad .$$

This proves uniqueness.

Now suppose we define ω^i_j by

$$\omega^i_j = \sum_k a^i_{jk} \theta^k \quad ,$$

where the a^i_{jk} are as defined above, and hence satisfy (b'). It is easy to check that equation (a') holds, and hence that equation (a) holds. Moreover, from equation (b') we have

$$\omega^i_j(X_k) - \omega^i_k(X_j) = a^i_{jk} - a^i_{kj} = b^i_{jk} \quad ;$$

it follows that

$$\sum_{\ell} \theta^{\ell} \wedge \omega_{\ell}^{i}(X_j, X_k) = \sum_{\ell} \theta^{\ell}(X_j)\omega_{\ell}^{i}(X_k) - \theta^{\ell}(X_k)\omega_{\ell}^{i}(X_j)$$

$$= \sum_{\ell} \delta_j^{\ell}\omega_{\ell}^{i}(X_k) - \delta_k^{\ell}\omega_{\ell}^{i}(X_j)$$

$$= \omega_j^i(X_k) - \omega_k^i(X_j)$$

$$= b_{jk}^i \quad ,$$

which is equivalent to equation (b). ∎

Although we have used Proposition 3 to motivate Proposition 4, the latter result seems to involve neither a Riemannian metric nor an orthonormal moving frame. However the two are, in a sense, really there, since there is a unique Riemannian metric on the domain of the moving frame $X_1, \ldots, X_n$ which makes it an orthonormal moving frame. Naturally, if we are already given a Riemannian metric $\langle \ , \ \rangle$ on M, then we will expect the 1-forms ω_j^i to have some significance for this metric only when the moving frame is orthonormal with respect to it.

Let us therefore consider an <u>orthonormal</u> moving frame $X_1, \ldots, X_n$ on a Riemannian manifold $(M, \langle \ , \ \rangle)$. The unique 1-forms ω_j^i given by Proposition 4 are called the <u>connection forms</u> for the moving frame $X_1, \ldots, X_n$. They satisfy $\omega_j^i = -\omega_i^j$ and the first structural equation, by their very definition. On the other hand, there is no reason to expect them to satisfy the second structural equation. Recognizing this, we define a matrix of 2-forms $\Omega = (\Omega_j^i)$ by

$$d\omega = \omega \wedge \omega + \Omega \qquad \text{i.e.,} \quad d\omega_j^i = \sum_k \omega_k^i \wedge \omega_j^k + \Omega_j^i \quad .$$

The 2-forms Ω^i_j are called the <u>curvature forms</u> for the orthonormal moving frame $X_1,\ldots,X_n$. The names "connection forms" and "curvature forms" are explained by the very next theorem. Let us set

$$\nabla_{X_i} X_j = \sum_{k=1}^{n} \mathbf{\Gamma}^k_{ij} X_k$$

$$R(X_i,X_j)X_k = \sum_{\ell=1}^{n} \mathbf{R}^{\ell}{}_{kij} X_\ell \qquad ,$$

where ∇ is the Levi-Civita connection for $<\ ,\ >$, and R is its curvature tensor. We use bold letters $\mathbf{\Gamma}$ and $\mathbf{R}$ to remind ourselves that these are not components with respect to a coordinate system; thus, for example, we do <u>not</u> necessarily have $\mathbf{\Gamma}^k_{ij} = \mathbf{\Gamma}^k_{ji}$ (but we clearly do have $\mathbf{R}^{\ell}{}_{kij} = \mathbf{R}^{\ell}{}_{kji}$.)

<u>5. THOEREM</u>. Let $X_1,\ldots,X_n$ be an <u>orthonormal</u> moving frame on a Riemannian manifold $(M, <\ ,\ >)$, and let $\theta^i, \omega^i_j, \Omega^i_j$ be the dual forms, connection forms, and curvature forms for this moving frame. Then we have the <u>structural equations of</u> $(M, <\ ,\ >)$:

$$d\theta^i = -\sum_k \omega^i_k \wedge \theta^k \qquad\qquad d\theta = -\omega \wedge \theta$$

$$d\omega^i_j = -\sum_k \omega^i_k \wedge \omega^k_j + \Omega^i_j \qquad\qquad d\omega = -\omega \wedge \omega + \Omega \ ,$$

where

$$\omega^i_j = \sum_k \mathbf{\Gamma}^i_{kj}\theta^k \ , \quad \Omega^i_j = \frac{1}{2}\sum_{k,\ell} \mathbf{R}^i{}_{jk\ell}\,\theta^k \wedge \theta^\ell = \sum_{k<\ell} \mathbf{R}^i{}_{jk\ell}\,\theta^k \wedge \theta^\ell \ .$$

<u>Proof</u>. By the uniqueness part of Proposition 4, we can prove the first structural equation by showing that if we <u>define</u> $\omega^i_j = \sum \mathbf{\Gamma}^i_{kj}\theta^k$, then the

ω^i_j do satisfy conditions (a) and (b) of that proposition. Now by Corollary 6-7 we have

$$0 = X_k < X_i, X_j > = < \nabla_{X_k} X_i, X_j > + < X_i, \nabla_{X_k} X_j >$$

$$= \Gamma^j_{ki} + \Gamma^i_{kj} \quad ;$$

this immediately implies that $\omega^i_j = -\omega^j_i$, which is condition (a).

As in the proof of Proposition 4, we have

$$\sum_\ell \theta^\ell \wedge \omega^i_\ell (X_j, X_k) = \omega^i_j(X_k) - \omega^i_k(X_j) \quad ,$$

while we also have

$$d\theta^i(X_j, X_k) = X_j(\theta^i(X_k)) - X_k(\theta^i(X_j)) - \theta^i([X_j, X_k]) \qquad \text{by Theorem I.7-13}$$

$$= 0 - 0 - \theta^i(\nabla_{X_j} X_k - \nabla_{X_k} X_j)$$

$$= \Gamma^i_{jk} - \Gamma^i_{kj}$$

$$= \omega^i_j(X_k) - \omega^i_k(X_j) \quad ;$$

this proves condition (b).

For the second structural equation we expand

$$\sum_i R^i{}_{jk\ell} X_i = R(X_k, X_\ell) X_j = \nabla_{X_k} \nabla_{X_\ell} X_j - \nabla_{X_\ell} \nabla_{X_k} X_j - \nabla_{[X_k, X_\ell]} X_j$$

to obtain

$$R^i{}_{jk\ell} = \sum_\mu (\Gamma^i_{k\mu} \Gamma^\mu_{\ell j} - \Gamma^i_{\ell\mu} \Gamma^\mu_{kj}) + X_k(\Gamma^i_{\ell j}) - X_\ell(\Gamma^i_{kj}) - \theta^i(\nabla_{[X_k, X_\ell]} X_j) \quad .$$

Comparing with

$$[d\omega_j^i + \sum_\mu \omega_\mu^i \wedge \omega_j^\mu](X_k,X_\ell) = X_k(\omega_j^i(X_\ell)) - X_\ell(\omega_j^i(X_k))$$

$$- \omega_j^i([X_k,X_\ell]) + \sum_\mu[\omega_\mu^i(X_k)\omega_j^\mu(X_\ell)$$

$$- \omega_\mu^i(X_\ell)\omega_j^\mu(X_k)] \quad ,$$

we see that this does indeed equal $R^i{}_{jk\ell} = \Omega_j^i(X_k,X_\ell)$. ∎

Notice that the results of Theorem 4 can also be written

$$\nabla_{X_k} X_j = \sum_i \omega_j^i(X_k)X_i$$

$$R(X_k,X_\ell)X_j = \sum_i \Omega_j^i(X_k,X_\ell)X_i \quad .$$

If we did not already have the ∇ operator and its curvature tensor, then we could, and E. Cartan did, use these equations to <u>define</u> R. Since the θ^i, hence the ω_j^i, and finally the Ω_j^i, all depend on the moving frame, it is then necessary to check that the resulting definition of R depends only on the values of X_j,X_k,X_ℓ at a given point. We save until the end of the chapter some remarks about what is involved in that. At the moment, we would like to point out that we are already in a position to prove the test case, and thus begin to justify the epithet "structural equations".

<u>6. THEOREM (THE TEST CASE; fourth version)</u>. Let $(M, \langle\ ,\ \rangle)$ be an n-dimensional Riemannian manifold for which the curvature tensor R is 0. Then M is locally isometric to $\mathbb{R}^n$ with its usual Riemannian metric.

Proof. Let $Y_1,\dots,Y_n$ be an orthonormal moving frame around $p \in M$, and let θ^i and ω^i_j be its dual forms and connection forms. By assumption, we have

$$d\omega^i_j = -\sum_k \omega^i_k \wedge \omega^k_j \quad .$$

Step 1. By Proposition 2, in a neighborhood of p there is a matrix $A = (A^i_j)$ of functions with $A(p)$ orthogonal and

$$(*) \qquad dA = A \wedge \omega \quad .$$

Define 1-forms φ^i by $\varphi^i = \sum_j A^i_j\theta^j$, which we can also write as

$$\varphi = A \wedge \theta \quad .$$

Step 2. We have

$$\begin{aligned} d\varphi &= (dA \wedge \theta) + (A \wedge d\theta) \\ &= (A \wedge \omega \wedge \theta) + (A \wedge d\theta) && \text{by } (*) \\ &= (A \wedge \omega \wedge \theta) + [A \wedge (-\omega \wedge \theta)] && \text{by the first structural equation} \\ &= 0 \quad . \end{aligned}$$

So there are functions x^i with $\varphi^i = dx^i$.

Step 3. We claim that $x^1,\dots,x^n$ is the desired coordinate system, i.e., that the natural frame $\partial/\partial x^1,\dots,\partial/\partial x^n$ is orthonormal. To prove this, we first note that the 1-forms φ^i satisfy $\varphi^i(X_j) = A^i_j$, so

$$X_j = \sum_i A^i_j \frac{\partial}{\partial x^i} \quad .$$

Since the X_j are orthonormal, it suffices to prove that (A^i_j) is always an orthogonal matrix. This is a consequence of (*) and the fact that $\omega = (\omega^i_j)$ is skew-symmetric. One argument for this conclusion has already been given on page 1-46, and here is another. Since

$$\begin{aligned} d(A \cdot A^T) &= (dA \wedge A^T) + (A \wedge dA^T) \\ &= (A \wedge \omega \wedge A^T) + (A \wedge (A \wedge \omega)^T) \qquad \text{by } (*) \\ &= (A \wedge \omega \wedge A^T) + (A \wedge \omega^T \wedge A^T) \\ &= (A \wedge \omega \wedge A^T) + (A \wedge - \omega \wedge A^T) = 0 \quad , \end{aligned}$$

and since $A \cdot A^T(0) = I$, it follows that $A \cdot A^T = I$ everywhere. ∎

We leave it to the reader to correlate the three Steps in this proof with those in previous versions. Note that the first structural equation involves the symmetry of the connection, as shown by the proof of Theorem 5, and that skew-symmetry of ω is equivalent to the definition of the Christoffel symbols, since it is the condition which determines the ω^i_j in Proposition 4.

In contrast to previous chapters, in this one we are going to give more than one proof of the Test Case. Although the proof just given is a natural use of the structural equations as integrability conditions, it does not illustrate the fundamental principle to be used in the method of the repère mobile, which is <u>to choose the moving frame most suitable to the particular problem</u>. When we return to the study of surfaces in 3-space, or submanifolds of Riemannian manifolds in general, we will see many instances of this principle. In our present setup there is one especially important moving frame, the investigation of which will take some time.

Let $X_{1p},\ldots,X_{np}$ be an orthonormal frame at M_p. In a sufficiently small neighborhood of p we can define a moving frame $X_1,\ldots,X_n$ by choosing $X_i(q)$ to be the parallel translate of X_{ip} along the unique geodesic from p to q. The moving frame $X_1,\ldots,X_n$ is said to be

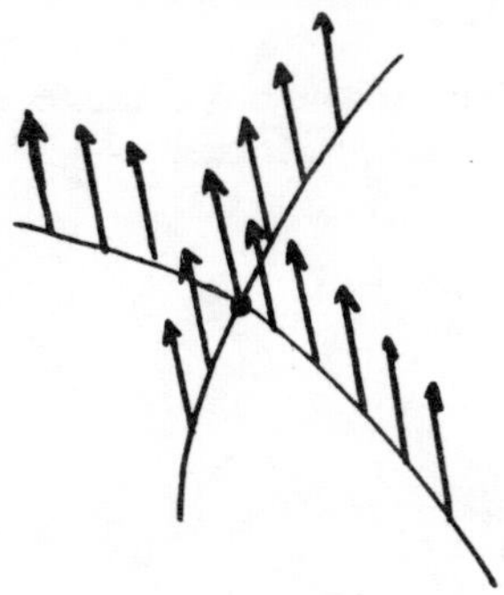

adapted to the frame $X_{1p},\ldots,X_{np}$. In order to explore the properties of this moving frame, it is convenient to introduce the map

$$\Phi\colon \mathbb{R} \times M_p \longrightarrow M$$

defined by

$$\Phi(t,X_p) = \exp(tX_p)$$

(actually, of course, Φ is usually not defined on all of $\mathbb{R} \times M_p$, but we will continue to write $\Phi\colon \mathbb{R} \times M_p \longrightarrow M$ for convenience). On M_p we introduce the coordinate system $t^1,\ldots,t^n$ by $t^i(\sum_j a^j X_{jp}) = a^i$, and we use t for the standard coordinate system on $\mathbb{R}$; we then let $(t,t^1,\ldots,t^n)$ denote the obvious coordinate system on $\mathbb{R} \times M_p$. We now begin to describe the dual forms θ^i and the connection forms ω^i_j in terms of their pull-backs to $\mathbb{R} \times M_p$.

7. PROPOSITION. When we write $\Phi^*\theta^i$ and $\Phi^*\omega^i_j$ in terms of $dt, dt^1,\ldots,dt^n$,

we have

$$\Phi^*\theta^i = t^i dt + \bar{\theta}^i$$

$$\Phi^*\omega^i_j = \bar{\omega}^i_j \qquad ,$$

where $\bar{\theta}^i$ and $\bar{\omega}^i_j$ are 1-forms which do not involve dt.

Proof. We can always write

$$\Phi^*\theta^i = f_i dt + \bar{\theta}^i$$

$$\Phi^*\omega^i_j = g_{ij} dt + \bar{\omega}^i_j \quad ;$$

it is only necessary to identify the f_i and g_{ij}. To do this, we fix $a^1,\ldots,a^n$ and consider the geodesic

$$\gamma(s) = \exp(s \sum_i a^i X_{ip}) , \qquad s \in (-\varepsilon,\varepsilon) \quad .$$

This can be written as $\gamma = \Phi \circ c$, where $c\colon (-\varepsilon,\varepsilon) \longrightarrow \mathbb{R} \times M_p$ is $c(s) = (s, \sum_i a^i X_{ip})$. So

$$\gamma^*\theta^i(s) = c^*\Phi^*\theta^i(s) = f_i(s,a^1,\ldots,a^n)dt$$

$$\gamma^*\omega^i_j(s) = c^*\Phi^*\omega^i_j(s) = g_{ij}(s,a^1,\ldots,a^n)dt.$$

On the other hand,

$$\gamma^*\theta^i(\left.\frac{d}{dt}\right|_s) = \theta^i(\left.\frac{d\gamma}{dt}\right|_s) = \theta^i(\sum_k a^k X_k(\gamma(s))) = a^i \quad ,$$

which shows that $f_i = t^i$. Also,

$$\gamma^* \omega^i_j \left(\frac{d}{dt}\bigg|_s\right) = \omega^i_j\left(\sum_k a^k X_k(\gamma(s))\right) = \sum_k \Gamma^i_{kj} a^k \quad ;$$

this sum vanishes, because X_j is parallel along γ, which means that

$$0 = \nabla_{\sum_k a^k X_k} X_j = \sum_k a^k \nabla_{X_k} X_j = \sum_k a^k \sum_i \Gamma^i_{kj} X_i \quad . \blacksquare$$

In our next result we will look at "partial derivatives" of the $\bar{\theta}^i$ and $\bar{\omega}^i_j$. If we are given an expression

$$\bar{\theta}^i = \sum_{j=1}^n g_j dt^j \quad ,$$

then we will use the symbol

$$\frac{\partial \bar{\theta}^i}{\partial t} \quad \text{for} \quad \sum_{j=1}^n \frac{\partial g_j}{\partial t} dt^j$$

and we will use similar symbols for the $\bar{\omega}^i_j$ [this definition depends on the particular coordinate system $(t, t^1, \ldots, t^n)$]. Note that in

$$d\bar{\theta}^i = \sum_{j=1}^n \left(\frac{\partial g_j}{\partial t} dt + \sum_k \frac{\partial g_j}{\partial t^k} dt^k\right) \wedge dt^j \quad ,$$

the terms involving dt are precisely

$$dt \wedge \frac{\partial \bar{\theta}^i}{\partial t} \quad .$$

8. PROPOSITION. We have the following "structural equations in polar coordinates":

$$\frac{\partial\bar\theta^i}{\partial t} = dt^i + \sum_k t^k \bar\omega_k^i \qquad \bar\theta^i(0,X) = 0 \quad \text{for all} \quad X \in M_p$$

$$\frac{\partial\bar\omega_j^i}{\partial t} = \sum_{k,\ell} (\mathbf{R}^i{}_{jk\ell}\circ\Phi) t^k \bar\theta^\ell \qquad \bar\omega_j^i(0,X) = 0 \quad \text{for all} \quad X \in M_p \quad .$$

Proof. By Proposition 7 we have

$$(1) \quad \Phi^*(d\theta^i) = d(\Phi^*\theta^i) = dt^i \wedge dt + dt \wedge \frac{\partial\bar\theta^i}{\partial t} + \text{terms not involving } dt$$

$$(2) \quad \Phi^*(d\omega_j^i) = d\Phi^*\omega_j^i = dt \wedge \frac{\partial\bar\omega_j^i}{dt} + \text{terms not involving } dt.$$

On the other hand, by the structural equations (Theorem 5) and Proposition 7 again, we have

$$(3) \quad \Phi^*(d\theta^i) = -\sum_k \Phi^*\omega_k^i \wedge \Phi^*\theta^k$$

$$= -\sum_k \bar\omega_k^i \wedge (t^k dt + \bar\theta^k)$$

$$(4) \quad \Phi^*(d\omega_j^i) = -\sum_k \Phi^*\omega_k^i \wedge \Phi^*\omega_j^k + \frac{1}{2}\Phi^*\Big(\sum_{k,\ell} \mathbf{R}^i{}_{jk\ell}\theta^k \wedge \theta^\ell\Big)$$

$$= -\sum_k \bar\omega_k^i \wedge \bar\omega_j^k + \frac{1}{2}\sum_{k,\ell} (\mathbf{R}^i{}_{jk\ell}\circ\Phi)(t^k dt + \bar\theta^k) \wedge (t^\ell dt + \bar\theta^\ell) \quad .$$

Comparing the coefficients of dt in (1) and (3) we obtain

$$dt^i - \frac{\partial\bar\theta^i}{\partial t} = -\sum_k t^k \bar\omega_k^i \quad ,$$

which gives us the first equation in the theorem. Similarly, from (2) and (4) we obtain

$$dt \wedge \frac{\partial \bar{\omega}^i_j}{\partial t} = \frac{1}{2} \sum_{k,\ell} (\mathbf{R}^i{}_{jk\ell} \circ \Phi)[dt \wedge (t^k \bar{\theta}^\ell - t^\ell \bar{\theta}^k)] \quad ;$$

together with the relation $\mathbf{R}^i{}_{jk\ell} = -\mathbf{R}^i{}_{j\ell k}$, this gives the second equation.

Since $\Phi(0,X) = p$ for all $X \in M_p$, we have $\Phi_{*(0,X)} = 0$, which gives the "initial conditions"

$$\bar{\theta}^i(0,X) = 0 \quad , \quad \bar{\omega}^i_j(0,X) = 0. \blacksquare$$

9. COROLLARY. The forms $\bar{\theta}^i$ satisfy the second order differential equation

$$\frac{\partial^2 \bar{\theta}^i}{\partial t^2} = \sum_{j,k,\ell} (\mathbf{R}^i{}_{jk\ell} \circ \Phi) t^j t^k \bar{\theta}^\ell$$

with the initial conditions

$$\bar{\theta}^i(0,X) = 0$$

$$\frac{\partial \bar{\theta}^i}{\partial t}(0,X) = dt^i \quad .$$

Proof. Differentiate the first equation in Proposition 8 and substitute in from the second. $\blacksquare$

It should not be hard to see that for all $X \in M_p$ we have

$$(*) \quad \Phi_*\left(\left.\frac{\partial}{\partial t^j}\right|_{(1,X)}\right) = \exp_*\left(\left.\frac{\partial}{\partial t^j}\right|_X\right) \quad ;$$

in this equation $\partial/\partial t^j$ denotes a tangent vector in $\mathbb{R} \times M_p$ as well as one in M_p, since we are using the same symbol t^j for a coordinate function on $\mathbb{R} \times M_p$ as on M_p. We have also written Φ_* instead of $\Phi_*(1,X)$, etc. Equation (*) shows that θ is determined by $\exp_*$ and $\bar{\theta}^i(1,X)$. This makes Corollary 9 particular significant, since it determines $\bar{\theta}^i(1,X)$ in terms of the functions $R^i{}_{jk\ell} \circ \Phi$. As a first illustration of this point, we bore ourselves to tears by twice again proving

10. THEOREM (THE TEST CASE; fifth version). Let $(M, \langle\ ,\ \rangle)$ be an n-dimensional Riemannian manifold for which the curvature tensor R is 0. Then M is locally isometric to $\mathbb{R}^n$ with its usual Riemannian metric.

Proof. Step 1. Let $X_1,\dots,X_n$ be the moving frame adapted to an orthonormal frame $X_{1p},\dots,X_{np}$ for M_p. From Proposition 8 we have

$$\frac{\partial \bar{\omega}^i_j}{\partial t} = 0 \qquad\qquad \bar{\omega}^i_j(0,X) = 0 \quad ,$$

which shows that $\bar{\omega}^i_j = 0$. This implies that $\omega^i_j = 0$, since (*) shows that Φ_* is a diffeomorphism at $(1,X)$ and

$$\bar{\omega}^i_j = \Phi^* \omega^i_j = \omega^i_j \circ \Phi_* \quad .$$

Step 2. Since $\omega^i_j = 0$, the first structural equation shows that $d\theta^i = 0$. So

$$\begin{aligned} 0 = d\theta^i(X_j,X_k) &= X_j(\theta^i(X_k)) - X_k(\theta^i(X_j)) - \theta^i([X_j,X_k]) \\ &= -\theta^i([X_j,X_k]) \quad . \end{aligned}$$

Thus $[X_j, X_k] = 0$ for all j,k, so there is a coordinate system $x^1, \ldots, x^n$ with $X_i = \partial/\partial x^i$.

Step 3. This is the desired coordinate system, since the X_i are obtained from the X_{ip} by parallel translation, and are consequently everywhere orthonormal. ∎

In this proof it is still possible to separate the argument into the standard three steps. But in the next proof everything happens at once.

11. THEOREM (THE TEST CASE; sixth version). Let $(M, \langle\ ,\ \rangle)$ be an n-dimensional Riemannian manifold for which the curvature tensor R is 0. Then M is locally isometric to $\mathbb{R}^n$ with its usual Riemannian metric.

Proof. Let $X_1, \ldots, X_n$ be the moving frame adapted to an orthonormal frame $X_{1p}, \ldots, X_{np}$ for M_p. From Corollary 9 we have

$$\frac{\partial^2 \bar{\theta}^i}{\partial t^2} = 0, \quad \bar{\theta}^i(0,X) = 0, \quad \frac{\partial \bar{\theta}^i}{\partial t}(0,X) = dt^i \quad ,$$

which implies that

$$\bar{\theta}^i(t,X) = t\,dt^i \quad , \quad \text{in particular } \bar{\theta}^i(1,X) = dt^i \quad .$$

So

$$\delta^i_j = \bar{\theta}^i(1,X)\left(\left.\frac{\partial}{\partial t^j}\right|_{(1,X)}\right) = \Phi^*\theta^i\left(\left.\frac{\partial}{\partial t^j}\right|_{(1,X)}\right) \quad \text{[for convenience we do not write } \Phi^*\theta^i(1,X)]$$

$$= \theta^i\left(\exp_*\left(\left.\frac{\partial}{\partial t^j}\right|_X\right)\right) \qquad \text{by } (*) \quad .$$

This shows that the value of the vector field X_j at $\exp X$ is

$$X_j(\exp X) = \exp_*\left(\left.\frac{\partial}{\partial t^j}\right|_X\right) .$$

Now the vector fields $\partial/\partial t^j$ on M_p are orthonormal with respect to the usual Riemannian metric $\langle\!\langle\, ,\, \rangle\!\rangle$ on M_p, so this equation shows that $\exp\colon (M_p, \langle\!\langle\, ,\, \rangle\!\rangle) \to (M, \langle\, ,\, \rangle)$ is an isometry. ∎

The reader may sort out for himself the vestigal forms in which the three Steps appear in this last proof. One thing does seem worth pointing out explicitly. In the two closely related proofs of Theorems 10 and 11 we use the device of expressing the integrability conditions $R = 0$ in terms of the map Φ. This is roughly equivalent to the method outlined in Problem I.6-8, where we solve a system of partial differential equations in $\mathbb{R}^n$ by reducing them to ordinary equations along lines through the origin.

The proof of Theorem 11, similar to the previous proof as it may be, is particularly important to us, for the methods used may be generalized to arbitrary Riemannian manifolds $(M, \langle\, ,\, \rangle)$. To do this we introduce 1-forms $\bar{\bar{\theta}}^i$ on M_p by $\bar{\bar{\theta}}^i(X) = \bar{\theta}^i(1,X)$. More precisely, if $X \in M_p$ and $v_X \in (M_p)_X$, then we have a tangent vector

$$(0,v_X) \in (\mathbb{R} \times M_p)_{(1,X)}$$

(recall that the tangent space of the product of two manifolds is isomorphic to the direct sum of the tangent spaces of the manifolds), so we may define

$$\bar{\bar{\theta}}^i(v_X)\left[= \bar{\bar{\theta}}^i(X)(v_X)\right] = \bar{\theta}^i(1,X)((0,v_X)) .$$

In particular, we have (leaving out the arguments for $\bar{\bar{\theta}}^i$ and $\bar{\theta}^i$)

$$(**) \qquad \bar{\bar{\theta}}^i\left(\left.\frac{\partial}{\partial t^j}\right|_X\right) = \bar{\theta}^i\left(\left.\frac{\partial}{\partial t^j}\right|_{(1,X)}\right) .$$

Now define a tensor $\boldsymbol{\langle\,,\,\rangle}$ of type $\binom{2}{0}$ on M_p by

$$\boldsymbol{\langle\,,\,\rangle} = \sum_{i=1}^{n} \bar{\bar{\theta}}^i \otimes \bar{\bar{\theta}}^i .$$

<u>12. THEOREM</u>. The map $\exp\colon (M_p, \boldsymbol{\langle\,,\,\rangle}) \longrightarrow (M, \langle\,,\,\rangle)$ is an isometry (in a neighborhood of $0 \in M_p$ on which $\exp$ is a diffeomorphism).

<u>Proof</u>. Since

$$\langle \sum_{j=1}^{n} a^j X_j, \sum_{j=1}^{n} b^j X_j \rangle = \sum_{i=1}^{n} a^i b^i$$

$$= \sum_{i=1}^{n} \theta^i\big(\sum_{j=1}^{n} a^j X_j\big)\cdot\theta^i\big(\sum_{j=1}^{n} b^j X_j\big) ,$$

we have $\langle\,,\,\rangle = \sum_i \theta^i \otimes \theta^i$. On the other hand, we also have

$$\boldsymbol{\Big\langle} \left.\frac{\partial}{\partial t^j}\right|_X , \left.\frac{\partial}{\partial t^k}\right|_X \boldsymbol{\Big\rangle} = \sum_{i=1}^{n} \bar{\bar{\theta}}^i\left(\left.\frac{\partial}{\partial t^j}\right|_X\right) \bar{\bar{\theta}}^i\left(\left.\frac{\partial}{\partial t^k}\right|_X\right)$$

$$= \sum_{i=1}^{n} \bar{\theta}^i\left(\left.\frac{\partial}{\partial t^j}\right|_{(1,X)}\right) \bar{\theta}^i\left(\left.\frac{\partial}{\partial t^k}\right|_{(1,X)}\right) \quad \text{by } (**)$$

$$= \sum_{i=1}^{n} \theta^i\left(\exp_*\left(\left.\frac{\partial}{\partial t^j}\right|_X\right)\right)\cdot\theta^i\left(\exp_*\left(\left.\frac{\partial}{\partial t^k}\right|_X\right)\right) ,$$

which means that

$$\boldsymbol{\langle\,,\,\rangle} = \exp^*\big(\sum_i \theta^i \otimes \theta^i\big) = \exp^* \langle\,,\,\rangle . \blacksquare$$

Recall (pg. 4D-15) that for a 2-dimensional subspace $W \subset M_q$ of the tangent space of a Riemannian manifold $(M, \langle\ ,\ \rangle)$ we define the sectional curvature $k(W)$ as

$$k(W) = \frac{\langle R(A,B)B, A \rangle}{\|A, B\|^2}, \quad A, B \text{ a basis for } W.$$

Let $p \in M$ be some fixed point. For every $X \in M_p$ and 2-dimensional subspace $V \subset M_p$, we will let $L(X,V)$ be the sectional curvature $k(W)$, where $W \subset M_{\exp X}$ is the parallel translate of V along the geodesic $t \longmapsto \exp tX$.

<u>13. COROLLARY (THE CURVATURE DETERMINES THE METRIC)</u>. Let M and M' be two Riemannian manifolds, and $T\colon M_p \to M'_{p'}$ an isometry for some $p \in M$ and $p' \in M'$. Suppose that $L(X,V) = L'(T(X),T(V))$ for all 2-dimensional subspaces $V \subset M_p$ and all sufficiently small X. Then there is an isometry from some neighborhood of $p \in M$ to a neighborhood of $p' \in M'$.

<u>Proof</u>. Choose an orthonormal frame $X_{1p}, \ldots, X_{np} \in M_p$, let $X_1, \ldots, X_n$ be the adapted moving frame in M, and let $X'_1, \ldots, X'_n$ be the moving frame in M' adapted to $T(X_{1p}), \ldots, T(X_{np})$. Let $\Phi\colon \mathbb{R} \times M_p \to M$ and $\Phi'\colon \mathbb{R} \times M'_{p'} \to M'$ be as defined previously; let $\bar\theta^i$ and $\bar\theta'^i$ be the corresponding forms on $\mathbb{R} \times M_p$ and $\mathbb{R} \times M'_{p'}$, and let $S\colon \mathbb{R} \times M_p \to \mathbb{R} \times M'_{p'}$ be $S(a,X) = (a, T(X))$. From the definition of $\mathbf{R}^i_{jkl}$ before Theorem 5 we see that

$$\mathbf{R}^i_{jk\ell} = \langle R(X_k, X_\ell)X_j, X_i \rangle .$$

The hypotheses of the theorem therefore imply that

$$R^{i}{}_{jij} \circ \Phi = R'^{i}{}_{jij} \circ (\Phi' \circ S) \quad \text{for all } i,j \quad .$$

From Proposition 4D-8 we deduce that

$$R^{i}{}_{jk\ell} \circ \Phi = R'^{i}{}_{jk\ell} \circ (\Phi' \circ S) \qquad \text{for all } i,j,k,\ell \qquad .$$

Now Corollary 9 (and the uniqueness of solutions of differential equations with given initial conditions) implies that

$$\bar{\theta}^{i} = \bar{\theta}'^{i} \circ S \ ,$$

and therefore that

$$\bar{\bar{\theta}}^{i} = \bar{\bar{\theta}}'^{i} \circ T \quad .$$

This means that T is an isometry of $(M_p, \langle\ ,\ \rangle)$ and $(M'_{p'}, \langle\ ,\ \rangle')$. The result then follows from Theorem 12. ∎

This corollary is the main assertion made by Riemann in his Habilitations lecture. We have proved a purely local result, but global results have also been obtained; see Ambrose, Parallel Translation of Riemannian curvature, Ann. of Math. 64(1956), 337-363.

The result of Corollary 13 is perhaps not what the reader may have understood by the assertion that "the curvature determines the metric", for it involves the parallel translation in the manifold, and not merely the curvature. Notice, however, that any rigorous statement about curvature determining the metric must involve a map $f: M \longrightarrow M'$, so that we know what it means to compare curvature in M with curvature in M'; in our case the map f is $\exp_{p'} \circ (\exp_p)^{-1}$. In this connection the following rather different question has always seemed to me the more interesting one. Suppose we have a diffeomorphism $f\colon M \longrightarrow M'$ such that for every 2-dimensional $V \in M_q$ we have $k(V) = k'(f_*(V))$; then is f an isometry? It is easy to see that as stated this is not true, because any diffeomorphism $f\colon \mathbb{R}^n \longrightarrow \mathbb{R}^n$ satisfies the hypothesis (when $\mathbb{R}^n$ has its usual Riemannian metric), but not necessarily the conclusion. It is also easy to obtain other examples. Consider the sphere $S^n \subset \mathbb{R}^{n+1}$ of radius a, with the induced Riemannian metric. For $n = 2$, we know that $k(S^2{}_p) = k(p) = 1/a^2$. If $n > 2$ and $\mathcal{O}$ is a neighborhood of 0 in a 2-dimensional subspace $V \subset M_p$, then $\exp(\mathcal{O})$ is isometric to a portion of S^2, so we have $K(V) = 1/a^2$ for all such V. Consequently, once again any diffeomorphism $f\colon S^n \longrightarrow S^n$ satisfies the hypothesis of our question, but not necessarily the conclusion. As we shall see in Addendum 1, there are also examples where all $k(V)$ are the same negative number.

In addition to these rather special counterexamples, it is simple to construct infinitely many other 2-dimensional examples. If we have a 2-dimensional manifold M such that the sets $k = \text{constant}$ give a foliation of M, then we can choose $f\colon M \longrightarrow M$ to be any diffeomorphism which

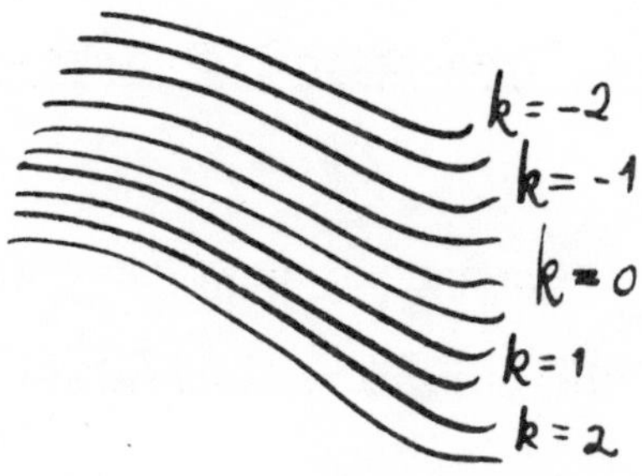

keeps each folium fixed as a set. This argument seems to have been made only recently (at least in print); however, there are specific classical examples of 2-manifolds M, M' which are not isometric under any map, but for which there is a diffeomorphism $f: M \longrightarrow M'$ with $k'(f(p)) = k(p)$ [see the references in the paper cited below]. Perhaps these examples are responsible for the fact that the higher dimensional cases remained unsettled for so long. Though it might be natural to assume that this case is even more hopeless, just the opposite is true: roughly speaking, when $n > 2$, a diffeomorphism which preserves sectional curvatures is an isometry, except for the special counterexamples mentioned in the previous paragraph. For details the reader is referred to a very recent paper,

R. S. Kulkarni, *Curvature and Metric*, Annals of Math. 91 (1970), pp. 211-331.

Before we proceed further with the study of moving frames, we pause to note that in the 2-dimensional case, Corollary 9 (and Corollary 13 which depends on it) have really been available to us since Chapter 3. Recall that we chose polar coordinate (ρ,ϕ) on M_p, and used them to introduce polar coordinates $(r,\varphi) = (\rho,\phi)\circ\exp^{-1}$ on a neighborhood of p [see page 3B-32]. If we write

$$\langle\ ,\ \rangle = dr \otimes dr + G\, d\varphi \otimes d\varphi \quad ,$$

and define g on $\mathbb{R}^2$ by

$$g = G\circ\exp\circ(\rho,\phi)^{-1} \quad ,$$

then we have the following formulas (collected together on page 3B-43):

$$\sqrt{g}(0,\phi) = 0$$

$$\frac{\partial\sqrt{g}}{\partial\rho}(0,\phi) = 1$$

$$\frac{\partial^2\sqrt{g}}{\partial\rho^2}(\rho,\phi) = -\sqrt{g}(\rho,\phi)\cdot k(\exp(0,\phi)) \quad .$$

These equations are easily seen to be equivalent to the equations in Corollary 9, and can be used the same way. For example, if $k = 0$ in a neighborhood of p, then

$$\frac{\partial^2 \sqrt{g}}{\partial \rho^2}(\rho,\phi) = 0 \quad ;$$

together with the initial conditions, this shows that $\sqrt{g} = \rho$, so

$$\langle\ ,\ \rangle = dr \otimes dr + r^2 d\varphi \otimes d\varphi \quad ,$$

which is exactly the expression in polar coordinates for the usual Riemannian metric on $\mathbb{R}^2$. Even the n-dimensional Test Case could be proved in this way, by considering $\exp(V)$ for various 2-dimensional $V \subset M_p$. However, the equations of Corollary 9 are generally most useful in higher dimensions. As an exercise in using them, Addendum 1 derives the form of the metric in an n-dimensional manifold of constant curvature.

Everything which we have done so far in this Chapter has involved the Levi-Civita connection associated to a metric $\langle\ ,\ \rangle$; moreover, we immediately interpreted the connection and curvature forms in terms of concepts defined in previous Chapters. But the method of the repère mobile is meant to treat arbitrary connections, and was used to define the curvature tensor before the ∇ operator had been invented. For the remainder of this chapter we will be concerned with this independent development of the theory of connections. We want to describe any connection in terms of "connection forms" ω^i_j, but we do not necessarily want the matrix $\omega = (\omega^i_j)$ to be skew-symmetric, so we do not have Proposition 4 to guide us. We do want to have the equations

$$\nabla_{X_k} X_j = \sum_i \omega^i_j(X_k) X_i \quad ,$$

which are a consequence of Theorem 5; these equations can also be written

$$\nabla X_j = \sum_i \omega^i_j \cdot X_i \quad .$$

For convenience, we let $\mathbf{X} = (X_1, \ldots, X_n)$ and abbreviate this equation as

$$\nabla \mathbf{X} = \mathbf{X} \cdot \omega \quad .$$

(recall the notation introduced on pp. 7-5 and 7-6).

Now consider another moving frame $\mathbf{X}' = \mathbf{X} \cdot a$. We want to have

$$\sum_i \omega'^i_j X'_i = \nabla X'_j = \nabla(\sum_\ell a^\ell_j X_\ell)$$

$$= \sum_\ell da^\ell_j X_\ell + \sum_\ell a^\ell_j \nabla X_\ell$$

[recall the formula for $\nabla_X fY$ and note that $X(f) = df(X)$] .

Using our matrix notation, this means that we want

$$\mathbf{X}' \cdot \omega' = \nabla \mathbf{X}' = \nabla(\mathbf{X} \cdot a)$$

$$= \mathbf{X} \cdot da + \nabla \mathbf{X} \cdot a$$

$$= \mathbf{X} \cdot da + \mathbf{X} \cdot (\omega \cdot a) \quad ,$$

so we want

$$\mathbf{X} \cdot (a\omega') = (\mathbf{X} \cdot a)\omega' = \mathbf{X}' \cdot \omega' = \mathbf{X} \cdot (da + \omega a)$$

for all moving frames $\mathbf{X}$. Thus we want the condition

$$(*) \qquad \omega' = a^{-1}da + a^{-1}\omega a \quad .$$

Hence we are lead to the following definition:*

> A (Cartan) connection on a manifold M is an assignment of a matrix $\omega = (\omega^i_j)$ of 1-forms to every moving frame $\mathbf{X}$ such that equation (*) holds between the 1-forms ω^i_j assigned to the moving frame $\mathbf{X}$ and the 1-forms ω'^i_j assigned to the moving frames $\mathbf{X}' = \mathbf{X}\cdot a$.

Given a Cartan connection, and a moving frame $\mathbf{X} = X_1,\dots,X_n$, the 1-forms ω^i_j which are assigned to $\mathbf{X}$ are called the connection forms for $\mathbf{X}$, and we define the dual forms θ^i by $\theta^i(X_j) = \delta^i_j$. From these forms we can define all the tensors which arise in Chapters 5 and 6. What follows is an outline of such a development of the theory of Cartan connections, independently of previous considerations.

We begin with a simple observation about the consistency of the transformation laws (*). Suppose $\mathbf{X}'' = \mathbf{X}'\cdot b = (\mathbf{X}\cdot a)\cdot b = \mathbf{X}\cdot(ab)$ is another moving frame, and that ω', ω'' satisfy

$$\omega' = a^{-1}da + a^{-1}\omega a$$
$$\omega'' = b^{-1}db + b^{-1}\omega' b \quad .$$

Then

$$\omega'' = b^{-1}db + b^{-1}(a^{-1}da + a^{-1}\omega a)b$$
$$= [b^{-1}a^{-1}(da)b + b^{-1}a^{-1}a(db)] + b^{-1}a^{-1}\omega ab$$
$$= (ab)^{-1}d(ab) + (ab)^{-1}\omega ab \quad .$$

*We are using the term "Cartan connection" as a convenient label, but the reader should be warned that in the literature this term is used for a more general concept.

This shows that if we are given connection forms for a certain set of moving frames whose domains cover M, and the various pairs of connection forms all satisfy (*), then there is a unique Cartan connection that assigns these forms to these particular moving frames. In view of this remark, it is very easy to determine a connection from a Riemannian metric.

14. PROPOSITION. On a Riemannian manifold $(M, \langle\ ,\ \rangle)$ there is a unique Cartan connection with the property that the connection forms ω^i_j for any orthonormal moving frame X satisfy

$$\omega^i_j = -\omega^j_i$$

$$d\theta^i = \sum_k \theta^k \wedge \omega^i_k \quad .$$

Proof. We already know, by Proposition 4, that for any moving frame there are unique 1-forms ω^i_j with this property. We just have to check that if $\mathbf{X}$ and $\mathbf{X}' = \mathbf{X}\cdot a$ are orthonormal moving frames, then

$$\omega' = a^{-1}da + a^{-1}\omega a \quad .$$

In view of uniqueness, we just have to show that if the forms ω^i_j satisfy the conditions of the theorem, and the forms ω'^i_j are defined by this formula, then they satisfy the same conditions,

$$\text{(a)}\quad \omega'^i_j = -\omega'^j_i$$

$$\text{(b)}\quad d\theta'^i = \sum_k \theta'^k \wedge \omega'^i_k \quad .$$

Since $\mathbf{X}$ and $\mathbf{X}'$ are orthonormal, the matrix a is everywhere orthogonal, $a\cdot a^T = I$. So if 0 denotes the zero matrix we have

$$0 = da\cdot a^T + a\cdot da^T \quad ,$$

or

$$da^T = - a^{-1}\cdot da\cdot a^T = - a^T\cdot da\cdot a^T \quad .$$

Consequently,

(1) $$(a^{-1}da)^T = (a^T da)^T = da^T\cdot a$$
$$= - a^T\cdot da\cdot a^T\cdot a = - a^T\cdot da = - a^{-1}da$$

(2) $$(a^{-1}\omega a)^T = (a^T\omega a)^T = a^T\omega^T a = - a^T\omega a = - a^{-1}\omega a \quad .$$

Clearly (1) and (2) imply (a).

To prove (b), we first note that

$$\theta'^i(X_j) = \theta'^i(\sum_k (a^{-1})^k_j X'_k) = (a^{-1})^i_j$$
$$= \sum_k (a^{-1})^i_k \theta^k(X_j) \quad ,$$

so $\theta'^i = \sum_k (a^{-1})^i_k \theta^k$, or

$$\theta' = a^{-1}\cdot\theta \quad .$$

Consequently,*

---

*To compute $d(a^{-1})$ we differentiate $a\cdot a^{-1} = I$ to obtain

$$da\cdot a^{-1} + a\cdot d(a^{-1}) = 0 \quad ,$$
$$d(a^{-1}) = - a^{-1}\cdot da\cdot a^{-1} \quad .$$

(3)
$$d\theta' = (-a^{-1}\cdot da\cdot a^{-1}) \wedge \theta + a^{-1}d\theta$$
$$= -a^{-1}da \wedge a^{-1}\theta + \left(a^{-1}\theta \wedge \omega\right).$$

On the other hand,

(4)
$$\theta' \wedge \omega' = (a^{-1}\theta) \wedge (a^{-1}da + a^{-1}\omega a)$$
$$= a^{-1}\theta \wedge a^{-1}da - \left(a^{-1}\omega \wedge \theta\right)$$
$$= a^{-1}\theta \wedge a^{-1}da - \left(a^{-1}\omega \wedge \theta\right).$$

[recall that $\omega \wedge \theta = (-1)^{k\ell}\theta \wedge \omega$ if ω is a matrix of k forms and θ if an n-tuple of ℓ forms; see page 7-6]

Clearly (3) and (4) imply (b). ∎

When we pass from the particular connection of Proposition 14 to a general Cartan connection, both structural equations need correction terms. If ω^i_j are the connection forms for a moving frame $\mathbf{X}$, with dual forms θ^i, we define 2-forms Θ^i and Ω^i_j by

$$d\theta = -\omega \wedge \theta + \Theta \qquad \text{i.e.,} \qquad d\theta^i = -\sum_k \omega^i_k \wedge \theta^k + \Theta^i$$

$$d\omega = -\omega \wedge \omega + \Omega \qquad \text{i.e.,} \qquad d\omega^i_j = -\sum_k \omega^i_k \wedge \omega^k_j + \Omega^i_j \quad .$$

We call the Θ^i and Ω^i_j the <u>torsion forms</u> and <u>connection forms</u> for the moving frame X. We now compute the transformation formulas for these forms.

15. PROPOSITION. If Θ'^i and Ω'^i_j are the torsion and connection forms for another moving frame $\mathbf{X}' = \mathbf{X}\cdot a$, then

$$\Theta' = a^{-1}\cdot\Theta$$
$$\Omega' = a^{-1}\Omega a \quad .$$

Proof. We have

(1) $\omega' = a^{-1}da + a^{-1}\omega a$

(2) $d\theta = -\omega \wedge \theta + \Theta$

(3) $d\theta' = -\omega' \wedge \theta' + \Theta'$

(4) $d\omega = -\omega \wedge \omega + \Omega$

(5) $d\omega' = -\omega' \wedge \omega' + \Omega'$,

and, as in the proof of Proposition 14,

(6) $\theta' = a^{-1}\theta$

(7) $d\theta' = -a^{-1}da \wedge a^{-1}\theta + a^{-1}d\theta$.

So

(8) $d\theta' = -a^{-1}da \wedge a^{-1}\theta + a^{-1}(-\omega \wedge \theta + \Theta)$ by (2), (7) .

We also have

(9) $d\theta' = -\omega' \wedge \theta' + \Theta'$ by (3)

$= -(a^{-1}da + a^{-1}\omega a) \wedge a^{-1}\theta + \Theta'$ by (1), (6)

$= -a^{-1}da \wedge a^{-1}\theta - a^{-1}(\omega \wedge \theta) + \Theta'$.

Comparison of (8) and (9) gives $a^{-1}\Theta = \Theta'$.

From (1) we obtain

(10) $d\omega' = [-a^{-1}da\, a^{-1} \wedge da] + (-a^{-1}da\, a^{-1}) \wedge \omega a + a^{-1}d\omega a - a^{-1}\omega \wedge da$

$=$ " + " $+ a^{-1}(-\omega \wedge \omega + \Omega)a -$ " by (4).

We also have

$$
\begin{aligned}
(11)\quad d\omega' &= -\,\omega' \wedge \omega' + \Omega' \qquad \text{by (5)} \\
&= -\,(a^{-1}da + a^{-1}\omega a) \wedge (a^{-1}da + a^{-1}\omega a) + \Omega' \qquad \text{by (1)} \\
&= -\,(a^{-1}da \wedge a^{-1}da) - (a^{-1}da\ a^{-1} \wedge \omega a) - (a^{-1}\omega \wedge da) \\
&\qquad - (a^{-1}\omega \wedge \omega a) + \Omega' \quad .
\end{aligned}
$$

Comparison of (10) and (11) gives $a^{-1}\Omega a = \Omega'$. ∎

The three relations

(1) $\omega' = a^{-1}da + a^{-1}\omega a$

(2) $\Theta' = a^{-1}\Theta$

(3) $\Omega' = a^{-1}\Omega a$

are precisely what enable us to define

(1) ∇Y for vector fields Y

(2) $T(X,Y)$ for tangent vectors X,Y

(3) $R(X,Y)Z$ for tangent vectors X,Y,Z.

We essentially know this already for the ∇ operator, which we consider first. Given a moving frame $\mathbf{X} = (X_1,\dots,X_n)$ with connection forms ω^i_j, we define

$$\nabla\mathbf{X} = \mathbf{X}\cdot\omega \qquad [\text{i.e. } \nabla X_j = \sum_i \omega^i_j X_j \quad \text{or} \quad \nabla_X X_j = \sum_i \omega^i_j(X) X_i]$$

and extend this to arbitrary vector fields $Y = \Sigma\, b^j X_j$ by defining

$$\nabla\cdot(\Sigma\, b^j X_j) = \sum_j db^j\cdot X_j + b^j \nabla X_j \quad .$$

For another moving frame $\mathbf{X}' = \mathbf{X}\cdot a$ with connection forms ω'^i_j we have

$$\begin{aligned}\nabla\mathbf{X}' &= \mathbf{X}'\cdot\omega' = (\mathbf{X}\cdot a)\cdot\omega' = (\mathbf{X}\cdot a)\cdot[a^{-1}da + a^{-1}\omega a] \\ &= \mathbf{X}\cdot[a\cdot(a^{-1}da + a^{-1}\omega a)] \\ &= \mathbf{X}\cdot da + \mathbf{X}\cdot(\omega\cdot a) = \mathbf{X}\cdot da + (\mathbf{X}\cdot\omega)\cdot a \\ &= \mathbf{X}\cdot da + \nabla\mathbf{X}\cdot a \quad ,\end{aligned}$$

which shows that the definition in terms of the two moving frames are consistent (written out, this equation becomes

$$\begin{aligned}\nabla(\sum_i a^i_j X_j) &= \sum_i da^i_j\cdot X_i + \sum_{k}(\sum_i \omega^k_i a^i_j)X_k \\ &= \sum_i da^i_j X_i + \sum_i a^i_j \nabla X_i \quad .)\end{aligned}$$

We next define

$$T(X_j,X_k) = \sum_i \Theta^i(X_j,X_k)\cdot X_i$$

and extend to arbitrary tangent vectors $X = \Sigma\, b^j X_j$, $Y = \Sigma\, c^k X_k$ by linearity,

$$T(X,Y) = \sum_{j,k} b^j c^k T(X_j,X_k) \quad .$$

For the moving frame $\mathbf{X}' = \mathbf{X}\cdot a$ we have

$$T(X'_\mu, X'_\nu) = \sum_\rho \Theta'^\rho(X'_\mu, X'_\nu)X'_\rho$$

$$= \sum_\rho \sum_\ell (a^{-1})^\rho_\ell \Theta^\ell(\sum_j a^j_\mu X_j, \sum_k a^k_\nu X_k) \sum_i a^i_\rho X_i$$

$$= \sum_{j,k} a^j_\mu a^k_\nu \sum_i \Theta^i(X_j, X_k)X_i$$

$$= \sum_{j,k} a^j_\mu a^k_\nu T(X_j, X_k) \quad ,$$

so the definitions in terms of the two moving frames are consistent. The tensor T is clearly alternating, since the Θ^i are 2-forms.

Finally, we define

$$R(X_k, X_\ell)X_j = \sum_i \Omega^i_j(X_k, X_\ell)X_i$$

and extend by linearity. For the moving frame $\mathbf{X}' = \mathbf{X}\cdot a$ we have

$$R(X'_\mu, X'_\nu)X'_\lambda = \sum_\rho \Omega'^\rho_\lambda(X'_\mu, X'_\nu)X'_\rho$$

$$= \sum_\rho \sum_{j,m} (a^{-1})^\rho_m \Omega^m_j a^j_\lambda(\sum_k a^k_\mu X_k, \sum_\ell a^\ell_\nu X_\ell) \sum_i a^i_\rho X_i$$

$$= \sum_{j,k,\ell} a^k_\mu a^\ell_\nu a^j_\lambda \Omega^i_j(X_k, X_\ell)X_i \quad ,$$

which again shows consistency. Clearly R is skew-symmetric in the first 2 arguments, since the Ω^i_j are 2-forms.

If we now define

$$\nabla_{X_k} X_j = \sum_{i=1}^{n} \Gamma^i_{kj} X_i$$

$$T(X_j, X_k) = \sum_{i=1}^{n} T^i_{jk} X_i$$

$$R(X_k, X_\ell) X_j = \sum_{i=1}^{n} R^i{}_{jk\ell} X_i \;,$$

then we have

$$\Gamma^i_{jk} = \omega^i_j(X_k) \qquad \text{or} \qquad \omega^i_j = \sum_k \Gamma^i_{kj} \theta^k$$

$$T^i_{jk} = \Theta^i(X_j, X_k) \qquad \text{or} \qquad \Theta^i = \frac{1}{2} \sum_{j,k} T^i_{jk} \theta^j \wedge \theta^k$$

$$R^i{}_{jk\ell} = \Omega^i_j(X_k, X_\ell) \qquad \text{or} \qquad \Omega^i_j = \frac{1}{2} \sum_{k,\ell} R^i{}_{jk\ell} \theta^k \wedge \theta^\ell \;.$$

Consequently, the "structural equations"

$$d\theta^i = -\sum_k \omega^i_k \wedge \theta^k + \Theta^i = -\sum_k \omega^i_k \wedge \theta^k + \frac{1}{2} \sum_{j,k} T^i_{jk} \theta^j \wedge \theta^k$$

$$d\omega^i_j = -\sum_k \omega^i_k \wedge \omega^k_j + \Omega^i_j = -\sum_k \omega^i_k \wedge \omega^k_j + \frac{1}{2} \sum_{k,\ell} R^i{}_{jk\ell} \theta^k \wedge \theta^\ell$$

$$\omega^i_j = \sum_k \Gamma^i_{kj} \theta^k$$

are purely a matter of definition. However, one can now easily reverse the computations in the proof of Theorem 5 to show that

$$T(X,Y) = \nabla_X Y - \nabla_Y X - [X,Y]$$

$$R(X,Y)Z = \nabla_X(\nabla_Y Z) - \nabla_Y(\nabla_X Z) - \nabla_{[X,Y]} Z \;,$$

thus verifying that T and R are the torsion and curvature tensors for ∇ as defined previously. We leave this computation to the reader, as well as the task of deriving the general form of the structural equations in polar coordinates.

Even though the structure equations are merely definitions in this approach, we can derive new relations from them.

<u>16. THEOREM</u>. We have the following relations between θ, ω, Θ and Ω:

(1) (Bianchi's 1st identity)

$$d\Theta + \omega \wedge \Theta = \Omega \wedge \theta$$

(2) (Bianchi's 2nd identity)

$$d\Omega + (\omega \wedge \Omega) - (\Omega \wedge \omega) = 0 .$$

(Notice that $\omega \wedge \Omega$ and $\Omega \wedge \omega$ are not equal up to sign, because Ω and ω are both matrices of forms and the order of matrix multiplication plays a role.)

<u>Proof</u>. We have

$$\begin{aligned} 0 = d(d\theta) &= d(-\omega \wedge \theta + \Theta) \\ &= -d\omega \wedge \theta + (\omega \wedge d\theta) + d\Theta \\ &= -(-\omega \wedge \omega + \Omega) \wedge \theta + [\omega \wedge (-\omega \wedge \theta + \Theta)] + d\Theta \\ &= -\Omega \wedge \theta + \omega \wedge \Theta + d\Theta \quad . \end{aligned}$$

Similarly,

$$\begin{aligned} 0 = d(d\omega) &= d(-\omega \wedge \omega + \Omega) \\ &= -d\omega \wedge \omega + (\omega \wedge d\omega) + d\Omega \\ &= (-\omega \wedge \omega + \Omega) \wedge \omega + [\omega \wedge (-\omega \wedge \omega + \Omega)] + d\Omega \\ &= -\Omega \wedge \omega + (\omega \wedge \Omega) + d\Omega \quad . \blacksquare \end{aligned}$$

It takes quite a bit of calculation to convince oneself that the equations in Theorem 16 really are the Bianchi identities. This calculation is left to those readers who have more endurance than the author; we will merely consider the special case of the first Bianchi identity for a connection without torsion. In this case $\Theta = 0$, so the identity is just $\Omega \wedge \theta = 0$. Thus we have

$$0 = (\Omega \wedge \theta)^i = \sum_j \Omega^i_j \wedge \theta^j = \frac{1}{2} \sum_{j,k,\ell} R^i_{jk\ell}\, \theta^k \wedge \theta^\ell \wedge \theta^j \ .$$

Applying this to (X_k, X_ℓ, X_j) we obtain the familiar formula

$$0 = \tfrac{1}{2}\{R^i_{jk\ell} - R^i_{jk\ell} + R^i_{\ell jk} - R^i_{\ell kj} + R^i_{k\ell j} - R^i_{kj\ell}\}$$

$$= R^i_{jk\ell} + R^i_{k\ell j} + R^i_{\ell jk} \ ,$$

(and thereby see why cyclic permutations of the indices should be involved). It turns out that taking the exterior derivative of the Bianchi identities does not give us any new relations, which indicates that these identities are the only general ones we should expect to find.

To complete the present Chapter we also derive the relations satisfied by the curvature tensor for the Levi-Civita connection.

17. THEOREM. Let $(M, \langle\ ,\ \rangle)$ be a Riemannian manifold, and consider the unique Cartan connection of Proposition 14. Then for every orthonormal moving frame we have

(1) $\Omega^i_j = -\Omega^j_i$.

Consequently, the curvature tensor R satisfies

(2) $< R(X,Y)Z,W > = - < R(X,Y)W,Z >$

(3) $< R(X,Y)Z,W > = < R(Z,W)X,Y >$.

<u>Proof</u>. Equation (1) is immediate from the fact that $\omega^i_j = -\omega^j_i$ (by assumption) and the second structural equation,

$$d\omega^i_j = -\sum_k \omega^i_k \wedge \omega^k_j + \Omega^i_j \quad .$$

Since

$$\begin{aligned} < R(X_k,X_\ell)X_j,X_i > &= < \sum_\mu R^\mu{}_{jk\ell} X_\mu , X_i > \\ &= R^i{}_{jk\ell} = \Omega^i_j(X_k,X_\ell) \quad , \end{aligned}$$

equation (1) implies (2). Then, as before, (3) follows from Proposition 4D-7. ▮

Addendum 1. Manifolds of Constant Curvature

A Riemannian manifold $(M, \langle\ ,\ \rangle)$ has <u>constant curvature</u> α if for all $p \in M$, and all 2-dimensional $V \subset M_p$ we have $k(V) = \alpha$. If M is 2-dimensional, this just means that $k(p) = \alpha$ for all $p \in M$. In this case, we can easily find the form of the metric $\langle\ ,\ \rangle$ from the considerations on page 7-29. We use geodesic polar coordinates $(r,\varphi) = (\rho,\phi)\circ\exp^{-1}$ so that the metric is

$$\langle\ ,\ \rangle = dr \otimes dr + G\, d\varphi \otimes d\varphi \quad .$$

If g on $\mathbb{R}^2$ is defined by $g = G\circ\exp\circ(\rho,\phi)^{-1}$, then we have seen that

$$\sqrt{g}(0,\phi) = 0$$

$$\frac{\partial\sqrt{g}}{\partial\rho}(0,\phi) = 1$$

$$\frac{\partial^2\sqrt{g}}{\partial\rho^2}(\rho,\phi) = -\,\alpha\sqrt{g}(\rho,\phi) \quad .$$

The general solution of the last equation is

$$\sqrt{g}(\rho,\phi) = c_1 \sin\sqrt{\alpha}\rho + c_2 \cos\sqrt{\alpha}\rho \qquad \alpha > 0$$

$$\sqrt{g}(\rho,\phi) = c_1 \sinh\sqrt{-\alpha}\rho + c_2 \cosh\sqrt{-\alpha}\rho \qquad \alpha < 0 \quad .$$

Taking into account the initial conditions, we find that

$$\sqrt{g} = \frac{\sin[h](\sqrt{|\alpha|}\rho)}{\sqrt{|\alpha|}} \quad ,$$

where we introduce the convention that

$$\sin[h] \text{ denotes } \begin{cases} \sin & \text{whenever we are dealing with } \alpha > 0 \\ \sinh & \text{" " " " " } \alpha < 0 \end{cases} \quad .$$

It follows that the metric is given by

$$< , > = dr \otimes dr + \frac{\sin[h]^2(\sqrt{|\alpha|}\,r)}{\sqrt{|\alpha|}}\, d\varphi \otimes d\varphi \quad .$$

To obtain the analagous results in the n-dimensional case, we will return to the equations of Corollary 9. However, we will need a preliminary result, which applies to spaces of constant curvature as a special case. A point p in an n-dimensional Riemannian manifold $(M, < , >)$ is called isotropic if the sectional curvatures $k(V)$ for all 2-dimensional $V \subset M_p$ have the same value α; this means that for all $X, Y \in M_p$ we have

$$< R(X,Y)Y,X > = \alpha \cdot \|X,Y\|^2 \quad .$$

Applying the formula on page I.9-12 (to the subspace of M_p spanned by X and Y), we can write this equation as

$$< R(X,Y)Y,X > = \alpha\left[< X,X > < Y,Y > - < X,Y >^2\right] \quad .$$

18. LEMMA. If $p \in M$ is isotropic, with all sectional curvatures equal to α, then for all $X, Y, Z, W \in M_p$ we have

$$< R(X,Y)Z,W > = \alpha\left[< X,W > < Y,Z > - < X,Z > < Y,W >\right] \quad .$$

Proof. Denote the right side of this equation by $\mathcal{R}(X,Y,Z,W)$. By hypothesis, $< R(X,Y)Y,X > = \mathcal{R}(X,Y,Y,X)$. It is easy to check that $\mathcal{R}$ has properties (1) - (3), and hence (4), of Proposition 4D-7. The desired result now follows from Proposition 4D-8. ∎

Although not necessary for our calculations, the following consequence of this Lemma is a standard result, and is pertinent to the topic of constant curvature manifolds.

19. THEOREM (SCHUR). If M is a Riemannian manifold of dimension $n \geq 3$ and all points of M are isotropic, then M has constant curvature.

Proof. By Lemma 18 we have, in a coordinate system x,

$$R_{hijk} = \alpha[g_{hj}g_{ik} - g_{hk}g_{ij}]$$

for some function α on M. Equivalently,

$$R^{h}{}_{ijk} = \alpha(\delta^h_j g_{ik} - \delta^h_k g_{ij}) .$$

Consequently, Ricci's Lemma (Proposition 5-3) and the relation $\delta^h_{j;\ell} = 0$ imply that

$$R^{h}{}_{ijk;\ell} = \frac{\partial \alpha}{\partial x^\ell}(\delta^h_j g_{ik} - \delta^h_k g_{ij}) .$$

From Bianchi's identity (Proposition 5-9(3')) we then obtain

$$0 = \frac{\partial \alpha}{\partial x^\ell}(\delta^h_j g_{ik} - \delta^h_k g_{ij}) + \frac{\partial \alpha}{\partial x^k}(\delta^h_k g_{i\ell} - \delta^h_\ell g_{ik}) + \frac{\partial \alpha}{\partial x^k}(\delta^h_\ell g_{ij} - \delta^h_j g_{i\ell}) .$$

In this identity set $h = k$ and sum over k, to obtain

$$0 = \frac{\partial \alpha}{\partial x^\ell}(g_{ij} - n g_{ij}) + \frac{\partial \alpha}{\partial x^j}(n g_{i\ell} - g_{i\ell}) + \frac{\partial \alpha}{\partial x^\ell} g_{ij} - \frac{\partial \alpha}{\partial x^j} g_{i\ell}$$

$$= (n-2)\cdot\left[\frac{\partial \alpha}{\partial x^j} g_{i\ell} - \frac{\partial \alpha}{\partial x^\ell} g_{ij}\right] .$$

Since $n \geq 3$ we have

$$\frac{\partial \alpha}{\partial x^j} g_{i\ell} = \frac{\partial \alpha}{\partial x^\ell} g_{ij} .$$

Hence

$$\delta^m_\ell \frac{\partial \alpha}{\partial x^j} = \sum_i g^{mi} g_{i\ell} \frac{\partial \alpha}{\partial x^j} = \sum_i g^{mi} g_{ij} \frac{\partial \alpha}{\partial x^\ell} = \delta^m_j \frac{\partial \alpha}{\partial x^\ell} .$$

Choosing $m = j \neq \ell$, we obtain $\partial\alpha/\partial x^\ell = 0$, for all ℓ. So α is constant. ∎

To obtain the metric in a space of constant curvature α, we now consider the equations of Corollary 9. By Lemma 18 we have

$$\mathsf{R}^i{}_{jk\ell} = \langle R(X_k,X_\ell)X_j,X_i \rangle = \alpha[\delta_{ki}\delta_{\ell j} - \delta_{kj}\delta_{\ell i}] ,$$

which implies that

$$\mathsf{R}^i{}_{jij} = -\mathsf{R}^i{}_{jji} = \alpha \ (j \neq i); \text{ all other } R^i{}_{jk\ell} = 0 .$$

So our equations become

$$(*) \quad \frac{\partial^2 \bar\theta^i}{\partial t^2} = -\alpha \sum_k t^k(t^k\bar\theta^i - t^i\bar\theta^k) .$$

To solve even these simplified equations requires quite a bit of trickiness. We first use the equations in Proposition 8 to obtain

$$\frac{\partial(\sum_i t^i\bar\theta^i)}{\partial t} = \sum_i t^i dt^i + \sum_{i,k} t^k t^i \bar\omega^i_k$$

$$= \sum_i t^i dt^i , \quad \text{using skew-symmetry of the } \bar\omega^i_k = \Phi^* \omega^i_k .$$

Since $\sum_i t^i\bar{\theta}^i(0,X) = 0$, we have

$$\sum_i t^i\theta^i = t \sum_i t^i dt^i \quad ,$$

and hence

$$(**)\ (\sum_i t^i\bar{\theta}^i)^2 = \sum_{i,j} t^i t^j \bar{\theta}^i\bar{\theta}^j = t^2 \sum_{i,j} t^i t^j dt^i dt^j = t^2(\sum_i t^i dt^i)^2 \quad .$$

We next obtain equations for the quantities $t^i\bar{\theta}^j - t^j\bar{\theta}^i$. By (*) we have

$$\frac{\partial^2(t^i\bar{\theta}^j - t^j\bar{\theta}^i)}{\partial t^2} = -\alpha \sum_k t^k t^i(t^k\bar{\theta}^j - t^j\bar{\theta}^k) + t^k t^j(t^k\bar{\theta}^i - t^i\bar{\theta}^k)$$

$$= -\alpha \sum_{k=1}^{n} (t^k)^2(t^i\bar{\theta}^j - t^j\bar{\theta}^i)$$

$$= -\alpha\rho^2(t^i\bar{\theta}^j - t^j\bar{\theta}^i) \quad ,$$

where we have set

$$\rho = \sqrt{\sum_{k=1}^{n} (t^k)^2} \quad .$$

Together with the initial conditions

$$(t^i\bar{\theta}^j - t^j\bar{\theta}^i)(0,X) = 0$$

$$\frac{\partial(t^i\bar{\theta}^j - t^j\bar{\theta}^i)}{\partial t}(0,X) = t^i dt^j - t^j dt^i \quad ,$$

we obtain

$$t^i\bar{\theta}^j - t^j\bar{\theta}^i = \frac{\sin[h](\rho\sqrt{|\alpha|}\,t)}{\rho\sqrt{|\alpha|}}\,(t^i dt^j - t^j dt^i) \quad .$$

Recall that for a form η, we use η^2 to denote the quadratic function $X \longmapsto \eta(X)\cdot\eta(X)$. Summing the squares of the equations just derived, we now obtain

$$\frac{\sin[h]^2(\rho\sqrt{|\alpha|}t)}{\rho^2|\alpha|}\sum_{i<j}(t^i dt^j - t^j dt^i)^2 = \frac{1}{2}\frac{\sin[h]^2(\rho\sqrt{|\alpha|}t)}{\rho^2|\alpha|}\sum_{i,j}(t^i dt^j - t^j dt^i)^2$$

$$= \frac{1}{2}\sum_{i,j}(t^i\bar{\theta}^j - t^j\bar{\theta}^i)^2 = \frac{1}{2}\left(\sum_{i,j}(t^i)^2(\bar{\theta}^j)^2 + (t^j)^2(\bar{\theta}^i)^2 - 2t^it^j\bar{\theta}^i\bar{\theta}^j\right)$$

$$= \sum_{i,j}(t^i)^2(\bar{\theta}^j)^2 - t^2\sum_{i,j}t^it^jdt^idt^j \qquad \text{by } (**)$$

$$= \rho^2\sum_{k=1}^n(\bar{\theta}^k)^2 - t^2(\sum_k t^kdt^k)^2 .$$

Since $\exp\colon (M_p, \Sigma\,\bar{\bar{\theta}}^k \otimes \bar{\bar{\theta}}^k) \longrightarrow (M, \langle\ ,\ \rangle)$ is an isometry, and $\bar{\bar{\theta}}$ is the value of $\bar{\theta}^i$ when $t = 1$, the expression for $\|\ \|^2$ in normal coordinates is obtained by seeing what $\Sigma(\bar{\theta}^k)^2$ becomes when we set $t = 1$ and $t^k = x^k$ in the above equation. We let r denote what ρ becomes when we perform this substitution; that is, we let

$$r = \sqrt{\sum_{k=1}^n (x^k)^2} .$$

We then obtain

$$\|\ \|^2 = \frac{1}{r^2}\left\{(\sum_k x^kdx^k)^2 + \frac{\sin[h]^2(\sqrt{|\alpha|}r)}{|\alpha|r^2}\cdot\sum_{i<j}(x^idx^j - x^jdx^i)^2\right\}.$$

Notice that we have

$$(\sum_k x^kdx^k)^2 = r^2[\sum_k(dx^k)^2] - \sum_{i<j}(x^idx^j - x^jdx^i)^2 ;$$

consequently, we can write

$$\| \ \|^2 = \sum_k (dx^k)^2 - \frac{|\alpha|r^2 - \sin[h]^2(\sqrt{|\alpha|}r)}{|\alpha|r^4} \sum_{i<j} (x^i dx^j - x^j dx^i)^2 .$$

From the Taylor series for sin it is easy to see that the coefficient of $\sum_{i<j} (x^i dx^j - x^j dx^i)^2$ is C^∞ (in fact it is analytic). This formula thus gives a direct verification of Riemann's assertion about the form of $\| \ \|^2$ in normal coordinates (pg. 4B-8). On the other hand, this form of the metric is not the one which Riemann mentioned in his lecture (pg. 4A-15). Riemann was probably led to this other form by the following considerations.

In a vector space V with a positive definite inner product $\langle \ , \ \rangle$ we define the <u>angle</u> $\angle(v,w)$ between two non-zero vectors $v, w \in V$ by

$$\angle(v,w) = \arccos \frac{\langle x,y\rangle}{\|x\| \cdot \|y\|} .$$

It is easy to say when two metrics give the same angle measurements.

<u>20. LEMMA</u>. Let $\langle \ , \ \rangle_1$ and $\langle \ , \ \rangle_2$ be two positive definite inner products on V. Then $\angle_1(v,w) = \angle_2(v,w)$ for all non-zero $v, w \in V$ if and only if there is a number $c > 0$ with $\langle \ , \ \rangle_2 = c \cdot \langle \ , \ \rangle_1$.

<u>Proof</u>. If $\langle \ , \ \rangle_2 = c \cdot \langle \ , \ \rangle_1$, then clearly $\angle_2(v,w) = \angle_1(v,w)$ for all $v, w \neq 0$.

Conversely, suppose $\angle_2(v,w) = \angle_1(v,w)$ for all $v, w \neq 0$. Let $v_1, \ldots, v_n$ be an orthonormal basis for V with respect to $\langle \ , \ \rangle_1$. Then for $i \neq j$ we have

$$\angle_2(v_i,v_j) = \angle_1(v_i,v_j) = 0 ,$$

so $< v_i, v_j > = 0$. Define c_i by $< v_i, v_i > = c_i$. Then

$$\angle_1(v_i, v_i + v_j) = \frac{< v_i, v_i + v_j >_1}{\sqrt{< v_i, v_i >_1}\sqrt{< v_i + v_j, v_i + v_j >_1}} = \frac{1}{\sqrt{2}} ,$$

$$\angle_2(v_i, v_i + v_j) = \frac{< v_i, v_i + v_j >_2}{\sqrt{< v_i, v_i >_2}\sqrt{< v_i + v_j, v_i + v_j >_2}} = \frac{c_i}{\sqrt{c_i}\sqrt{c_i + c_j}} .$$

It follows that $c_i = c_j$. ▮

Now a diffeomorphism $f: (M_1 < , >_1) \to (M_2, < , >_2)$ between Riemannian manifolds is called conformal if each f_{*p} is angle-preserving, and the two Riemannian manifolds are then called conformally equivalent. By Lemma 20, a diffeomorphism f is conformal if and only if

$$< , >_1 = \lambda f^* < , >_2$$

for some positive function $\lambda: M_1 \to \mathbb{R}$. In particular, if a Riemannian manifold $(M, < , >)$ is conformal to $\mathbb{R}^n$ with its usual Riemannian metric, then around each point $p \in M$ we can choose a coordinate system x such that $g_{ij} = \lambda\delta_{ij}$ for some positive function $\lambda: M \to \mathbb{R}$. Such a coordinate system x is called isothermic.

In 1822 Gauss showed that isothermic coordinates can be found at any point of an arbitrary surface. His proof depends on a trick that worked only for analytic (C^ω) manifolds, and uses a little knowledge of complex function theory. Later proofs, depending on theorems about partial differential equations, settled the differentiable case. We shall not present any proofs here; the problems are more analytic than geometric. We have mentioned the result simply because it must have been known to Riemann, and probably guided him in determining the metric for spaces of constant curvature. Once we move to dimension $n > 2$, it is no longer true that all n-manifolds are locally conformally equivalent to $\mathbb{R}^n$. However, it is reasonable to suppose that manifolds of constant curvature, which are presumably the next simplest class of manifolds after flat ones, might be conformally equivalent to $\mathbb{R}^n$.

We are therefore led to consider metrics on (a subset of) $\mathbb{R}^n$ of the form

$$g_{ij} = \frac{\delta_{ij}}{F^2} \qquad F \text{ nowhere } 0 .$$

Then $g^{ij} = \delta^{ij}/F^2$. We also have

$$\frac{\partial g_{ij}}{\partial x^k} = \frac{-2\delta_{ij}}{F^3}\frac{\partial F}{\partial x^k} = \frac{-2\delta_{ij}}{F^2}\frac{\partial \log F}{\partial x^k} .$$

Setting $\log F = f$, and using the formulas on pages I.9-37 and 9-39, we obtain

$$\Gamma^i_{ii} = \frac{-\partial f}{\partial x^i}, \quad \Gamma^i_{jj} = \frac{\partial f}{\partial x^i}, \quad \Gamma^i_{ij} = \Gamma^i_{ji} = \frac{-\partial f}{\partial x^j} \quad (i \neq j); \quad \text{all other } \Gamma^i_{jk} = 0 .$$

From the formula on page 5-8 we obtain

$$R^i{}_{jij} = -R^i{}_{jji} = \frac{\partial f}{\partial x^j \partial x^j} + \frac{\partial^2 f}{\partial x^i \partial x^i} - \sum_{r \neq i,j} \left(\frac{\partial f}{\partial x^r}\right)^2 \qquad i \neq j$$

$$R^i{}_{jj\ell} = \frac{-\partial^2 f}{\partial x^i \partial x^\ell} - \frac{\partial f}{\partial x^i}\frac{\partial f}{\partial x^\ell} \qquad i,j,\ell \text{ distinct}$$

$$R^i{}_{ji\ell} = \frac{\partial^2 f}{\partial x^j \partial x^\ell} + \frac{\partial f}{\partial x^j}\frac{\partial f}{\partial x^\ell} \qquad i,j,\ell \text{ distinct}$$

all other $R^i{}_{jk\ell} = 0$.

Now by Lemma 18, the metric has constant curvature α if and only if

$$R_{ijk\ell} = \alpha(g_{ik}g_{j\ell} - g_{i\ell}g_{jk}) \quad ,$$

or equivalently

$$R^i{}_{jk\ell} = \alpha(\delta^i_k g_{j\ell} - \delta^i_\ell g_{jk})$$

$$= \frac{\alpha}{F^2}(\delta^i_k \delta_{j\ell} - \delta^i_\ell \delta_{jk}) \; ,$$

and hence

$$R^i{}_{jij} = - R^i{}_{jji} = \frac{\alpha}{F^2} \qquad (j \neq i) \; ; \quad \text{all other} \quad R^i{}_{jk\ell} = 0 \; .$$

So the metric has constant curvature α if and only if

$$\frac{\partial^2 f}{\partial x^j \partial x^\ell} + \frac{\partial f}{\partial x^j}\frac{\partial f}{\partial x^\ell} = 0 \qquad j \neq \ell$$

$$\frac{\partial^2 f}{\partial x^i \partial x^i} + \frac{\partial^2 f}{\partial x^j \partial x^j} - \sum_{r \neq i,j} \left(\frac{\partial f}{\partial x^r}\right)^2 = \frac{\alpha}{F^2} \qquad i \neq j \; .$$

Since

$$\frac{\partial^2 f}{\partial x^j \partial x^\ell} + \frac{\partial f}{\partial x^j}\frac{\partial f}{\partial x^\ell} = \frac{1}{F}\frac{\partial^2 F}{\partial x^j \partial x^\ell} \qquad \text{for all} \quad j, \ell \; ,$$

these equations hold if and only if

$$(1) \qquad \frac{\partial^2 F}{\partial x^j \partial x^\ell} = 0 \; , \qquad j \neq \ell$$

$$(2) \quad F\left(\frac{\partial^2 F}{\partial x^i \partial x^i} + \frac{\partial^2 F}{\partial x^j \partial x^j}\right) = \alpha + \sum_{r=1}^{n} \left(\frac{\partial F}{\partial x^r}\right)^2 \qquad i \neq j \; .$$

Equation (1) implies that $F = G_1 + \dots + G_n$, where G_j depends only on x^j. Using equation (2) for i, i and then for j, j, we obtain

$$\frac{\partial^2 G_i}{\partial x^i \partial x^i} = \frac{\partial^2 G_j}{\partial x^j \partial x^j} .$$

So we must have

$$G_i = c x_i^{\,2} + b_i x_i + a_i \quad ,$$

for some c. From (2) we then obtain

$$\alpha = \sum_{r=1}^{n} (4 a_r c - b_r^{\,2}) \quad .$$

There are two important special cases. If we choose $c = \alpha/4$, $b_i = 0$, $a_i = 1/n$ we obtain a metric $\langle\ ,\ \rangle = \sum g_{ij} dx^i \otimes dx^j$ with

$$g_{ij} = \frac{\delta_{ij}}{\left[1 + \frac{\alpha}{4} \sum_i (x^i)^2\right]^2} .$$

This is the form which Riemann mentions (pg. 4A-15). Notice that for $\alpha < 0$, this metric is defined only on $M = \{a \in \mathbb{R}^n : \Sigma(a^i)^2 < -4/\alpha\}$. Nevertheless, $(M, \langle\ ,\ \rangle)$ is complete. To see this, we compute that

$$\gamma(t) = 2 \frac{\sinh \frac{\sqrt{|\alpha|}\, t}{2}}{\sqrt{|\alpha|} \cosh \frac{\sqrt{|\alpha|}\, t}{2}}$$

is a geodesic through 0, parameterized by arc length, and defined for all t. Since the metric $\langle\ ,\ \rangle$ is radially symmetric around 0, there are geodesics through 0 in all directions, which are defined for all t.

When $\alpha > 0$, the same metric is defined on all of $\mathbb{R}^n$, but it is <u>not</u> complete; a computation shows that the geodesics through 0 cannot be extended past $(-2\pi/\sqrt{\alpha}, 2\pi/\sqrt{\alpha})$; on this time interval they already extend infinitely far in both directions in $\mathbb{R}^n$. On the other hand, we have previously mentioned that the n-sphere of radius $\sqrt{\alpha}$ is a Riemannian manifold of constant curvature α; since this sphere is compact, the manifold is complete. Some theorems from Volume 3 will throw more light on these matters.

The other important case occurs when $c = a_i = 0$ and $b_i = \delta_{in}$, so that the metric is

$$\sum_{i=1}^{n} \frac{dx^i \otimes dx^i}{(x^n)^2} ,$$

with constant curvature $\alpha = -1$. The 2-dimensional case, in particular, gives the nicest example of a non-Euclidean geometry (see Problem I.9-41).

Addendum 2. E. Cartan's Treatment of Normal Coordinates

In this Addendum we use moving frames to prove Riemann's claims about the form of the metric in normal coordinates, and in fact obtain stronger results; a special case of our result is represented by the first form of the metric for spaces of constant curvature which we found in Addendum 1. Throughout, we will be working with a moving frame $X_1,\dots,X_n$ adapted to an orthonormal basis $X_{ip},\dots,X_{np}$ for M_p; the forms θ^i, $\bar{\theta}^i$ are as defined previously, and $x^1,\dots,x^n$ denotes the Riemannian normal coordinate system determined by $X_{1p},\dots,X_{np}$.

21. PROPOSITION. The quadratic form $\|\ \|^2 - \sum_i (dx^i)^2$ can be written as a quadratic form in the differentials $x^r dx^s - x^s dx^r$.

Proof. Consider the functions A^i_{jk} on $\mathbb{R} \times M_p$ which satisfy the equations

$$(1)\qquad \frac{\partial^2 A^i_{jk}}{\partial t^2} = \sum_r (\mathbf{R}^i{}_{rjk} \circ \Phi) t t^r + \sum_{r,s,\ell} (\mathbf{R}^i{}_{rs\ell} \circ \Phi) A^\ell_{jk} t^r t^s$$

$$A^i_{jk}(0,X) = 0$$

$$\frac{\partial A^i_{jk}}{\partial t}(0,X) = 0 \ .$$

We claim that

$$(2)\qquad \bar{\theta}^i = t\,dt^i + \sum_{j,k} A^i_{jk} t^j dt^k \ .$$

To prove this, we simply note that if we define $\bar{\theta}^i$ by this formula, then an easy calculation shows that the $\bar{\theta}^i$ satisfies the equations and initial conditions of Corollary 9; since the solutions are unique, the result follows.

Now note, by skew-symmetry of $\mathbf{R}^i{}_{jk\ell}$ in the last two indices, that $A^i_{jk} + A^i_{kj}$ satisfies a linear second order differential equation, with initial conditions

$$A^i_{jk} + A^i_{kj}(0,X) = 0$$

$$\frac{\partial}{\partial t}(A^i_{jk} + A^i_{kj})(0,X) = 0 \quad .$$

It follows that

$$(3) \qquad A^i_{jk} = - A^i_{kj} \quad .$$

[The motivation for this proof is the following. The equations in Proposition 8 clearly imply that $\overline{\theta}^i - t dt^i = 0$ at $(t,0)$. This means (by Lemma I.3-2) that we can write $\overline{\theta}^i$ as in (2). Then the equations in Corollary 9 imply that the A^i_{jk} satisfy (1).]

Next consider the functions B^r_{sjk} which satisfy the equations

$$(4) \qquad \frac{\partial^2 B^r_{sjk}}{\partial t^2} = (\mathbf{R}^r{}_{sjk} \circ \Phi)t + \sum_{\mu,\ell} \mathbf{R}^r{}_{s\mu\ell} A^\ell_{jk} t^\mu$$

$$B^r_{sjk}(0,X) = 0$$

$$\frac{\partial B^r_{sjk}}{\partial t}(0,X) = 0 \quad .$$

We claim that

$$(5) \qquad A^r_{jk} = \sum_s B^r_{sjk} t^s \quad .$$

This is proved by checking that if we define A^r_{jk} by (5), then equations (4) imply equations (1).

Using the skew symmetry of $\mathcal{R}^r{}_{sjk}$ in j and k and the skew-symmetry of A^ℓ_{jk} in j and k [equation (3)] , we easily see that

$$(6) \qquad B^r_{sjk} = -B^r_{skj} \quad .$$

Also, using skew-symmetry of $\mathcal{R}^r{}_{sjk}$ in r and s, we see that

$$(7) \qquad B^r_{sjk} = -B^s_{rjk} \quad .$$

We know that the expression for $\| \|^2 = \Sigma(\theta^i)^2$ in the Riemannian normal coordinate system $x^1,\ldots,x^n$ is what $\Sigma(\bar{\theta}^i)^2$ becomes when we set $t = 1$ and $t^i = x^i$. So by (2) we have

$$\| \|^2 = \sum_i (dx^i)^2 + \sum_i(\sum_{j,k} A^i_{jk}x^j dx^k)^2 + \sum_r(\sum_{j,k} A^r_{jk}dx^r x^j dx^k) \quad .$$

Now (3) implies that the first triple sum is a quadratic form in the $x^r dx^s - x^s dx^r$. The second triple sum can be written

$$\sum_{r,s,i,j} B^r_{sjk}x^s dx^r x^j dx^k \quad ;$$

using (6) and (7), we see that this can also be written as a quadratic form in the $x^r dx^s - x^s dx^r$. ∎

For further developments, see E. Cartan's Lecons sur la Geometrie des Espaces de Riemann, pp. 242 ff.

Chapter 8. Connections in Principal Bundles

The method of moving frames turns out to be surprisingly powerful, but it leads us to a definition of a connection which does not have the "invariance" property of the Koszul definition (although it is a big improvement over the classical definition, simply because the transformation law can be stated so much more elegantly). At the same time, we should note a certain deficiency in our proofs of the Test Case. We certainly have enough of them (six so far), and they use the integrability conditions $R = 0$ in many different ways. In the first proof we use the classical integrability theorem, and in the second and third proofs we essentially reprove this theorem. In the fourth proof we use the differential form version of the Frobenius Theorem, and in the fifth and sixth proofs we use the proof of the integrability theorem outlined in Problem I.6-8. However, in all this time we have never used the distribution formulation of the Frobenius Integrability Theorem (I.6-5), although this is the most geometric version of all.

These two phenomena will eventually turn out to be closely related, but for the present we will concentrate on the first problem. The main step in the solution of this problem was accomplished by Ehresmann* in 1950. With the advantages of hindsight we can reconstruct the solution in a way that makes it seem natural and almost obvious.

It will be helpful to begin by reconsidering the classical and modern definitions of vector fields. As we pointed out long ago, the snazziest modern definition of tangent vectors is essentially the same as the classical definition, as n-tuples of numbers which "transform" according to certain rules. On the other hand, the modern definition of a vector field represents

---

*Ehresmann, Les connexions infinitesimales dans un espace fibre differentiable. Colloque de Topologie, Bruxelles (1950), 29-55.

a definite improvement over the classical definition. Instead of dealing with n-tuples of functions transforming according to certain rules, we make the set of tangent vectors into a new manifold, the tangent bundle, and define vector fields to be sections of this vector bundle. The idea behind the modern treatment of connections is to obtain a bundle whose sections are just the moving frames on M. To do this, we will imitate the construction of the tangent bundle in a rather straight forward way.

Recall that a frame u for M_p is just an ordered basis $u = (u_1, \dots, u_n)$ for M_p. Let $F(M)$ denote the set of all frames u for all tangent spaces M_p. We call $F(M)$ the bundle of frames of M, and define $\pi: F(M) \longrightarrow M$ to be the map which takes a basis u for M_p to $\pi(u) = p$. If (x,U) is a coordinate system on M and $p \in U$, then every frame $u = (u_1, \dots, u_n)$ for M_p can be written uniquely as

$$u_j = \sum_{j=1}^{n} x_j^i(u) \left.\frac{\partial}{\partial x^i}\right|_{\pi(u)} .$$

The matrix $(x_j^i(u))$ is non-singular, and any non-singular matrix can occur, so the map

$$u \longmapsto (x^i(\pi(u)),\ x_j^i(u)) \in \mathbb{R}^n \times GL(n,\mathbb{R})$$

is a one-one map $x_\#$ from $\pi^{-1}(U)$ onto $x(U) \times GL(n,\mathbb{R})$. It is easy to see that if (y,V) is another coordinate system, then $y_\# \circ (x_\#)^{-1}$ is C^∞, from $x_\#(\pi^{-1}(U \cap V))$ to $y_\#(\pi^{-1}(U \cap V))$. This means that we can make $F(M)$ into a C^∞ manifold in such a way that each $x_\#$ is a diffeomorphism; with this C^∞ structure on $F(M)$, the map π is clearly C^∞. Notice, finally, that we have a C^∞ map $F(M) \times GL(n,\mathbb{R}) \longrightarrow F(M)$, given by $(u,A) \longmapsto u \cdot A$, where $(u \cdot A)_i = \sum_j A_i^j u_j$; we have $u \cdot A = A$ only when $A = I$, and the set of all

$u \cdot A$ for $A \in GL(n,\mathbb{R})$ is just $\pi^{-1}(\pi(u))$.

Although we have called $F(M)$ the "bundle of frames", it is clearly not a vector bundle; each fibre $\pi^{-1}(p)$ is diffeomorphic to $GL(n,\mathbb{R})$, rather than to some $\mathbb{R}^N$. However, $F(M)$ is another special sort of "bundle" which we will now define.

Let M be a C^∞ manifold, and G a Lie group. A (C^∞) principal bundle over M, with group G, is a triple $(P,\pi,\cdot)$ where

(1) P is a C^∞ manifold (the total space of the principal bundle)

(2) $\pi\colon P \longrightarrow M$ is a C^∞ map (the projection map of the bundle) onto M (the base space of the principal bundle), satisfying

$$\pi(u \cdot a) = \pi(a) \quad \text{for all} \quad u \in P \quad \text{and} \quad a \in G$$

(3) the map $\cdot$ (the action of G) is a C^∞ map $(u,a) \longmapsto u \cdot a$ from $P \times G$ to P with

$$u \cdot (ab) = (u \cdot a) \cdot b \quad \text{for all} \quad u \in P \quad \text{and} \quad a,b \in G$$

such that the following "local triviality" condition holds: for each $p \in M$ there is a neighborhood U of p and a diffeomorphism $t\colon \pi^{-1}(U) \longrightarrow U \times G$ of the form $t(u) = (\pi(u),\varphi(u))$ where φ satisfies $\varphi(u \cdot a) = \varphi(u)a$ [the latter product being the product in G].

From the condition $\pi(u \cdot a) = \pi(a)$ we see that $\{u \cdot a\colon a \in G\} \subset \pi^{-1}(\pi(u))$.

Using the property $\varphi(u\cdot a) = \varphi(u)a$ of the map φ, we see that we actually have $\{u\cdot a: a \in G\} = \pi^{-1}(\pi(u))$; for if $v \in P$ satisfies $\pi(v) = \pi(u)$, and $\varphi(v) = \varphi(u)a$ for $a \in G$, then $\varphi(v) = \varphi(u\cdot a)$ and $\pi(v) = \pi(u.a)$, so $v = u\cdot a$. Notice also that if $u\cdot a = u$ for some $u \in P$, then $a = e$.

Each "fibre" $\pi^{-1}(p)$ of P is clearly diffeomorphic to G. If $p = \pi(u)$ for $u \in P$, and $i\colon \pi^{-1}(p) \longrightarrow P$ is the inclusion, then the image of the tangent space $i_*(\pi^{-1}(p)_u)$ is a subspace V_u of P_u, called the vertical subspace at u; tangent vectors in this subspace are called vertical tangent vectors at u. Clearly $Y \in V_u$ is vertical if and only if $\pi_* Y = 0$.

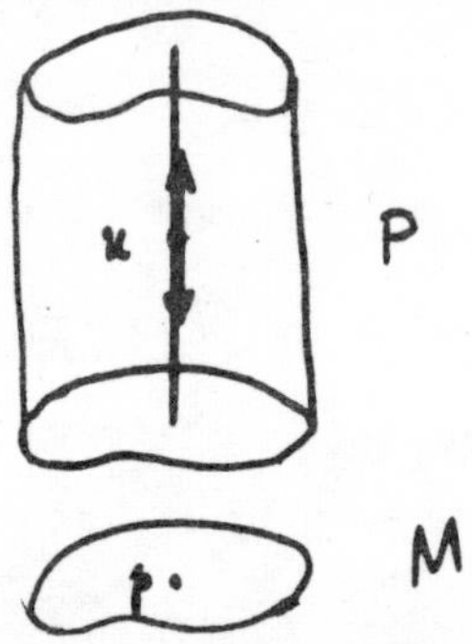

The simplest example of a principal bundle is $M \times G$, with $\pi\colon M \times G \longrightarrow M$ the projection on the first factor, and $(p,a)\cdot b = (p,ab)$. This is called the trivial principal bundle with group G. So far, we have given only one other example of a principal bundle, the bundle of frames $F(M)$, with group $GL(n,\mathbb{R})$. However, we can use the construction of this bundle to acquire many other examples. If $\pi\colon E \longrightarrow M$ is any C^∞ vector bundle over M, we can let $F(E)$ be the collection of all frames u for the vector space $\pi^{-1}(p)$, for all $p \in M$; the projection map $\varpi\colon F(E) \longrightarrow M$ takes a frame u for $\pi^{-1}(p)$ into $\varpi(u) = p$. Consider, in particular, the Möbius strip as a 1-dimensional vector bundle $\pi\colon E \longrightarrow S^1$ over S^1. A frame in a 1-dimensional vector space is just a non-zero vector, so $F(E)$ consists of the Möbius strip with the zero-section deleted. This space is connected (cut a paper Möbius strip along the center if you don't believe it); more generally, a vector bundle $\pi: E \longrightarrow M$ over a connected space M is orientable if and only if $F(E)$ is disconnected.

Examples of a different sort are obtained if we begin with a vector bundle $\pi: E \longrightarrow M$ equipped with a Riemannian metric, and define $O(E)$ to be the set of all orthonormal frames u for $\pi^{-1}(p)$, for all $p \in M$. This is a principal bundle over M with group $O(n)$. In the case of a 1-dimensional bundle $\pi: E \longrightarrow M$, every fibre of $O(E)$ has exactly 2 points, and $O(1) = Z_2$; the action of the non-zero element $O(1)$ on $O(E)$ interchanges these 2 points in each fibre. For the Möbius strip, the principal bundle $O(E)$ looks like the picture below: more precisely, the total space of $O(E)$

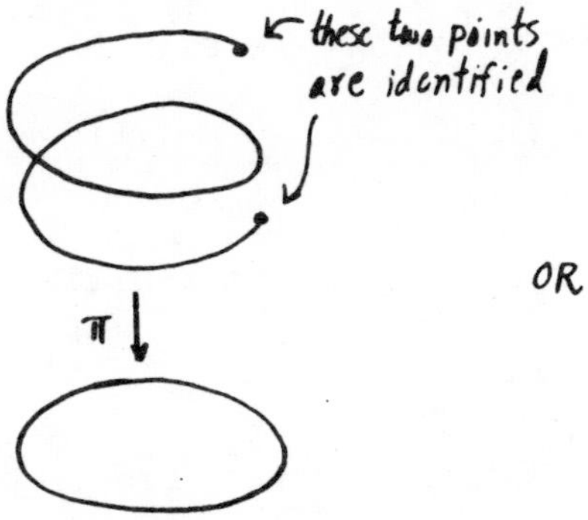

OR

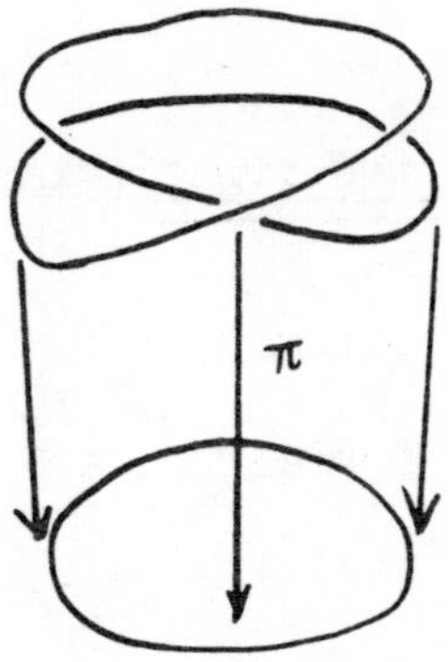

is a circle S^1, and the projection map $S^1 \longrightarrow S^1$ is given by $\theta \longmapsto 2\theta$.

In the case of an oriented bundle $\pi: E \longrightarrow M$, these constructions can be modified to give principal bundles $SF(E)$ and $SO(E)$, the bundle of positively oriented frames, and positively oriented orthonormal frames, respectively. For 1-dimensional bundles, the total space $SF(E)$ looks like $M \times (0,\infty)$, while $SO(E)$ looks just like M. For 2-dimensional bundles, $SO(E)$ has a circle above each point of M; the action of $\theta \in SO(2) = S^1$ rotates each circle through an angle of θ.

Although examples in which the group G is discrete are the least interesting for us, we will nevertheless give two more, as examples of principal bundles are hard to come by. In the first example, the group is the integers $\mathbb{Z}$, the total space and base space are $\mathbb{R}$ and $\mathbb{Z}$, respectively, and the map $\pi\colon \mathbb{R} \longrightarrow \mathbb{Z}$ is $\pi(\theta) = (\cos\theta, \sin\theta)$. In the second example

we define $\pi\colon S^2 \longrightarrow P^2$ to be $\pi(p) = \{p, -p\}$; the group is $\mathbb{Z}_2$ and the action of the non-zero element is to take $p \in S^2$ to $-p$.

Before discussing principal bundles in particular, we need some generalities about Lie groups acting on manifolds. Consider a Lie group G, a C^∞ manifold M, and a C^∞ map $(p,a) \longmapsto p\cdot a$ from $M \times G$ to M. We say that G acts on M on the right (via this map) if

(1) the map $R_a\colon M \longrightarrow M$ defined by $R_a(p) = p\cdot a$ is a diffeomorphism for all $a \in G$

(2) $p\cdot(ab) = (p\cdot a)\cdot b$ for all $p \in M$ and $a,b \in G$.

Condition (2) can also be written as $R_{ab} = R_b \circ R_a$; since each R_a is a diffeomorphism, it follows easily that R_e is the identity map of M. We say that G acts effectively if e is the only element a with R_a the identity map of M, and we say that G acts without fixed point if the following stronger condition holds: if $p.a = p$ for some $p \in M$, then $a = e$.

Now let $\mathfrak{g}$ be the Lie algebra of a Lie group G which acts on M on the right. For every $X \in \mathfrak{g}$, we have the curve $t \longmapsto \exp tX$ in G; for each $p \in M$ this gives rise to a curve $c_p(t) = p\cdot(\exp tX) = R_{\exp tX}(p)$. We denote $c_p'(0)$ by $\sigma(X)(p)$; we thus have a vector field $\sigma(X)$ on M, and hence a map $\sigma\colon \mathfrak{g} \longrightarrow$ (vector fields on M). The 1-parameter group of diffeomorphisms generated by $\sigma(X)$ is $\varphi_t(p) = p\cdot(\exp tX)$, by the very definition of $\sigma(X)$. It is important to note that we can also describe $\sigma(X)$ as follows. For $p \in M$, let $\sigma_p\colon G \to M$ be $\sigma_p(a) = p\cdot a$. Then

$$\sigma(X)(p) = \sigma_{p*}(X).$$

To discuss this operation σ we will also need to introduce an important map, the "adjoint map"

$$\mathrm{Ad}(a) = (L_a R_a{}^{-1})_* = (R_a{}^{-1} L_a)_*\colon \mathfrak{g} \to \mathfrak{g} \quad ,$$

where L_a and R_a now denote left and right translations in G. Thus $\mathrm{Ad}(a)$ is the differential at e of the map $b \mapsto aba^{-1} = L_a R_a{}^{-1}(b) = R_a{}^{-1} L_a(b)$. Usually $\mathrm{Ad}(a)(X)$ is denoted simply by $\mathrm{Ad}(a)X$. If $\tilde{X}$ is the left invariant vector field on G with $\tilde{X}(e) = X \in \mathfrak{g}$, then

$$\mathrm{Ad}(a)X = (R_a{}^{-1})_*(L_{a*}\tilde{X})(e) = [(R_a{}^{-1})_*\tilde{X}](e) \quad ,$$

since $L_{a*}\tilde{X} = \tilde{X}$. Consider, in particular, the special case where $G = GL(n,\mathbb{R})$, so that $\mathfrak{g} = \mathfrak{gl}(n,\mathbb{R})$ is the set of all $n \times n$ matrices. For any $n \times n$ matrix M and any $A \in GL(n,\mathbb{R})$ we have

$$\mathrm{Ad}(A)M = AMA^{-1} \quad ,$$

since $L_{A*} = L_A$ and $R_{A*} = R_A$, because L_A and R_A are linear functions (compare Problem I.10-19).

<u>1. PROPOSITION</u>. Let G act on the right on M. Then

(1) The map $\sigma: \mathfrak{g} \longrightarrow$ (vector fields on M) is linear.

(2) $\sigma([X,Y]) = [\sigma(X),\sigma(Y)]$.

(3) If G acts effectively and $X \neq 0$, then $\sigma(X)$ is not the zero vector field.

(4) If G acts without fixed point and $X \neq 0$, then $\sigma(X)$ is nowhere 0.

<u>Proof</u>. Linearity is clear from the equation $\sigma(X)(p) = \sigma_{p*}(X)$.

To prove (2) we note that since the bracket of two vector fields is the same as the Lie derivative (Theorem I.5-10), we have

(1) $$[X,Y] = [\tilde{X},\tilde{Y}](e) = \lim_{h\to 0} \frac{1}{h}[Y - (R_{\exp tX})_*\tilde{Y}(e)]$$

$$= \lim_{h\to 0} \frac{1}{h}[Y - Ad(\exp -tX)Y].$$ (Compare Problem I.10-19.)

On the other hand, if R_a now denotes the map $p \longmapsto p\cdot a$ from M to G, then

(2) $$R_{\exp tX} \circ \sigma_{p\cdot(\exp -tX)}(a) = p\cdot([\exp -tX]\ a \exp tX).$$

Since $\varphi_t(p) = p.(\exp tX)$ is the 1-parameter group of diffeomorphisms generated by $\sigma(X)$, on M we have

$$[\sigma(X),\sigma(Y)](p) = \lim_{h\to 0} \frac{1}{h}[\sigma(Y)(p) - [R_{\exp tX}]_*\sigma(Y)(p)]$$

$$= \lim_{h\to 0} \frac{1}{h}[\sigma_{p*}Y - \sigma_p(Ad(\exp -tX)Y)] \qquad \text{by (2)}$$

$$= \sigma_{p*}(\lim_{h\to 0} \frac{1}{h}[Y - Ad(\exp -tX)Y]) = \sigma_{p*}([X,Y]) \qquad \text{by (1)}$$

$$= \sigma([X,Y]) \ .$$

To prove (3), suppose $\sigma(X) = 0$. Then for every p the 1-parameter group of diffeomorphisms $\varphi_t(p) = p\cdot(\exp tX)$ must be $\varphi_t(p) = p$ (for the unique integral curve c of $\sigma(X)$ through p is clearly $c(p) = p$). If G acts effectively, this means that all $\exp tX = 0$, so $X = 0$.

To prove (4), suppose $\sigma(X)(p) = 0$ for some p. Then $p\cdot(\exp tX) = p$ for all t. If G acts without fixed point, then all $\exp tX = e$, so $X = 0$. ∎

Let us now apply this construction to a principle bundle $\pi\colon P \longrightarrow M$, with group G. The map $\cdot$ from $P \times G \longrightarrow P$ is an action of G on P to the right (the map $u \longmapsto u\cdot a$ is a diffeomorphism by condition (3) of the definition) and G acts without fixed point (we have already pointed out that this follows from condition (3).) Therefore we have the <u>fundamental vector field</u> $\sigma(X)$ <u>corresponding to</u> X for all $X \in \mathfrak{g}$; for every $u \in P$, the map $X \longmapsto \sigma(X)(u)$ is an isomorphism, since G acts without fixed point. Since the maps $R_a\colon P \longrightarrow P$ take fibres to themselves, the set of all $\sigma(X)(u)$ is precisely the set of vertical vectors at u.

2. PROPOSITION. For all $X \in \mathfrak{g}$ and $a \in G$, the vector field $(R_a)_*\sigma(X)$ is the fundamental vector field

$$(R_a)_*\sigma(X) = \sigma(\mathrm{Ad}(a^{-1})X) .$$

Proof. Since $\varphi_t(p) = p\cdot\exp tX = R_{\exp tX}(p)$ is the 1-parameter group of diffeomorphisms generated by $\sigma(X)$, it follows from Lemma I.5-11 that $(R_a)_*\sigma(X)$ generates the 1-parameter group of diffeomorphisms

$$\psi_t = R_a \circ R_{\exp tX} \circ R_a{}^{-1} = R_{a^{-1}\exp tX\, a} .$$

Now $\{a^{-1} \exp tX\, a\}$ is the 1-parameter group of diffeomorphisms of G generated by $\mathrm{Ad}(a^{-1})X$. So ψ_t generates $\sigma(\mathrm{Ad}(a^{-1})X)$. ∎

The vector fields $\sigma(X)$ are rather difficult to picture, especially since most principal bundles themselves are impossible to visualize. Nevertheless they will be very important, as we shall see upon returning to the structure which began our whole discussion, the principal bundle of frames $F(M)$. A section s of this bundle over an open set $U \subset M$ is a C^∞ map $s\colon U \longrightarrow F(M)$ with $\pi \circ s =$ identity map of U. Clearly, a section s is just what we used to call a moving frame on U. Note that, unlike a vector bundle, which always has a section defined on all of M, namely the 0-section, a principal bundle need not have such a section. In fact, if the principal bundle $\pi\colon P \longrightarrow M$ over M with group G has a section $s\colon M \longrightarrow P$, then the bundle is trivial — we let $(p,a) \in M \times G$ correspond to $s(p)\cdot a \in P$.

We can use some of this new language to give an alternative, but completely equivalent, definition of a Cartan connection: A Cartan connection is an assignment of an $n \times n$ matrix-valued 1-form $\omega_s [= (\omega_{s\,j}^{\;i})]$ to every section $s\colon U \longrightarrow M$ in such a way that

$$(*) \qquad \omega_{s\cdot a} = a^{-1}da + a^{-1}\omega_s a$$

for every C^∞ function $a\colon U \longrightarrow GL(n,\mathbb{R})$. (Here $s\cdot a$ is the section $(s\cdot a)(p) = s(p)\cdot a(p)$, the $\cdot$ denoting the action of $GL(n,\mathbb{R})$ on $F(M)$.) This formulation of the definition suggests how we may obtain a definition of a connection which has all the advantages of the Cartan definition but which is also "invariant". We ask if there is an $n \times n$ matrix-valued 1-form ω on the manifold $F(M)$ such that for each section (moving frame) s we have

$$\omega_s = s^*(\omega) \quad ;$$

notice that previously we used ω alone to denote the connection form for some moving frame, but from now on we will have to be careful to use subscripts to distinguish the forms on M from the form ω which we hope to find on $F(M)$. Since the forms ω_s for a Cartan connection satisfy (*), our question is then the following:

> If we are given a collection of ω_s satisfying (*), is there an ω on $F(M)$ such that each $\omega_s = s^*(\omega)$? More generally, which $n \times n$ matrix-valued 1-forms ω on $F(M)$ have the property that for every section $s\colon U \longrightarrow F(M)$ and $GL(n,\mathbb{R})$-valued function a on U we have

$$(**) \qquad (s\cdot a)^*(\omega) = a^{-1}da + a^{-1}s^*(\omega)\cdot a \quad ?$$

In order to answer these questions, we need to know something about the section $s.a$. If a has the constant value $A \in GL(n,\mathbb{R})$, then $s\cdot a = R_A \circ s$, where $R_A(u) = u\cdot A$, the dot denoting the action of $GL(n,\mathbb{R})$ on $F(M)$. So for any tangent vector $X_p \in M_p$ we have $(s.a)_*(X_p) = R_{A*}(s_*X_p)$; when a is not constant there is a correction term.

<u>3. PROPOSITION</u>. Let s be a section of $F(M)$, over some open set U, and let $a\colon U \longrightarrow GL(n,\mathbb{R})$ be C^∞. Then for any tangent vector X_p at a point $p \in U$ we have

$$(s\cdot a)_*(X_p) = R_{a(p)*}(s_*X_p) + \sigma(X_p(a))(s(p)\cdot a(p)) \quad .$$

[Note that $X_p(a)$ is an $n \times n$ matrix, and hence may be considered as an element of $\mathfrak{gl}(n,\mathbb{R})$, so $\sigma(X_p(a))$ is a vector field on $F(M)$.]

Proof. For convenience, let

$$m: F(M) \times GL(n,\mathbb{R}) \longrightarrow F(M)$$

be $m(u,A) = u\cdot A$. Remember that the tangent space W of $F(M) \times GL(n,\mathbb{R})$ at (u,A) is isomorphic to the direct sum $F(M)_u \oplus GL(n,\mathbb{R})_A$, so we can consider every element of W as a pair $(Y_1,Y_2) = Y_1 \oplus Y_2$, where $Y_1 \in F(M)_u$ and $Y_2 \in GL(n,\mathbb{R})_A$. Note that if

c_1 is an integral curve in $F(M)$ for Y_1
c_2 " " " " $GL(n,\mathbb{R})$ for Y_2 ,

then

$t \longmapsto m(c_1(t),A) = c_1(t)\cdot A$ is an integral curve for $X \oplus 0$
$t \longmapsto m(u,c_2(t)) = u\cdot c_2(t)$ " " " " " $0 \oplus Y$.

Now let c be an integral curve for X_p. Since $s\cdot a = m\circ(s,a)$, we have

$$(s\cdot a)_*(X_p) = m_*(s_*(X_p),a_*(X_p)) = m_*(s_*(X_p) \oplus a_*(X_p))$$
$$= \frac{d}{dt}\bigg|_{t=0} s(c(t))\cdot a(p) + \frac{d}{dt}\bigg|_{t=0} s(p)\cdot a(c(t)) \ .$$

The first term on the right can be written as

$$\left.\frac{d}{dt}\right|_{t=0} R_{a(p)}(s(c(t))) = R_{a(p)*}\left(\left.\frac{ds(c(t))}{dt}\right|_{t=0}\right)$$

$$= R_{a(p)*}(s_* X_p) \quad .$$

To identify the second term, recall that for all $u \in F(M)$ we have

$$\sigma(M)(u) = \sigma_{u*}(M) \quad , \quad \text{where } \sigma_u(A) = u\cdot A \text{ for } A \in GL(n,\mathbb{R}) \quad .$$

We write

$$s(p)\cdot a(c(t)) = s(p)\cdot a(p)\cdot[a(p)^{-1}\cdot a(c(t))] \quad .$$

The term in brackets gives a curve $\gamma(t) = a(p)^{-1}\cdot a(c(t))$ in $GL(n,\mathbb{R})$ with $\gamma(0) = I$; then $\gamma'(0) \in GL(n,\mathbb{R})_I = \mathfrak{gl}(n,\mathbb{R})$ is the tangent vector corresponding to the matrix

$$\left.\frac{da(c(t))}{dt}\right|_{t=0} = X_p(a) \quad ,$$

when we identify the $n \times n$ matrices with $\mathfrak{gl}(n,\mathbb{R})$. Consequently,

$$\left.\frac{d}{dt}\right|_{t=0} s(p)\cdot a(c(t)) = \left.\frac{d}{dt}\right|_{t=0} s(p)\cdot a(p)\cdot[a(p)^{-1}a(c(t))]$$

$$= \sigma_{s(p)\cdot a(p)*}\left(\left.\frac{d}{dt}\right|_{t=0} a(p)^{-1}a(c(t))\right)$$

$$= \sigma_{s(p)\cdot a(p)*}(X_p(a))$$

$$= \sigma(X_p(a))(s(p)\cdot a(p)) \quad . \blacksquare$$

<u>Remark</u>. Proposition 3 can actually be formulated for any principal bundle $\pi\colon P \longrightarrow M$ over M, with group G. If s is a section over U and $a\colon U \longrightarrow G$ is C^{∞}, then for any tangent vector X_p at $p \in U$ we have

$$(s\cdot a)_*(X_p) = R_{a(p)*}(s_*X_p) + \sigma\left(L_{a(p)^{-1}*}a_*(X_p)\right)(s(p)\cdot a(p)).$$

With Proposition 3 at hand, let us reconsider our question. We want to know which $n \times n$ matrix-valued 1-forms ω on $F(M)$ have the property that for each section s we have

$$(**) \qquad (s\cdot a)^*(\omega) = a^{-1}da + a^{-1}s^*(\omega)a \quad .$$

This is equivalent to saying that for every $X_p \in M_p$ we have

$$\omega((s\cdot a)_*X_p) = a^{-1}(p)X_p(a) + a^{-1}(p)\omega(s_*X_p)a(p) \quad .$$

According to Proposition 3 this is equivalent to

$$\omega(R_{a(p)*}(s_*X_p)) + \omega(\sigma(X_p(a))(s(p)\cdot a(p))) = a^{-1}(p)X_p(a) + a^{-1}(p)\omega(s_*X_p)a(p) \quad .$$

We can extract 2 separate equations from this, as follows:

(1) First suppose that $a(p) = I$. Then we obtain

$$\omega(s_*X_p) + \omega(\sigma(X_p(a))(s(p)) = X_p(a) + \omega(s_*X_p) \quad ,$$

or

$$\text{(I)} \qquad \omega(\sigma(X_p(a)).(s(p)) = X_p(a) \qquad \text{for } a(p) = I \quad .$$

(2) Now suppose a has the constant value A. Then we obtain

$$\text{(II)} \qquad \sigma(R_{A*}(s_* X_p)) = A^{-1}\omega(A_* X_p)A \quad .$$

Conversely, it is not hard to show that if ω satisfies (I) and (II), then ω satisfies (**).

Now the equations (I) and (II) can be simplified considerably, resulting in equations that do not involve s at all. In (I), the matrix $X_p(a)$ can obviously be any $n \times n$ matrix, since we just have to satisfy $a(p) = I$. Since we can also choose any s, we see that (I) is equivalent to

$$\text{(I')} \qquad \omega(\sigma(M)(u)) = M \qquad \text{for all } M \in \mathfrak{gl}(n,\mathbb{R}), \text{ and } u \in F(M).$$

We claim that (II) is equivalent to

$$\text{(II')} \qquad \omega(R_{A*}Y) = A^{-1}\omega(Y)A \qquad \text{for all } Y \in F(M)_u \quad .$$

To see this we note that by choosing s appropriately we can make $s_* X_p$ be any vector in $u = s(p)$ that is not vertical; since the vertical vectors are all of the form $\sigma(M)$, equation (II') for vertical vectors follows from (I') and Proposition 2, remembering that for $GL(n,\mathbb{R})$ we have $Ad(A)M = AMA^{-1}$.

Summing up, we see that a matrix-valued 1-form ω on $F(M)$ satisfies

$$(s\cdot a)^*(\omega) = a^{-1}da + a^{-1}s^*(\omega)a$$

[and consequently the assignment of $s^*\omega$ to s is a Cartan connection] if and only if

$$\omega(\sigma(M)) = M \qquad \text{for all } M \in \mathfrak{gl}(n,\mathbb{R})$$

$$\omega(R_{A*}(Y)) = A^{-1}\omega(Y)A = \mathrm{Ad}(A^{-1})\omega(Y) \qquad \text{for all } Y \in F(M)_u .$$

We leave it to the reader to show that if we are given a Cartan connection $\{\omega_s\}$, then there is a unique such ω on $F(M)$ with $\omega_s = s^*(\omega)$. (Define $\omega(Y_u) = \omega_s(X_p)$ whenever $Y_u = s_* X_p$, and use Proposition 3 and the transformation rules for a Cartan connection to verify that ω is well-defined.) We are consequently ready for the final definition of a connection; since all our conditions make sense in any principal bundle, our new definition is not only more abstract, more elegant, and more incomprehensible, but also more general.

> An (Ehresmann) connection in a principal bundle $\pi: P \longrightarrow M$ over M with group G is a C^∞ $\mathfrak{g}$-valued 1-form ω on P such that
>
> (1) $\omega(\sigma(X)) = X$ for all $X \in \mathfrak{g}$
>
> (2) $\omega(R_{a*}Y) = \mathrm{Ad}(a^{-1})\omega(Y)$ for all $a \in G$, and every vector field Y on E.

If ω is an Ehresmann connection, then for every $u \in P$, the map $\omega(u): P_u \longrightarrow \mathfrak{g}$ is onto, by (1), so its kernel $H_u = \ker \omega(U)$ is a subspace of P_u having the same dimension as M. This subspace is called the horizontal subspace at u (determined by the connection), and tangent vectors in H_u are called horizontal. Thus every Ehresmann connection ω on P gives rise to a certain distribution H on P.

4. PROPOSITION. If H is the distribution on P determined by an Ehresmann connection ω, then

(1) $P_u = V_u \oplus H_u$

(2) $H_{u\cdot a} = (R_a)_* H_u$

(3) H is a C^∞ distribution.

Conversely, if H is a distribution on P satisfying (1) - (3), then H is the distribution determined by a unique Ehresmann connection ω.

Proof. Condition (1) is obvious from the definition of H_u as $\ker \omega(u)$ (and the fact that $\omega(u)$ is onto $\mathfrak{g}$).

If $Y \in H_u$, then

$$\begin{aligned} \omega(u\cdot a)(R_{a*}Y) &= \omega(R_{a*}Y) = \mathrm{Ad}(a^{-1})\omega(Y) \qquad \text{by condition (2) in the definition of a connection} \\ &= 0 , \end{aligned}$$

so $R_{a*}Y \in H_{u\cdot a}$. Since R_{a*} is one-one, and $\dim H_u = \dim H_{u\cdot a}$, it follows that $H_{u\cdot a} = (R_a)_* H_u$.

To prove that H is a C^∞ distribution, choose vector fields $Y_1,\ldots,Y_n,\ldots,Y_{n+k}$ which span P_v for all v in a neighborhood of u. Let $X_1,\ldots,X_k$ be a basis for $\mathfrak{g}$, so that we can write $\omega = \Sigma\, \omega^j \cdot X_j$ for ordinary C^∞ 1-forms ω^j on P. Let $\overline{Y}_i$ be the vector field

$$\overline{Y}_i = Y_i - \sum_j \omega^j(Y_i)\sigma(X_j).$$

The C^∞ vector fields $\overline{Y}_i$ are clearly horizontal and span the distribution H in a neighborhood of u.

Conversely, given H, we (must) define ω by $\omega(Y) = 0$ for Y vertical and $\omega(\sigma(X)) = X$ for $X \in \mathfrak{g}$. Then ω is C^∞, since $\omega(Y)$ is C^∞ when Y is a horizontal vector field, or when $Y = \sigma(X)$, and these vector fields span the set of all vector fields, over the C^∞ functions. Condition (1) for a connection holds by definition of ω. To prove condition (2), we need only prove it for horizontal Y and vertical Y. If Y is horizontal, then $(R_a)_*Y$ is also, by condition (2) on H, so we have

$$\omega((R_a)_*Y) = 0 = \mathrm{Ad}(a^{-1})\omega(Y) \quad .$$

When Y is vertical, we may assume that $Y = \sigma(X)$ for $X \in \mathfrak{g}$. Then $(R_a)_*Y = \sigma(\mathrm{Ad}(a^{-1})X)$ by Proposition 2. So

$$\omega((R_a)_*Y) = \mathrm{Ad}(a^{-1})X = \mathrm{Ad}(a^{-1})\omega(Y) \quad ,$$

as desired. ∎

Often, a connection is <u>defined</u> to be a distribution H satisfying (1) - (3) of Proposition 4, and ω is defined as in the second part of the proof — it is then called the "connection form" for the connection H. Using the decomposition given by (1) of Proposition 4, we can write, for any tangent vector $Y \in P_u$, a unique expression

$$Y = v(Y) + h(Y)$$

where $v(Y)$, the **vertical component** of Y, is vertical and $h(Y)$, the **horizontal component** of Y, is horizontal. As we noted in the proof of Proposition 4,

$$h(Y) = Y - \sum_j \omega^j(Y)\sigma(X_j) \quad .$$

From this formula it is clear that $h(Y)$, and hence $v(Y)$, is C^∞ if Y is C^∞.

From the decomposition $P_u = V_u \oplus H_u$ and the fact that V_u is the kernel of $\pi_*: P_u \longrightarrow M_{\pi(u)}$, it is also clear that $\pi_*: H_u \longrightarrow M_{\pi(u)}$ is an isomorphism for each $u \in P$. Consequently, for every vector field X on M there is a unique vector field X^* on P such that X^* is everywhere horizontal and $\pi_*(X^*{}_u) = X_{\pi(u)}$ for all $u \in P$; this vector field X^* is called the lift of X. There are two simple propositions about lifts.

5. PROPOSITION. If X is a C^∞ vector field on M, then X^* is a C^∞ vector field on P, and for all $a \in G$ we have $R_{a*}(X^*) = X^*$. Conversely, if Y is a vector field on P such that $R_{a*}(Y) = Y$ for all $a \in G$, then $Y = X^*$ for a unique vector field X on M.

Proof. Using local triviality, we can choose a C^∞ vector field X' on some $\pi^{-1}(U) \subset P$ such that $\pi_*(X'{}_u) = X_{\pi(u)}$ for all $u \in \pi^{-1}(U)$. Then $X^* = h(X')$ is also C^∞. The other parts are left to the reader. ∎

6. PROPOSITION. If X^* and Y^* are the lifts of vector fields X and Y on M, then

(1) $X^* + Y^*$ is the lift of $X + Y$

(2) for every $f: M \longrightarrow \mathbb{R}$ we have $(fX)^* = (f\circ\pi)\cdot X^*$

(3) $h([X^*,Y^*]) = [X,Y]^*$.

Proof. The first 2 parts are trivial. For the third we note that X^* and X are π-related, so by Proposition I.6-3,

$$\pi_*(h[X^*,Y^*]_u) = \pi_*([X^*,Y^*]_u) = [X,Y]_{\pi(u)} \quad . \blacksquare$$

Our aim now is to develop, and generalize, the material of the last 3 Chapters, from the point of view of Ehresmann connections. The topics to be discussed include parallel translation, covariant derivatives, the curvature and torsion tensors, the structural equations, and the Bianchi identities. Presumably, somewhere along the way an elegant proof of the Test Case will also turn up. We begin with a discussion of parallel translation.

A C^1 curve $\gamma\colon [0,1] \longrightarrow P$ is called <u>horizontal</u> if all tangent vectors $c'(t)$ are horizontal vectors. If γ is merely piecewise C^1, then we also

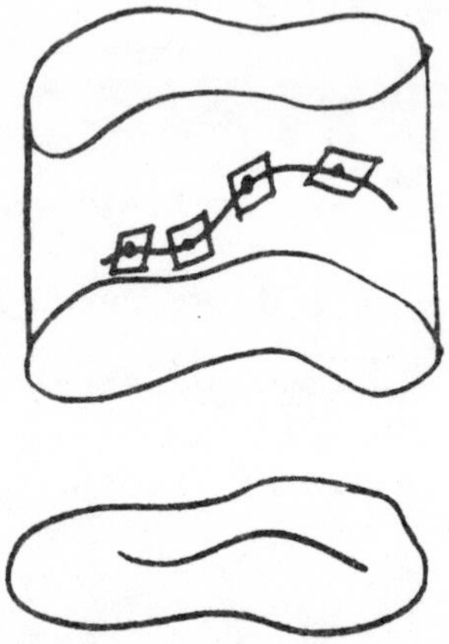

require $c'(t+)$ and $c'(t-)$ to be horizontal at every point t where c is not C^1. If $c\colon [0,1] \longrightarrow M$ is a piecewise C^1 curve in M, we now define a <u>lift</u> of c to be a horizontal curve $c^*\colon [0,1] \longrightarrow M$ such that c^* <u>covers</u> c, i.e. such that $\pi \circ c^* = c$. Notice that if c^* is a lift of c, then so is $R_a \circ c^*$, by Proposition 4(2).

<u>7. PROPOSITION</u>. Let $c\colon [0,1] \longrightarrow M$ be a piecewise C^1 curve, and choose $u_0 \in P$ with $\pi(u_0) = c(0)$. Then there is a unique lift c^* of c with $c^*(0) = u_0$.

<u>Proof</u>. Using local triviality of the principal bundle, it is easy to show that there is a curve $\gamma\colon [0,1] \longrightarrow P$ with $\gamma(0) = u_0$ and $\pi \circ \gamma = c$. A lift c^* of c must be of the form $c^*(t) = \gamma(t) \cdot a(t)$ for some $a\colon [0,1] \longrightarrow G$

with $a(0) = e$. Using the method of proof for Proposition 3 (see also the Remark following it), we find that

$$c^{*\prime}(t) = R_{a(t)*}(\gamma'(t)) + \sigma\left(L_{a(t)^{-1}*}a'(t)\right)(c^*(t)) \quad .$$

Consequently,

$$\omega(c^{*\prime}(t)) = \mathrm{Ad}(a(t)^{-1})\omega(\gamma'(t)) + L_{a(t)^{-1}*}a'(t) \quad .$$

Now c^* is horizontal if and only if $\omega(c^{*\prime}(t)) = 0$; using the above equation, and remembering the definition of Ad, this means that

$$L_{a(t)^{-1}*}a'(t) = -L_{a(t)^{-1}*}R_{a(t)*}\omega(\gamma'(t)) \quad ,$$

$$a'(t) = -R_{a(t)*}\omega(\gamma'(t)) \quad .$$

In this equation, $\omega(\gamma'(t))$ is a given curve in $\mathfrak{g}$. If we introduce a coordinate system on G, this equation becomes a differential equation for a, and we know that a unique solution exists locally. So we can always find a lift c^* defined in a neighborhood of any $t \in [0,1]$, with $c^*(t) = u$ for any given $u \in \pi^{-1}(c(t))$; moreover, this lift is unique.

We now have to show that the lift can be defined on all of $[0,1]$; clearly we just have to show that a lift on $[0,t_0)$ can be extended past t_0. To do this, pick a lift $\bar{c}^*$ defined in a neighborhood of t_0 (with any old initial condition $\bar{c}^*(t_0)$). Choose $t_1 < t_0$ so that $\bar{c}^*$ is defined at t_1, and then choose $a \in G$ with $c^*(t_1) = \bar{c}^*(t_1)\cdot a$. Clearly we can extend c^* past t_0 by letting it be $R_a \circ \bar{c}^*$. ∎

Notice that the last part of this argument is just that used in the proof of Proposition I.5-17. The first part of the argument is much simpler when we are looking for a lift in a neighborhood of a point t with $c'(t) \neq 0$. For then (a portion of) c is the integral curve of a vector field X on M, and c^* is just an integral curve of the lift X^*.

Using Proposition 7, we can now define <u>parallel translation</u> of fibres of the principal bundle $\pi: P \to M$ along any curve $c: [0,1] \to M$. For any $u \in \pi^{-1}(0)$ we let $\tau_t(u) \in \pi^{-1}(c(t))$ be $c^*(t)$, where c^* is the lift of c with $c^*(0) = u$. In this way we obtain a map

$$\tau_t: \pi^{-1}(c(0)) \to \pi^{-1}(c(t)) \quad .$$

It is clear that $\tau_t \circ R_a = R_a \circ \tau_t$, since $R_a \circ c^*$ is again a lift. The map τ_t is a diffeomorphism whose inverse is just the parallel translation along the reversed portion of c from t to 0.

Consider, in particular, the principal bundle of frames $\pi: F(M) \to M$. Every frame $u \in F(M)$ determines an isomorphism from $\mathbb{R}^n$ to $M_{\pi(u)}$; namely, we send $e_i \in \mathbb{R}^n$ to $u_i \in M_{\pi(u)}$. This isomorphism will be denoted by the same letter, $u: \mathbb{R}^n \to M_{\pi(u)}$. It is easy to check that for $\xi \in \mathbb{R}^n$ we have

$$(u \cdot a)(\xi) = u(a \cdot \xi) \quad ,$$

where the product $a \cdot \xi$ of an $n \times n$ matrix a and a vector $\xi \in \mathbb{R}^n$ is defined in the footnote on page 7-5. Now suppose $c: [0,1] \to M$ is a piecewise C^1 curve, $X_p \in M_p$ is a tangent vector at $p = c(0)$, and $c^*: [0,1] \to F(M)$ is a lift of c with $c^*(0) = u$. There is a unique $\xi \in \mathbb{R}^n$ with $c^*(0)(\xi) = u(\xi) = X_p$; we let

$$\tau_t(X_p) = c^*(t)(\xi) \quad ,$$

thus defining parallel translation of vectors. To check that this parallel translation is well-defined, we consider any other lift, which must be of the form $\bar{c}^* = R_a \circ c^*$. Then

$$X_p = c^*(0)(\xi) = c^*(0) \cdot a(a^{-1} \cdot \xi)$$

$$= \bar{c}^*(0)(a^{-1} \cdot \xi) \quad ,$$

so the definition of τ_t with respect to $\bar{c}^*$ gives

$$\tau_t(X_p) = \bar{c}^*(t)(a^{-1} \cdot \xi) = \bar{c}^*(t) \cdot a(a^{-1} \cdot \xi) = c^*(t)(\xi)$$

$$= \text{old } \tau_t(X_p).$$

If we choose $c^*(0) = u$ so that $u((1,0,\ldots,0)) = X_p$, we see that we can parallel translate X_p by making it the first vector in a basis, parallel translating the basis, and then taking the first vector in the translated basis. It is also easy to see that τ_t is a vector space isomorphism.

Having defined parallel translation of vectors, we can now define covariant differentiation of vector fields by the formula on page 6-11,

$$(*) \qquad \nabla_{X_p} Y = \lim_{h \to 0} \frac{1}{h}(\tau_h{}^{-1} Y_{c(h)} - Y_p) \quad ,$$

where $c(0) = p$ and $c'(0) = Y_p$. It is not yet clear that this covariant differentiation is the same as the one we obtained on page 7-37 from the corresponding Cartan connection $\{s^*\omega\}$; to prove this we will need a Lemma that is also used frequently later on. Given the vector field Y, we consider the function $f_Y \colon F(M) \to \mathbb{R}^n$ whose value at a frame v is just the set of components of $Y_{\pi(v)}$ with respect to v — in symbols,

$$f_Y(v) = v^{-1}(Y_{\pi(v)}) \quad .$$

8. LEMMA. Let Y be a vector field on M, and let $X_p \in M_p$. Then for any $u \in F(M)$ with $p = \pi(u)$ we have

$$\nabla_{X_p} Y = u(X^*_u(f_Y)) \quad ,$$

where $X^*_u \in P_u$ is the unique horizontal vector with $\pi_*(X^*_u) = X_p$ [notice that since $f\colon P \longrightarrow R^n$, the value $X^*_u(f_Y)$ of the vector X^*_u on f_Y is an element of R^n, so $u(X^*_u(f_Y)) \in M_p$ makes sense].

Proof. Let c be a curve with $c(0) = p$ and $c'(0) = X_p$, and let c^* be the lift of c with $c^*(0) = u$, so that $c^{*\prime}(0) = X^*_u$. Recall the definition of the parallel translation $\tau_h^{-1}(Y_{c(h)})$ of $Y_{c(h)}$ along the reversed part of c from 0 to h: we choose $\xi \in R^n$ with

$$c^*(h)(\xi) = Y_{c(h)} \qquad \text{or} \qquad c^*(h)^{-1}(Y_{c(h)}) = \xi$$

and then define

$$\tau_h^{-1}(Y_{c(h)}) = c^*(0)(\xi) = u(\xi) \quad .$$

Consequently,

$$u \circ c^*(h)^{-1}(Y_{c(h)}) = \tau_h^{-1}(Y_{c(h)}) \quad .$$

So we have

$$\nabla_{X_p} Y = \lim_{h\to 0} \frac{1}{h} [\tau_h^{-1}(Y_{c(h)}) - Y_p]$$

$$= \lim_{h\to 0} \frac{1}{h} [u\circ c^*(h)^{-1}(Y_{c(h)}) - u\circ u^{-1}(Y_p)]$$

$$= u(\lim_{h\to 0} \frac{1}{h}[c^*(h)^{-1}(Y_{c(h)}) - Y_p])$$

$$= u(\lim_{h\to 0} \frac{1}{h}[f_Y(c^*(h)) - f_Y(u)])$$

$$= u(X^*_u(f)). \blacksquare$$

9. PROPOSITION. The ∇ defined by (*) is a Koszul connection; that is,

(1) $\nabla_{X_p+X'_p} Y = \nabla_{X_p} Y + \nabla_{X'_p} Y$

(2) $\nabla_{X_p}(Y_1 + Y_2) = \nabla_{X_p} Y_1 + \nabla_{X_p} Y_2$

(3) $\nabla_{aX_p} Y = a\nabla_{X_p} Y$ for all $a \in R$

(4) $\nabla_{X_p}(fY) = f(p)\cdot\nabla_{X_p} Y + X_p(f)\cdot Y_p$

(5) if X and Y are C^∞ vector fields, then so is $p \mapsto \nabla_{X_p} Y$.

Proof. Equations (2), (3), and (4) are easy to prove from the definition. Condition (5) follows from Lemma 8, since we can choose a C^∞ section $p \mapsto u(p)$ in a neighborhood of p. Equation (1) also follows from the Lemma, since we have $(X_p + X'_p)^* = X^*_u + X'^*_u$. $\blacksquare$

Now let us compare this Koszul connection ∇ with the one obtained from the Cartan connection corresponding to an Ehresmann connection ω on $F(M)$.

We can write $\omega = (\omega^i_j)$, where each ω^i_j in the matrix is an ordinary 1-form on $F(M)$. Equivalently, we can write

$$\omega = \sum_{i,j} \omega^i_j \cdot E^j_i \quad ,$$

where E^j_i is the matrix with zeros everywhere except for a 1 in the ith row and j^{th} column, so that

$$(E^j_i)^\alpha_\beta = \delta^\alpha_i \delta^j_\beta \quad .$$

Let (x,U) be a coordinate system, and let $s\colon U \longrightarrow F(M)$ be the "natural section"

$$s(q) = \left(\left.\frac{\partial}{\partial x^1}\right|_q , \ldots , \left.\frac{\partial}{\partial x^n}\right|_q \right) \quad .$$

Then the Cartan connection corresponding to ω assigns $(s^* \omega^i_j)$ to this moving frame. So the operation $\bar{\nabla}$ defined on page 7-37 is determined on U by

$$\bar{\nabla}_{\frac{\partial}{\partial x^i}} \frac{\partial}{\partial x^j} = \sum_k s^* \omega^k_j \left(\frac{\partial}{\partial x^i} \right) \frac{\partial}{\partial x^k} = \sum_k \omega^k_j \left(s_* \frac{\partial}{\partial x^i} \right) \frac{\partial}{\partial x^k} \quad .$$

10. PROPOSITION. These two connections are the same, $\bar{\nabla} = \nabla$.

Proof. It obviously suffices to show that $\bar{\nabla}_{\partial/\partial x^i} \partial/\partial x^j = \nabla_{\partial/\partial x^i} \partial/\partial x^j$, since $\bar{\nabla}$ amd ∇ are both Koszul connections. Let $f = f_{\partial/\partial x^j}$. By Proposition 8, it suffices to show that

$$\sum_k \omega_j^k\left(s_*\left.\frac{\partial}{\partial x^i}\right|_p\right)\left.\frac{\partial}{\partial x^k}\right|_p = s(p)\left(\left(\frac{\partial}{\partial x^i}^*\right)_{s(p)}(f)\right)$$

$$= \sum_k\left[\left(\frac{\partial}{\partial x^i}^*\right)_{s(p)}(f^k)\right]\cdot\left.\frac{\partial}{\partial x^k}\right|_p \quad .$$

So we need to show that

$$\omega_j^k\left(s_*\left.\frac{\partial}{\partial x^i}\right|_p\right) = \left(\frac{\partial}{\partial x^i}^*\right)_{s(p)}(f^k) \quad .$$

Now, by definition of f we have

$$(*) \qquad \sum_k f^k(u)\cdot u_k = \left.\frac{\partial}{\partial x^j}\right|_{\pi(u)} \qquad \text{for all } u = (u_1,\dots,u_n)\in F(M) \quad .$$

In particular,

$$f^k(s(q)) = \delta_j^k \quad .$$

Writing equation (*) for $u = s(p)\cdot\exp tM$, and differentiating with respect to t, we obtain

$$\sum_k \left.\frac{d}{dt}\right|_{t=0} f^k(s(p)\cdot\exp tM)\cdot\left.\frac{\partial}{\partial x^k}\right|_p$$

$$+ \sum_k f^k(s(p))\cdot\left.\frac{d}{dt}\right|_{t=0}(s(p)\cdot\exp tM) = 0 ,$$

and hence

$$\sum_k \sigma(M)_{s(p)}(f^k)\cdot\left.\frac{\partial}{\partial x^k}\right|_p = -\sum_k \delta_j^k(s(p)\cdot M)_k = -(s(p)\cdot M)_j \quad .$$

In particular,

$$\sum_k \sigma(E^{\nu}_{\mu})_{s(p)}(f^k)\cdot\frac{\partial}{\partial x^k}\bigg|_p = -(s(p)\cdot E^{\nu}_{\mu})_j = -\sum_\beta \frac{\partial}{\partial x^\beta}\bigg|_p (E^{\nu}_{\mu})^{\beta}_j$$

$$= -\sum_\beta \frac{\partial}{\partial x^\beta}\bigg|_p \delta^{\beta}_{\mu}\delta^{\nu}_{j} = -\delta^{\nu}_{j}\frac{\partial}{\partial x^\mu}\bigg|_p \quad .$$

So

$$\sigma(E^{\nu}_{\mu})_{s(p)}(f^k) = \begin{cases} -\delta^{\nu}_{j} & k = \mu \\ 0 & k \neq \mu \end{cases} \quad .$$

Now the lift $(\partial/\partial x^i)^*$ of $\partial/\partial x^i$ at points $s(q)$ is given by

$$\frac{\partial}{\partial x^i}^* = h\left(s_*\left(\frac{\partial}{\partial x^i}\right)\right) = s_*\left(\frac{\partial}{\partial x^i}\right) - \sum_{\mu,\nu} \omega^{\mu}_{\nu}\left(s_*\frac{\partial}{\partial x^i}\right)\sigma(E^{\nu}_{\mu}) \quad .$$

So

$$\left(\frac{\partial}{\partial x^i}^*\right)_{s(p)}(f^k) = \frac{\partial(f^k\circ s)}{\partial x^i}(p) - \sum_{\mu,\nu}\omega^{\mu}_{\nu}\left(s_*\frac{\partial}{\partial x^i}\bigg|_p\right)\sigma(E^{\nu}_{\mu})_{s(p)}(f^k)$$

$$= \frac{\partial(\delta^k_j)}{\partial x^i} - \sum_\nu \omega^{k}_{\nu}\left(s_*\frac{\partial}{\partial x^i}\bigg|_p\right)(-\delta^{\nu}_{j})$$

$$= \omega^k_j\left(s_*\frac{\partial}{\partial x^i}\bigg|_p\right) \quad . ∎$$

There is no need to repeat here the definition of $\nabla_X A$ for arbitrary tensor fields A; this can be defined in either of the two ways used in Chapter 6, and we will use any results from that Chapter which we require.

We return for a moment to a connection ω on a general principal bundle $\pi: P \longrightarrow M$ with group G. Consider a k-form α on P, with values in a vector space V. We define a V-valued $(k+1)$-form $D\alpha$, the <u>covariant differential</u> of α, by

$$D\alpha(Y_1,\ldots,Y_{k+1}) = (d\alpha)(hY_1,\ldots,hY_{k+1}) \quad ,$$

where d is the ordinary differential and hY is the horizontal component of Y. In particular, we define the <u>curvature form</u> Ω of ω by $\Omega = D\omega$. Thus Ω is a $\mathfrak{g}$-valued 2-form on P.

11. PROPOSITION. For all $a \in G$, we have $R_a^{\ *}\Omega = \mathrm{Ad}(a^{-1})\Omega$. In other words, for all tangent vectors $Y_1, Y_2 \in P_u$ we have

$$R_a^{\ *}\Omega(Y_1,Y_2) = \mathrm{Ad}(a^{-1})\Omega(Y_1,Y_2)$$

[this makes sense since $\Omega(Y_1,Y_2) \in \mathfrak{g}$].

Proof. We have

$$\begin{aligned}
R_a^{\ *}\Omega(Y_1,Y_2) &= R_a^{\ *}(D\omega)(hY_1,hY_2)\\
&= d(R_a^{\ *}\omega)(hY_1,hY_2)\\
&= d(\mathrm{Ad}(a^{-1})\omega)(hY_1,hY_2)\\
&= \mathrm{Ad}(a^{-1})[D\omega(hY_1,hY_2)] \qquad \text{since } \mathrm{Ad}(a^{-1})\colon \mathfrak{g} \to \mathfrak{g} \text{ is linear}\\
&= \mathrm{Ad}(a^{-1})\Omega(Y_1,Y_2) \qquad . \blacksquare
\end{aligned}$$

We will eventually see that this Ω does indeed correspond to the Ω in Chapter 7, but we first introduce the analogue of $\boldsymbol{\Theta}$. This analogue cannot be defined for connections in all principal bundles, but only for connections ω in the bundle of frames $\pi\colon F(M) \to M$. On this bundle we have a certain $\mathbb{R}^n$-valued 1-form θ, defined by

$$\theta_u(Y_u) = u^{-1}(\pi_* Y_u) \quad .$$

We will call this 1-form the canonical form or the dual form of the principal bundle $F(M)$. To see the appropriateness of the latter term, consider a section $s\colon U \to F(M)$ given by $s = (X_1,\ldots,X_n)$. For any tangent vector $Y_p \in M_p$ we have

$$s^*\theta(Y_p) = \theta_{s(p)}(s_* Y_p) = s(p)^{-1}(Y_p) \quad ,$$

so for the ith component θ^i of θ we have

$$\begin{aligned} s^*\theta^i(Y_p) &= i^{\text{th}} \text{ component of } s(p)^{-1}(Y_p) \\ &= i^{\text{th}} \text{ component of } Y_p \text{ with respect to the basis } X_1(p),\ldots,X_n(p). \end{aligned}$$

Consequently, the $s^*\theta^i$ are just the dual forms for the moving frame $(X_1,\ldots,X_n)$.

We now define the torsion form $\boldsymbol{\Theta}$ of a connection ω on $F(M)$ by $\boldsymbol{\Theta} = D\theta$ (this depends on ω, since h, and hence D, depends on ω).

12. PROPOSITION. For all $A \in GL(n,\mathbb{R})$ we have

$$R_A{}^*\theta = A^{-1}\cdot\theta \;, \qquad R_A{}^*\boldsymbol{\Theta} = A^{-1}\cdot\boldsymbol{\Theta} \quad .$$

In other words, for all tangent vectors $Y_1, Y_2 \in P_u$ we have

$$R_A^*\theta(Y_1) = A^{-1}\cdot\theta(Y_1)$$

$$R_A^*\Theta(Y_1,Y_2) = A^{-1}\cdot\Theta(Y_1,Y_2)$$

[these equations make sense since $\theta(Y_1)$ and $\Theta(Y_1,Y_2) \in \mathbb{R}^n$].

Proof. For θ we have

$$\begin{aligned} R_A^*\theta(Y_1) &= \theta_{u\cdot A}(R_{A*}Y_1) = (u\cdot A)^{-1}(\pi_*Y_1) \\ &= A^{-1}\cdot u^{-1}(\pi_*Y_1) \\ &= A^{-1}\cdot\theta(Y_1) \quad . \end{aligned}$$

Then for Θ we have

$$\begin{aligned} R_A^*\Theta(Y_1,Y_2) &= R_A^*(d\theta)(hY_1,hY_2) \\ &= d(R_A^*\theta)(hY_1,hY_2) \\ &= d(A^{-1}\cdot\theta)(hY_1,hY_2) \\ &= A^{-1}\cdot d\theta(hY_1,hY_2) \\ &= A^{-1}\cdot\Theta(Y_1,Y_2) \quad . \blacksquare \end{aligned}$$

A connection ω in the bundle of frames $F(M)$ also allows us to define certain special vector fields in $F(M)$. For $\xi \in \mathbb{R}^n$ we define the basic vector field $B(\xi)$ corresponding to ξ by letting $B(\xi)_u$ be the unique horizontal vector at u such that $\pi_*(B(\xi)_u) = u(\xi)$. In particular, $B(e_i)_u$ is the unique horizontal vector at u which covers u_i.

13. PROPOSITION. For all $\xi \in \mathbb{R}^n$ we have

(1) $\theta(B(\xi)) = \xi$

(2) $R_{A*}B(\xi) = B(A^{-1}\cdot\xi)$ for all $A \in GL(n,\mathbb{R})$.

Moreover, if $\xi \neq 0$, then $B(\xi)$ is nowhere 0. Consequently, if $\xi_1,\ldots,\xi_n$ is a basis for $\mathbb{R}^n$, then $B(\xi_1)_u,\ldots,B(\xi_n)_u$ is a basis for H_u.

Proof. (1) and (2) are left to the reader. For the third assertion, note that if $B(\xi)_u = 0$, then

$$0 = \pi_*(B(\xi)_u) = u(\xi) \quad ,$$

so $\xi = 0$. The final assertion follows immediately. ∎

To prove the structural equations in our new setup, we need two lemmas; the first involves fundamental vector fields and basic vector fields, but the second, which holds for connections in any bundle, involves fundamental vector fields and arbitrary horizontal vector fields.

14. LEMMA. Consider the basic vector fields determined by a connection on the bundle of frames $F(M)$. For every $M \in \mathfrak{gl}(n,\mathbb{R})$ and $\xi \in \mathbb{R}^n$ we have

$$[\sigma(M),B(\xi)] = B(M\cdot\xi) \quad .$$

Proof. Since $\varphi_t(u) = u\cdot\exp tM = R_{\exp tM}(u)$ is the 1-parameter group of diffeomorphisms generated by $\sigma(M)$, we have

$$\begin{aligned}[\sigma(M),B(\xi)] &= \lim_{h\to 0} \frac{1}{h}\,[B(\xi) - R_{\exp tM*}B(\xi)] \\ &= \lim_{h\to 0} \frac{1}{h}\,[B(\xi) - B([\exp -tM]\cdot\xi)] && \text{by Proposition 13} \\ &= B(\lim_{h\to 0} \frac{1}{h}(\xi - (\exp -tM)\cdot\xi)) && \text{since } \xi \mapsto B(\xi) \text{ is linear onto } H_u \\ &= B(M\cdot\xi)\ . \blacksquare\end{aligned}$$

<u>15. LEMMA</u>. Consider a connection on any principal bundle P over M with group G. For any $X \in \mathfrak{g}$ and any horizontal vector field Y on P, the vector field $[\sigma(X),Y]$ is also horizontal.

<u>Proof</u>. We have

$$[\sigma(X),Y] = \lim_{h\to 0} \frac{1}{h}\,[Y - R_{\exp tX*}(Y)]\ .$$

Since Y is horizontal, so is each $R_{\exp tX*}(Y)$. $\blacksquare$

<u>16. THEOREM</u>. Let ω be a connection on a principal bundle P over M. If P is the bundle of frames, with the dual form θ, and the torsion form Θ determined by ω, then we have the <u>first structural equation</u>:

$$d\theta(Y_1,Y_2) = -\{\omega(Y_1)\cdot\theta(Y_2) - \omega(Y_2)\cdot\theta(Y_1)\} + \Theta(Y_1,Y_2) \qquad \text{for all } Y_1,Y_2 \in P_u$$

[where $\omega(Y_1)\cdot\theta(Y_2)$ is the action of the matrix $\omega(Y_1)$ on $\theta(Y_2) \in \mathbb{R}^n$].

If P is any principal bundle, and Ω is the curvature form of ω, then we have the <u>second structural equation</u>:

$$d\omega(Y_1,Y_2) = -[\omega(Y_1),\omega(Y_2)] + \Omega(Y_1,Y_2) \qquad \text{for all } Y_1,Y_2 \in P_u\ .$$

Proof. Since each Y_i is the sum of a vertical and a horizontal vector, and since both sides of the first structural equation are skew-symmetric and bilinear, we can prove this equation by considering 3 cases.

Case 1. Y_1 and Y_2 are horizontal. Then $\omega(Y_i) = 0$, so the equation reduces to the definition of Θ as $D\theta$.

Case 2. Y_1 and Y_2 are vertical. Then the right side is 0. If we extend Y_1 and Y_2 to vertical vector fields $\tilde{Y}_1, \tilde{Y}_2$, then the left side is the value at u of

$$Y_1(\theta(Y_2)) - Y_2(\theta(Y_1)) - \theta([Y_1,Y_2]) \ ,$$

which is 0, since $[Y_1,Y_2]$ is also vertical.

Case 3. Y_1 is vertical and Y_2 is horizontal. Let $Y_1 = \sigma(M)_u$ for $M \in \mathfrak{gl}(n,\mathbb{R})$ and let $Y_2 = B(\xi)_u$ for $\xi \in \mathbb{R}^n$. Then

$$- \{\omega(Y_1)\cdot\theta(Y_2) - \omega(Y_2)\cdot\theta(Y_1)\} + \Theta(Y_1,Y_2) = - M\cdot\xi + 0 + 0 \ ,$$

while

$$\begin{aligned} d\theta(Y_1,Y_2) &= \sigma(M)(\theta(B(\xi)))(u) - B(\xi)(\theta(\sigma(M)))(u) - \theta([\sigma(M),B(\xi)](u) \\ &= 0 - 0 - \theta(B(M\cdot\xi))(u) \qquad\qquad \text{by Lemma 14} \\ &= -M\cdot\xi \ . \end{aligned}$$

This proves the first structural equation.

The second structural equation will be proved similarly.

Case 1. Y_1 and Y_2 are horizontal. The proof is as before.

Case 2. Y_1 and Y_2 are vertical. Let $Y_i = \sigma(X_i)_u$ for $X_i \in \mathfrak{g}$. Then

$\Omega(Y_1,Y_2) = 0$, while

$$d\omega(Y_1,Y_2) = \sigma(X_1)(\omega(\sigma(X_2)))(u) - \sigma(X_2)(\omega(\sigma(X_1)))(u) - \omega([\sigma(X_1),\sigma(X_2)])(u)$$

$$= 0 - 0 - \omega(\sigma([X_1,X_2]))(u) \qquad \text{by Proposition 2}$$

$$= - [X_1,X_2] = - [\omega(Y_1),\omega(Y_2)] \quad .$$

<u>Case 3</u>. Y_1 <u>is vertical and</u> Y_2 <u>is horizontal</u>. Then the right side is 0. If we extend Y_2 to a horizontal vector field $\tilde{Y}_2$ and let $Y_1 = \sigma(X)_u$ for $X \in \mathfrak{g}$, then the left side is the value at u of

$$\sigma(X)(\omega(\tilde{Y}_2)) - \tilde{Y}_2(\omega(\sigma(X))) - \omega([\sigma(X),\tilde{Y}_2])$$

$$= 0 \quad , \qquad \text{by Lemma 15.} \blacksquare$$

The structural equations for a connection ω on the bundle of frames $F(M)$ can also be written in terms of ordinary forms. With respect to the standard basis $e_1,\ldots,e_n$ of $\mathbb{R}^n$ we can write the $\mathbb{R}^n$-valued forms θ and Θ as

$$\theta = \sum_i \theta^i \cdot e_i \qquad\qquad \Theta = \sum_i \Theta^i \cdot e_i \quad ,$$

for certain ordinary forms θ^i and Θ^i. Similarly, with respect to the basis E_i^j of $\mathfrak{gl}(n,\mathbb{R})$ introduced previously, we can write

$$\omega = \sum_{i,j} \omega_j^i \cdot E_i^j \qquad\qquad \Omega = \sum_{i,j} \Omega_j^i \cdot E_i^j \quad .$$

It is easy to see that the structural equations can then be written

$$d\theta^i = -\sum_j \omega^i_j \wedge \theta^i + \Theta^i$$

$$d\omega^i_j = -\sum_k \omega^i_k \wedge \omega^k_j + \Omega^i_j \quad .$$

Instead of introducing these forms explicitly, it will be conveient to write the structural equations in the abbreviated form

$$d\theta = -\omega \wedge \theta + \Theta$$

$$d\omega = -\omega \wedge \omega + \Omega \quad .$$

For a section $s\colon U \longrightarrow F(M)$ we obtain

$$d(s^*\theta^i) = -\sum_j s^*\omega^i_j \wedge s^*\theta^i + s^*\Theta^i$$

$$d(s^*\omega^i_j) = -\sum_k s^*\omega^i_k \wedge s^*\omega^k_j + s^*\Omega^i_j \quad .$$

Since $s^*\theta^i$ are just the dual forms to the moving frame s, this shows that $s^*\Theta^i$ and $s^*\Omega^i_j$ are precisely the torsion and curvature forms* for the moving frame s which are determined by the Cartan connection which assigns $s^*\omega$ to s. Propositions 11 and 12 correspond to Proposition 15 of Chapter 7. Recall that the formulas of this Proposition allowed us to define the tensors T and R. Essentially equivalent, but much neater, definitions will now be given directly from Propositions 11 and 12.

Let ω be a connection on $F(M)$ with torsion form Θ and curvature from Ω. For $X_1, X_2 \in M_p$, we let

$$T(X_1,X_2) = u(\Theta(\bar{X}_1,\bar{X}_2)) \in M_p$$

where $\bar{X}_i \in F(M)_u$ are any vectors with $\pi_*(\bar{X}_i) = X_i$. This definition is

*In Ehresmann's original treatment, the torsion and curvature forms Θ and Ω on $F(M)$ were defined by the structural equations. The definition in terms of D was given by Ambrose and Singer, A Theorem on Holonomy, Trans. Amer. Math. Soc. 75 (1953), 428-443. This paper also introduced much of the convenient terminology, like "horizontal vectors."

independent of the choice of u, and of $\bar{X}_i \in F(M)_u$. To see this, first note that if $\bar{X}_1$ is replaced by $\bar{\bar{X}}_1$ with $\pi_*(\bar{\bar{X}}_1) = X_1$, then $\bar{X}_1 - \bar{\bar{X}}_1$ is vertical, so

$$\Theta(\bar{X}_1 - \bar{\bar{X}}_1, \bar{X}_2) = D\theta(\bar{X}_1 - \bar{\bar{X}}_1, \bar{X}_2) = 0;$$

similarly, $\bar{X}_2$ may be replaced by any $\bar{\bar{X}}_2$ with $\pi_*(\bar{\bar{X}}_2) = X_2$, without changing the value of $u(\Theta(\bar{X}_1, \bar{X}_2))$. If we change u to $u\cdot A$, then we can pick $R_{A*}\bar{X}_i$ for the new X_i, and we have

$$u\cdot A(\Theta(R_{A*}\bar{X}_1, R_{A*}\bar{X}_2)) = (u\cdot A)(A^{-1}\cdot\Theta(\bar{X}_1, \bar{X}_2)) \qquad \text{by Proposition 12}$$

$$= u(\Theta(\bar{X}_1, \bar{X}_2)) \quad .$$

Since Θ is a form, it is clear that T is skew-symmetric.

Similarly, for $X_1, X_2, X_3 \in M_p$ we let

$$R(X_1, X_2)X_3 = (\Omega(\bar{X}_1, \bar{X}_2)\cdot(u^{-1}X_3));$$

here $\Omega(\bar{X}_1, \bar{X}_2) \in \mathfrak{gl}(n, \mathbb{R})$ is an $n \times n$ matrix, so it acts on $u^{-1}X_3 \in \mathbb{R}^n$. Just as before, we see that the definition does not depend on the choice of $\bar{X}_1$ or $\bar{X}_2$. If we change u to $u\cdot A$, then

$$u\cdot A(\Omega(R_{A*}\bar{X}_1, R_{A*}\bar{X}_2)\cdot([u\cdot A]^{-1}X_3)$$

$$= u\cdot A([A^{-1}\Omega(\bar{X}_1, \bar{X}_2)A]\cdot(A^{-1}\cdot u^{-1}(X_3)))$$

$$= u(\Omega(\bar{X}_1, \bar{X}_2)\cdot(u^{-1}X_3)) \quad .$$

Since Ω is a 2-form, it is clear that R is skew-symmetric in X_1 and X_2.

In Chapter 7, we mentioned that the structural equations could be used to prove that the torsion and curvature tensors T and R, defined for a Cartan connection in terms of Θ^i and Ω^i_j, were just those derived from the ∇ which could also be defined for the Cartan connection. Here we will actually carry out this proof for an Ehresmann connection.

<u>17. PROPOSITION</u>. For any vector fields X, Y, Z on M we have

$$T(X,Y) = \nabla_X Y - \nabla_Y X - [X,Y]$$
$$R(X,Y)Z = \nabla_X \nabla_Y Z - \nabla_Y \nabla_X Z - \nabla_{[X,Y]} Z \quad .$$

<u>Proof</u>. Recall that we defined $f_Y \colon F(M) \longrightarrow \mathbb{R}^n$ by $f_Y(u) = u^{-1}(Y_{\pi(u)})$. If Y^* is the lift of Y, we can write $f_Y(u) = \theta(Y^*_u)$. So Lemma 8 gives

$$\nabla_X Y(p) = u(X^*_u(\theta(Y^*))) \qquad \text{for } u \in \pi^{-1}(p) \quad .$$

Consequently,

$$\begin{aligned} T(X_p, Y_p) &= u(\Theta(X^*_u, Y^*_u)) = u(D\theta(X^*_u, Y^*_u)) \\ &= u(d\theta(X^*_u, Y^*_u)) \\ &= u[X^*_u(\theta(Y^*)) - Y^*_u(\theta(X^*)) - \theta([X^*,Y^*](u)] \\ &= \nabla_{X_p} Y - \nabla_{Y_p} X - [X,Y]_p \quad , \end{aligned}$$

since $\pi_*([X^*,Y^*]) = [X,Y]$.

To prove the second equality, we note that since X^* and Y^* are horizontal, the second structural equation gives

$$\begin{aligned}\Omega(X^*{}_u, Y^*{}_u) &= d\omega(X^*, Y^*)(u) \\ &= X^*(\omega(Y^*))(u) - Y^*(\omega(X^*))(u) - \omega([X^*, Y^*])(u) \\ &= -\omega([X^*, Y^*])(u).\end{aligned}$$

If we set the vertical component of $[X^*, Y^*]$ equal to

$$v[X^*, Y^*](u) = \sigma(M)_u \qquad \text{for } M \in \mathfrak{gl}(n, \mathbb{R}) ,$$

then we obtain

$$\Omega(X^*{}_u, Y^*{}_u) = -M ,$$

so

$$\text{(1)} \qquad \begin{aligned} R(X_p, Y_p)Z_p &= u(\Omega(X^*{}_u, Y^*{}_u)\cdot(u^{-1}Z_p)) \\ &= u(-M\cdot f_Z(u)).\end{aligned}$$

On the other hand, since we also have $f_Z(u) = \theta(Z^*{}_u)$, we obtain

$$\text{(2)} \qquad \begin{aligned} &\nabla_X\nabla_Y Z(p) - \nabla_Y\nabla_X Z(p) - \nabla_{[X,Y]}Z(p) \\ &\quad = u(X^*{}_u(Y^* f_Z) - Y^*{}_u(X^* f_Z) - (h[X^*, Y^*]_u) f_Z) \\ &\quad = u((v[X^*, Y^*]_u) f_Z) .\end{aligned}$$

The expressions in (1) and (2) are equal, since

$$\sigma(M)_u f_Z = \lim_{h \to 0} \frac{1}{h}[f_Z(u \cdot \exp tM) - f_Z(u)]$$

$$= \lim_{h \to 0} \frac{1}{h}[(\exp tM)^{-1} \cdot f_Z(u) - f_Z(u)]$$

$$= - M \cdot f_Z(u) \quad . \blacksquare$$

One of the steps in this proof is of sufficient importance to be stated explicitly, along with a counterpart, in the following corollary of the structural equations.

18. PROPOSITION. If ω is a connection on the bundle of frames $F(M)$, and B_1, B_2 are basic vector fields, then the horizontal component of $[B_1, B_2](u)$ is the value of $B(-\Theta(B_{1u}, B_{2u}))$ at u.

If ω is a connection on any principal bundle P over M, and Y_1, Y_2 are horizontal vector fields, then the vertical component of $[Y_1, Y_2](u)$ is the value of $\sigma(-\Omega(Y_{1u}, Y_{2u}))$ at u.

Proof. Since B_1 and B_2 are horizontal, the first structural equation gives

$$\Theta(B_{1u}, B_{2u}) = d\theta(B_1, B_2)(u)$$

$$= B_1(\theta(B_2))(u) - B_2(\theta(B_1))(u) - \theta([B_1, B_2])(u)$$

$$= - \theta(h[B_1, B_2](u)) \qquad \text{since } \theta = 0 \text{ on vertical vectors.}$$

Since we can always write $h[B_1, B_2](u)$ as $B(\xi)_u$ for some $\xi \in \mathbb{R}^n$, the first result follows from Proposition 13(1).

Since Y_1 and Y_2 are horizontal, the second structural equation gives

$$\Omega(Y_{1u}, Y_{2u}) = d\omega(Y_1, Y_2)(u) = - \omega([Y_1, Y_2](u)) \quad ,$$

which gives the second result. ∎

Suddenly, we are ready for

<u>19. THEOREM (THE TEST CASE ; seventh version)</u>. Let $(M, \langle\ ,\ \rangle)$ be an n-dimensional Riemannian manifold for which the curvature tensor R (for the Levi-Civita connection) is 0. Then M is locally isometric to $\mathbb{R}^n$ with its usual Riemannian metric.

<u>Proof</u>. On the bundle of frames $F(M)$ we have a connection ω with $\Theta = 0$ and $\Omega = 0$, for which parallel translation preserves the inner product $\langle\ ,\ \rangle$.

<u>Step 1</u>. The second part of Proposition 18 shows that the bracket of two horizontal vector fields is again horizontal. Thus, <u>the distribution</u> H <u>is integrable</u>. At a point $p \in M$ choose an <u>orthonormal</u> frame $u \in \pi^{-1}(p)$, and let N be the (n-dimensional) integral manifold of H through u. Clearly, N is locally the image of a section $s: U \longrightarrow F(M)$ with $s(p) = u$.

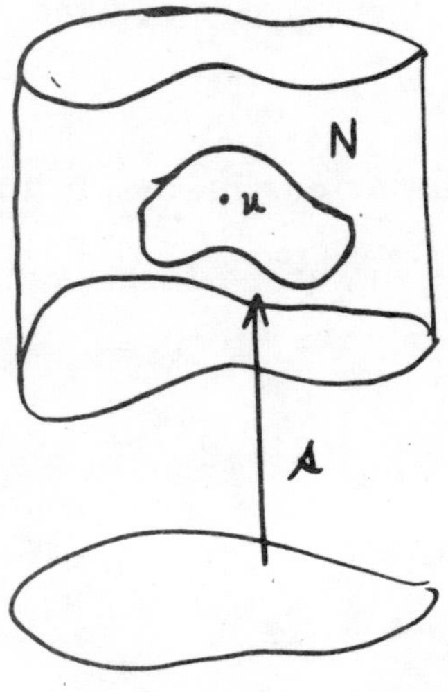

<u>Step 2</u>. Consider the basic vector fields $B(e_i)$, for $e_1, \ldots, e_n$ the standard basis of $\mathbb{R}^n$. Since they are horizontal, the bracket $[B(e_i), B(e_j)]$ is horizontal. But the first part of Proposition 18 shows that

the horizontal component is 0, so $[B(e_i),B(e_j)] = 0$. If we let $X_i(q) = \pi_*(B(e_i)(s(q)))$, then we also have $[X_i,X_j] = 0$, since $B(e_i)$ and X_i are π-related. But, by definition of $B(\xi)$, we have

$$\begin{aligned}\pi_* B(e_i)(s(q)) &= s(q)(e_i) \\ &= i^{th} \text{ vector of the frame } s(q).\end{aligned}$$

Thus, $s = (X_1,\ldots,X_n)$ where $[X_i,X_j] = 0$; hence there is a coordinate system $x^1,\ldots,x^n$ with $X_i = \partial/\partial x^i$.

Step 3. We claim this is the desired coordinate system. We just have to show that $s(q)$ is always orthonormal, so it suffices to show that $s(q)$ is the parallel translate of $u = s(p)$ along any curve c in U from p to q. This is obvious: to translate u along c, we choose a lift c^* of c with $c^*(0) = u$, and then the translate of u is $c^*(1)$; but clearly $c^* = s \circ c$. ∎

To wrap things up, we present the new version of the Bianchi identities.

20. THEOREM. For a connection ω on the bundle of frames $F(M)$, with dual form θ, torsion form $\boldsymbol{\Theta}$, and curvature form Ω, we have

(1) (Bianchi's 1st identity) $D\boldsymbol{\Theta} = \Omega \wedge \theta$. In other words, for $X,Y,Z \in F(M)_u$ we have

$$D\Theta(X,Y,Z) = \frac{(2+1)!}{2!1!} \cdot \frac{1}{3!} [\Omega(X,Y)\cdot\theta(Z) - \Omega(Y,X)\cdot\theta(Z)$$
$$+ \Omega(Y,Z)\cdot\theta(X) - \Omega(Z,Y)\cdot\theta(X)$$
$$+ \Omega(Z,X)\cdot\theta(Y) - \Omega(X,Z)\cdot\theta(Y)]$$
$$= \Omega(X,Y)\cdot\theta(Z) + \Omega(Y,Z)\cdot\theta(X) + \Omega(Z,X)\cdot\theta(Y) \quad .$$

For a connection ω on any principal bundle, with curvature from Ω, we have

(2) (Bianchi's 2nd identity) $D\Omega = 0$.

Proof. Applying d to the first structural equation, $d\theta = -\omega \wedge \theta + \Theta$, we obtain

$$0 = -(d\omega \wedge \theta) + (\omega \wedge d\theta) + d\Theta \; .$$

So

$$\begin{aligned} D\Theta(X,Y,Z) &= d\Theta(hX,hY,hZ) \\ &= (d\omega \wedge \theta)(hX,hY,hZ) - 0 \\ &= (\Omega \wedge \theta)(X,Y,Z) \quad , \end{aligned}$$

since $d\omega(hA,hB) = \Omega(A,B)$ and $\theta(hA) = \theta(A)$ for all $A,B \in F(M)_u$.

Applying d to the second structural equation, $d\omega = -\omega \wedge \omega + \Omega$, we obtain

$$0 = -(d\omega \wedge \omega) + (\omega \wedge d\omega) + d\Omega \; .$$

So

$$\begin{aligned} D\Omega(X,Y,Z) &= d\Omega(hX,hY,hZ) \\ &= (d\omega \wedge \omega)(hX,hY,hZ) - (\omega \wedge d\omega)(hX,hY,hZ) \\ &= 0 \; , \end{aligned}$$

since $\omega(hA) = 0$ for all $A \in F(M)_u$. ∎

SUMMARY

The following diagram summarizes the relationship between the various definitions of a connection:

(1) n^3 functions Γ^k_{ij} assigned to each coordinate system (classical),

(2) ∇ operator on vector fields (Koszul),

(3) n x n matrix of 1-forms (ω^i_j) assigned to each moving frame (E. Cartan),

(4) n x n matrix-valued 1-form ω on $F(M)$ (Ehresmann),

(5) distribution H on $F(M)$.

The relationship between (1) and (2) is immediate, as is the relationship between (4) and (5). In Chapter 7 we saw how to pass between (1)-(2) and (3), and in this Chapter we saw how to pass between (4)-(5) and (3). We can also go directly from (5) to (1), since ∇ is defined by parallel translation. It remains to indicate how one passes directly from (1)-(2) to (5).

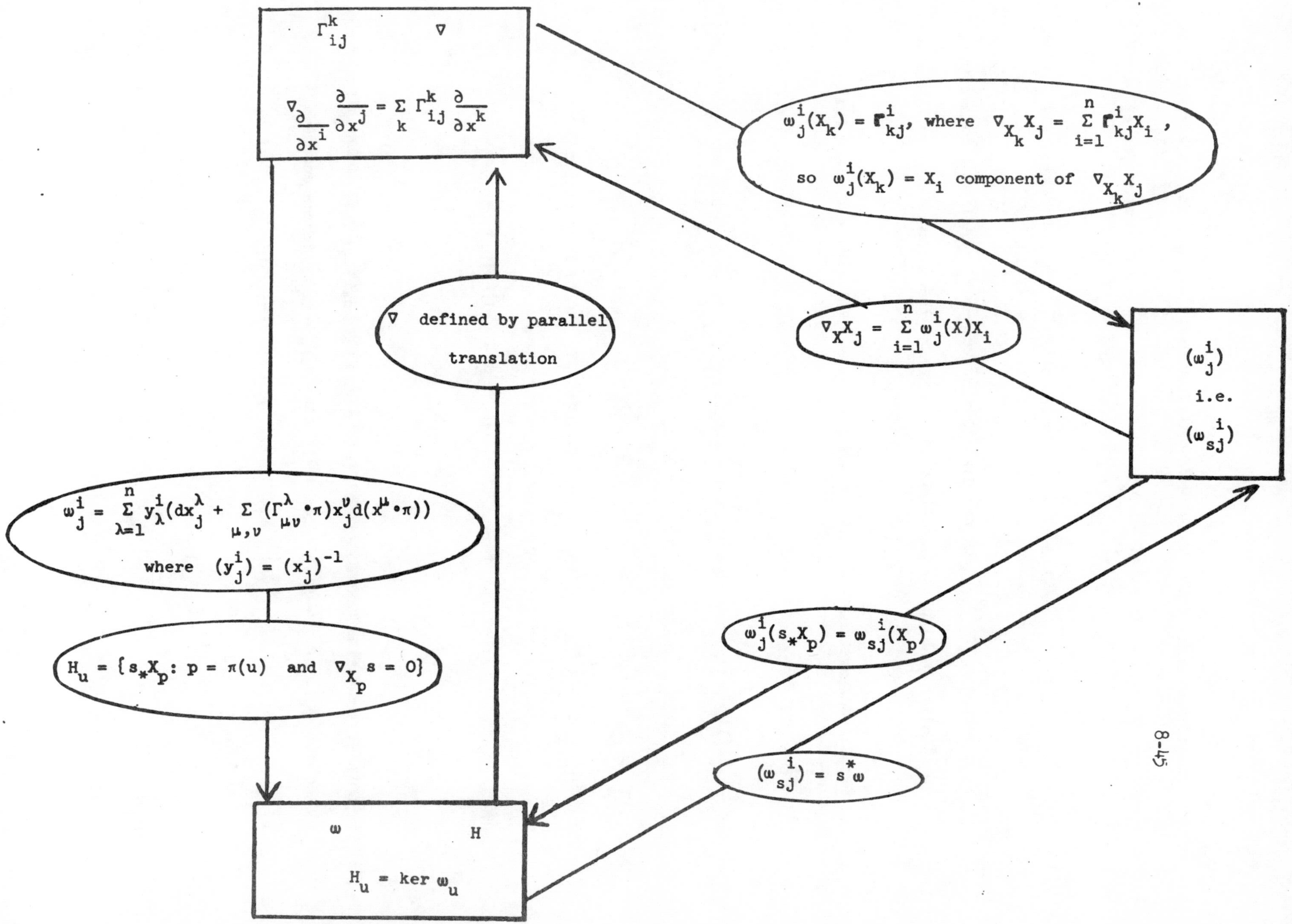
Γ^k_{ij} ∇
$\nabla_{\frac{\partial}{\partial x^i}} \frac{\partial}{\partial x^j} = \sum_k \Gamma^k_{ij} \frac{\partial}{\partial x^k}$
$\omega^i_j(X_k) = \Gamma^i_{kj}$, where $\nabla_{X_k} X_j = \sum_{i=1}^n \Gamma^i_{kj} X_i$,
so $\omega^i_j(X_k) = X_i$ component of $\nabla_{X_k} X_j$
$\nabla_X X_j = \sum_{i=1}^n \omega^i_j(X) X_i$
(ω^i_j)
i.e.
$(\omega_{s\,j}^{\,i})$
∇ defined by parallel translation
$\omega^i_j = \sum_{\lambda=1}^n y^i_\lambda (dx^\lambda_j + \sum_{\mu,\nu} (\Gamma^\lambda_{\mu\nu} \circ \pi) x^\nu_j d(x^\mu \circ \pi))$
where $(y^i_j) = (x^i_j)^{-1}$
$H_u = \{s_* X_p : p = \pi(u)$ and $\nabla_{X_p} s = 0\}$
$\omega^i_j(s_* X_p) = \omega_{s\,j}^{\,i}(X_p)$
$(\omega_{s\,j}^{\,i}) = s^* \omega$
ω H
$H_u = \ker \omega_u$

21. PROPOSITION. Let ∇ be a Koszul connection on M. For $u \in F(M)$, let

$$H_u = \{s_*X_p : p = \pi(u) \text{ and } \nabla_{X_p} s = 0\} ,$$

[where $\nabla_{X_p} s$ denotes $(\nabla_{X_p} X_1, \ldots, \nabla_{X_p} X_n)$ if $s = (X_1, \ldots, X_n)$]. Then H_u is a subspace of P_u, and the distribution $u \longmapsto H_u$ defines a connection on F(M) for which the covariant derivative is the given ∇.

Proof. Suppose that $s \cdot a$ is another section with $a(p) = I$, and that $\nabla_{X_p} s \cdot a = 0$. Then

$$(1) \qquad 0 = \nabla_{X_p} s \cdot a = s \cdot X_p(a) + (\nabla_{X_p} s) \cdot a$$
$$= s \cdot X_p(a) ,$$

which implies that $X_p(a)$ is the zero matrix. So by Proposition 3 we have

$$(s \cdot a)_*(X_p) = s_*X_p + \sigma(X_p(a))(s(p))$$
$$= s_*X_p \quad .$$

Thus H_u contains exactly one vector X^*_u with $\pi_*(X^*_u) = X_p$. Now given vectors $A_u, B_u \in H_u$, let $\pi_* A_u = X_p$ and $\pi_* B_u = Y_p$. We can choose s with $s(p) = u$ and $\nabla_{X_p} s = \nabla_{Y_p} s = 0$; then we also have $\nabla_{X_p + Y_p} s = 0$, so we must have

$$A_u + B_u = s_*(X_p + Y_p) = s_*X_p + s_*Y_p = A_u + B_u \quad .$$

Thus H_u is a subspace*. It has the same dimension as M, and contains no vertical vectors except 0, so we have $F(M)_u = V_u \oplus H_u$. It is clear that $R_{A*}H_u = H_{u\cdot A}$ for all $A \in GL(n,\mathbb{R})$, since $s\cdot A$ also satisfies $\nabla_{X_p} s\cdot A = 0$.

To prove that H is a C^∞ distribution, we consider a C^∞ vector field X on M, and a C^∞ section s. We claim that there is a function $a\colon U \longrightarrow GL(n,\mathbb{R})$, defined in some open set U, such that $\nabla_{X_p} s\cdot a = 0$ for all $p \in U$. This is because equation (1) shows that this condition is equivalent to

$$0 = s\cdot X_p(a) + (\nabla_{X_p} s)\cdot a \quad ,$$

and this is a differential equation for a. For each $A \in GL(n,\mathbb{R})$ and $p \in U$, let $a_{p,A}$ be the solution with $a_{p,A}(p) = s(p)\cdot A$. For every $u \in F(M)$ we have $u = s(\pi(u))\cdot A(u)$ for some $A(u) \in GL(n,\mathbb{R})$ which depends differentially on u. Then the vector field

$$X^*(u) = [s\cdot a_{\pi(u),A(u)}]_*(X_{\pi(u)}) \in F(M)_u$$

is C^∞, since $a_{p,A}$ is C^∞ in p and A.

Thus the distribution H is a connection on $F(M)$. For a curve c in M, consider a curve c^* in $F(M)$ with $\pi\circ c^* = c$ and all $c^{*\prime}(t)$

*We can also define H_u simply as $s_*(M_p)$, where s is a section with $s(p) = u$ and $(\nabla s)(p) = 0$.

horizontal. By definition of the connection, this means that for some $X_p \in M_p = M_{c(t)}$ we have

$$c^{*\prime}(t) = s_*(X_p) \quad \text{where} \quad \nabla_{X_p} s = 0 \ .$$

Of course, X_p must be $\pi_* s_* X_p = \pi_* c^{*\prime}(t) = c'(t)$. So $\nabla_{c'(t)} s = 0$. From this it is easy to see that the components of c^* are vector fields along c which are parallel along c, with respect to the original ∇. Hence parallel translation defined with respect to the connection H is the same as parallel translation for ∇. This implies that the covariant derivatives are also the same. ∎

Proposition 21 gives the most geometric way of going from ∇ to H, and hence to ω. We can also give a computational description of ω, in terms of a coordinate system (x,U) on M. Recall that $x_\# = (x^i \circ \pi, x^i_j)$ is a coordinate system on $\pi^{-1}(U)$, where we write a frame u as

$$u_j = \sum_{i=1}^{n} x^i_j(u) \cdot \left.\frac{\partial}{\partial x^i}\right|_{\pi(u)} .$$

At each u, the non-singular matrix $(x^i_j(u))$ has an inverse matrix $(y^i_j(u)) = (x^i_j(u))^{-1}$.

<u>22. PROPOSITION</u>. Let Γ^k_{ij} be the n^3 functions assigned to the coordinate system (x,U) by a classical connection on M. For the corresponding Ehresmann connection $\omega = (\omega^i_j) = \sum_{i,j} \omega^i_j \cdot E^j_i$ we have

$$\omega^i_j = \sum_{\lambda=1}^{n} y^i_\lambda \Big(dx^\lambda_j + \sum_{\mu,\nu} (\Gamma^\lambda_{\mu\nu} \circ \pi) x^\nu_j d(x^\mu \circ \pi)\Big) \qquad \text{on } \pi^{-1}(U) \ .$$

Proof. Let $s = (\partial/\partial x^1,\ldots,\partial/\partial x^n)$ be the natural section. We know that the matrix of 1-forms $(\omega_s{}^i_j)$ for the corresponding Cartan connection is given by

$$\Gamma^i_{\mu j} = (\omega_s{}^i_j)\left(\frac{\partial}{\partial x^\mu}\right) ,$$

and consequently the corresponding Ehresmann connection satisfies

$$(1) \qquad \Gamma^i_{\mu j}(p) = \omega^i_j\left(s_*\left(\left.\frac{\partial}{\partial x^\mu}\right|_p\right)\right) .$$

If we write

$$s_*\left(\left.\frac{\partial}{\partial x^\mu}\right|_p\right) = \Sigma\, a^i \left.\frac{\partial}{\partial(x^i\circ\pi)}\right|_{s(p)} + \Sigma\, b^i_j \left.\frac{\partial}{\partial x^i_j}\right|_{s(p)} ,$$

then

$$a^i = s_*\left(\left.\frac{\partial}{\partial x^\mu}\right|_p\right)(x^i\circ\pi) = \left.\frac{\partial(x^i\circ\pi\circ s)}{\partial x^\mu}\right|_p = \left.\frac{\partial x^i}{\partial x^\mu}\right|_p = \delta^i_\mu$$

$$b^i_j = s_*\left(\left.\frac{\partial}{\partial x^\mu}\right|_p\right)(x^i_j) = \left.\frac{\partial(x^i_j\circ s)}{\partial x^\mu}\right|_p = \left.\frac{\partial\delta^i_j}{\partial x^\mu}\right|_p = 0 ,$$

so

$$s_*\left(\left.\frac{\partial}{\partial x^\mu}\right|_p\right) = \left.\frac{\partial}{\partial(x^\mu\circ\pi)}\right|_{s(p)} .$$

A similar calculation shows that for any $A \in GL(n,\mathbb{R})$ we have

$$(2)\quad \left.\frac{\partial}{\partial(x^\mu\circ\pi)}\right|_{s(p)\cdot A} = (s\cdot A)_*\left(\left.\frac{\partial}{\partial x^\mu}\right|_p\right) = R_{A*}\left(\left.\frac{\partial}{\partial x^\mu}\right|_p\right) .$$

Now any $u \in \pi^{-1}(U)$ can be written $u = s(\pi(u))\cdot(x^i_j(u)) = s(p)\cdot A$, say. So for the matrix (ω^i_j) we have

$$\begin{aligned}
(\omega^i_j)\left(\left.\frac{\partial}{\partial(x^\mu\circ\pi)}\right|_u\right) &= (\omega^i_j)\left(R_{A*}\left.\frac{\partial}{\partial x^\mu}\right|_p\right) && \text{by (2)}\\
&= A^{-1}(\omega^i_j)\left(\left.\frac{\partial}{\partial x^\mu}\right|_p\right)A && \text{by definition of an Ehresmann connection}\\
&= A^{-1}(\Gamma^i_{\mu j}(p))A && \text{by (1)} ,
\end{aligned}$$

so that

$$\begin{aligned}
\omega^i_j\left(\left.\frac{\partial}{\partial(x^\mu\circ\pi)}\right|_u\right) &= \sum_{\lambda,\nu} (A^{-1})^i_\lambda \Gamma^\lambda_{\mu\nu}(p)A^\nu_j\\
&= \sum_{\lambda,\nu} y^i_\lambda(u)\Gamma^\lambda_{\mu\nu}(\pi(u))x^\nu_j(u) .
\end{aligned}$$

This accounts for the coefficient of $d(x^\mu\circ\pi)$ in the desired expression for ω^i_j .

Now let us write

$$\sigma(E^\alpha_\beta)(u) = \sum_{\lambda,j} a^\lambda_j \left.\frac{\partial}{\partial x^\lambda_j}\right|_u$$

(clearly each $\partial/\partial x^\lambda_j$ is vertical, since $\pi_*(\partial/\partial x^\lambda_j)(f) = \partial(f\circ\pi)/\partial x^\lambda_j = 0$, as $f\circ\pi$ is constant on fibres; hence the $\partial/\partial x^\lambda_j$ span V_u). Then

$$a_j^\lambda = \sigma(E_\beta^\alpha)(x_j^\lambda)(u)$$

$$= \sigma_{u*}(E_\beta^\alpha)(x_j^\lambda)$$

$$= \lim_{h \to 0} \frac{x_j^\lambda(u \cdot \exp hE_\beta^\alpha) - x_j^\lambda(u)}{h}$$

$$= x_j^\lambda(u \cdot E_\beta^\alpha) = \lambda^{th} \text{ component with respect to } s(\pi(u)) \text{ of } (u \cdot E_\beta^\alpha)_j$$

$$= \text{" " " " " " ." } \sum_\gamma u_\gamma (E_\beta^\alpha)_j^\gamma$$

$$= \text{" " " " " " " } u_\beta \delta_j^\alpha$$

$$= x_\beta^\lambda \delta_j^\alpha \; .$$

So

$$\sigma(E_\beta^\alpha)(u) = \sum_\lambda x_\beta^\lambda \left. \frac{\partial}{\partial x_\alpha^\lambda} \right|_u \; .$$

Hence

$$\delta_\beta^i \delta_j^\alpha = (E_\beta^\alpha)_j^i = \omega_j^i(\sigma(E_\beta^\alpha)(u))$$

$$= \omega_j^i\left(\sum_\lambda x_\beta^\lambda \left.\frac{\partial}{\partial x_\alpha^\lambda}\right|_u\right)$$

$$= \sum_\lambda x_\beta^\lambda \omega_j^i\left(\left.\frac{\partial}{\partial x_\alpha^\lambda}\right|_u\right) \; .$$

This is easily seen to account for the other term, $\sum_\lambda y_\lambda^i dx_j^\lambda$, in the expression for ω_j^i . ∎

Addendum 1. The Tangent Bundle of F(M).

The existence of a connection on the bundle of frames $F(M)$ has some interesting implications.

23. PROPOSITION. If there is a connection on $F(M)$, then the tangent bundle of $F(M)$ is trivial.

Proof. The $n^2 + n$ vector fields $\sigma(E^i_j)$ and $B(e_i)$ are everywhere linearly independent. ▮

24. COROLLARY. If there is a connection on $F(M)$, then M is paracompact.

Proof. Since the tangent bundle of $F(M)$ is trivial, there is clearly a Riemannian metric on $F(M)$. So $F(M)$ is metrizable (Theorem I.9-7), so every component of $F(M)$ is σ-compact (Theorem I.A-1). Since $\pi: F(M) \longrightarrow M$ is a continuous map onto M, it follows that each component of M is σ-compact, so M is metrizable (Theorem I.A-1). ▮

Of course, a non-paracompact manifold M may still have a connection in some principal bundle $\pi: P \longrightarrow M$ with group G. For example, the trivial bundle $M \times G \longrightarrow M$ has an obvious connection (the horizontal vectors Y are those with $\pi_{2*}Y = 0$, where $\pi_2: M \times G \longrightarrow G$ is projection on the second coordinate). If M is paracompact, then there is a connection in any principal bundle $\pi: P \longrightarrow M$; this can be proved using partitions of unity, noting that any convex combination of Ehresmann connections is also a connection (a convex combination of connections ω_i makes sense, since the values of ω_i are in the vector space $\mathfrak{g}$). For the special case of a Riemannian manifold $(M, \langle\ ,\ \rangle)$, we have a specific connection in $F(M)$, the Levi-Civita connection. It is easy to see that the Levi-Civita connection for a Riemannian metric can also be defined for an indefinite Riemannian metric (a non-degenerate inner product in each tangent space); it is the unique symmetric connection for which parallel translation is an isometry, and the Γ_{ij}^{k} are given by exactly the same formula as the Christoffel symbols. So Corollary 24 implies

<u>25. COROLLARY</u>. If M has an indefinite Riemannian metric, then M is paracompact.

As far as I know, there is no simpler way to prove this result (an indefinite Riemannian metric does not give rise to a metric on M, for there are paths of negative length; and the inf of the lengths of paths between two points may be $-\infty$).

Addendum 2. Complete Connections

If ω is a connection on $F(M)$, we define a **geodesic**, as usual, to be a curve c such that dc/dt is parallel along c. The connection ω is **complete** if every geodesic segment can be extended to $\mathbb{R}$. Unlike the Levi-Civita connection for a Riemannian metric, a general connection on $F(M)$ may not be complete even if M is compact. To construct an example, we begin with $R^+ = \{x \in \mathbb{R}: x > 0\}$. The bundle of frames $F(R^+)$ can obviously be identified with $R^+ \times (\mathbb{R} - \{0\})$. Consider the connection for which the horizontal distribution H has the lines $y = x + b$ as integral manifolds. For a curve c in $\mathbb{R}^+$ to be a

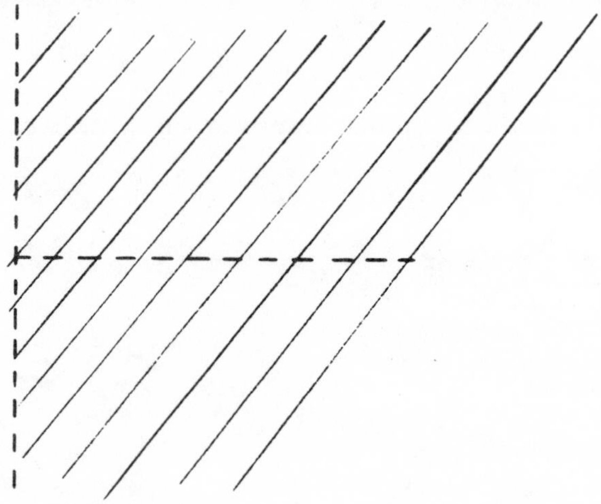

geodesic, the vectors dc/dt must lie along one of these lines, so we must have

$$\frac{dc}{dt} = c(t) + \alpha \qquad \text{for some } \alpha,$$

and consequently

$$c(t) = \beta e^t - \alpha \qquad \text{for some } \alpha, \beta \;.$$

Now identify x with $2x$, for all $x \in \mathbb{R}^+$; the resulting manifold is a circle S^1, and we have a map $\pi\colon R^+ \to S^1$ given by $\pi(x) =$ equivalence class of x. The distribution on $F(R^+)$ gives rise to an obvious distribution on $F(S^1)$, which determines a connection on $F(S^1)$. Since the curve $c: [0,1) \longrightarrow R^+$ defined by $c(t) = 1 - e^t$ is a geodesic, the curve $\pi \circ c$ is a geodesic in S^1. It cannot be defined past 1, since $\pi(c(t))$ goes around S^1 infinitely often as $t \longrightarrow 1$. (A similar example can be constructed if we start with the Levi-Civita connection on $\mathbb{R}$ for the metric $\langle\ ,\ \rangle = e^x dx \otimes dx$, and identify x with $x+1$.)

There are other anomalies for general connections. For example, even though a connection ω is complete, it may not be possible to join every pair of points with a geodesic. If we consider a Lie group G, it is easy to see that there is a unique connection ω on $F(G)$ which makes all left invariant vector fields parallel. Then geodesics through e are just 1-parameter subgroups. The connection is complete, since 1-parameter subgroups can be extended to all of $\mathbb{R}$, but (Problem I.10-29) it is not necessarily true that every element lies on a 1-parameter subgroup. It is not known whether every two points can be joined by a geodesic in a compact manifold with a complete connection.

Finally, it should be pointed out that results about geodesics for the Levi-Civita connection of a Riemannian metric need not hold for the Levi-Civita connection of an indefinite Riemannian metric. For example, there is an indefinite metric whose Levi-Civita connection is complete but for which not every pair of points can be joined by a geodesic (see J. W. Smith, Lorentz structures on the plane, Trans. Amer. Math. Soc. 95(1960)pp. 226-237).

Addendum 3. Connections in Vector Bundles

Suppose $\pi\colon E \longrightarrow M$ is a C^∞ vector bundle over M. We have already seen how to form a principal bundle $\varpi\colon F(E) \longrightarrow M$, in which* every $u \in \varpi^{-1}(p)$ is a frame for $\pi^{-1}(p)$. We can then define a connection in the original vector bundle E to be a connection in the principal bundle $F(E)$. However, there is a more direct way of defining a connection in E, which makes certain things work out more simply, since the bundle of frames contains a lot of superfluous stuff in it.

For a vector bundle $\pi\colon E \longrightarrow M$ (as for a principal bundle) we define the vertical subspace $V_e \subset E_e$ at any point $e \in E$ to be the image $i_*(\pi^{-1}(p)_e)$, where $p = \pi(e)$; vectors in V_e are called vertical, and V_e is clearly the kernel of $\pi_*\colon E_e \longrightarrow M_p$. For every frame $u \in \varpi^{-1}(p)$ and every $\xi \in \mathbb{R}^n$ we also have $u(\xi) \in \pi^{-1}(p)$ defined by $u(\xi) = \sum \xi^i \cdot u^i$. Finally, for every non-zero $\alpha \in \mathbb{R}$, we let $\bar{\alpha}\colon E \longrightarrow E$ be the diffeomorphism defined by $\bar{\alpha}(e) = \alpha \cdot e$. We define a connection in E to be a distribution H such that

(1) $E_e = H_e \oplus V_e$ for all $e \in E$.

(2) $\bar{\alpha}_*(H_e) = H_{\bar{\alpha}(e)} = H_{\alpha \cdot e}$ for all non-zero $\alpha \in E$.

(3) H is a C^∞ distribution .

The subspace H_e is called the horizontal subspace at e, and vectors in H_e are called horizontal. Applying (2) with any $\alpha \neq 1$, and using a local trivialization, we see that if $s\colon M \longrightarrow E$ is the zero section, then

---

*This bundle if a special case of the "associated principal bundle", which is used in the theory of fibre bundles.

$H_{s(p)}$ must be $s_*(M_p)$.

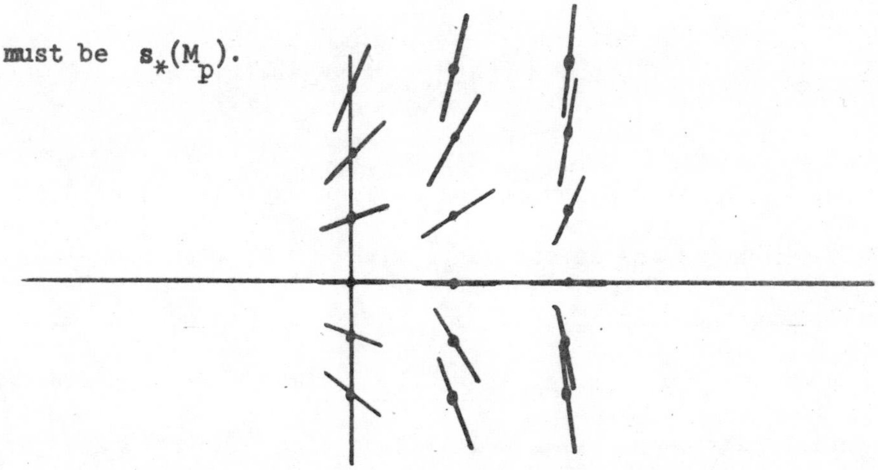

We will first show how such a connection arises from a connection ω on the principal bundle $F(M)$. Let H be the distribution on $F(M)$ determined by ω. Given $e \varepsilon \pi^{-1}(p)$, choose any frame $u \varepsilon \omega^{-1}(p)$; then there is a unique $\xi \varepsilon \mathbb{R}^n$ with $u(\xi) = e$. Now we can define a map $\varphi_\xi : F(E) \longrightarrow E$ by $\varphi_\xi(v) = v(\xi)$ for all $v \varepsilon F(E)$; this map takes fibres of $F(E)$ to fibres of E. We let $H_e = \varphi_{\xi *}(H_u)$. This is well-defined, for if we choose $u \cdot A$ instead of u, then we must choose $A^{-1} \cdot \xi$ instead of ξ; since

$$\varphi_{A^{-1} \cdot \xi}(v) = v(A^{-1} \cdot \xi) = v \cdot A^{-1}(\xi) \quad ,$$

we have $\varphi_{A^{-1} \cdot \xi} = \varphi_\xi \circ R_{A^{-1}}$, so

$$(\varphi_{A^{-1} \cdot \xi})_*(H_{u \cdot A}) = \varphi_{\xi *} R_{A^{-1} *}(H_{u \cdot A})$$

$$= \varphi_{\xi *} H_u \, , \qquad \text{by Proposition 4.}$$

We leave it to the reader to verify that these well-defined H_e do satisfy conditions (1),(2),(3).

We also want to show that every connection H in E does arise in this way from some connection H in $F(E)$. To do this, we first consider parallel translation. If $c: [0,1] \longrightarrow M$ is a curve, then the parallel translation

$$\tau_t\colon \pi^{-1}(c(0)) \longrightarrow \pi^{-1}(c(p))$$

determined by H is defined in the obvious way: $\tau_t(e) = c^\dagger(t)$ where $c^\dagger\colon [0,1] \longrightarrow E$ is the unique curve with $c^\dagger(0) = e$ such that each $c^{\dagger\prime}(t)$ is horizontal; the existence of $c^\dagger$ in E is proved similarly to the existence of $c*$ in a principal bundle. Clearly τ_t is a diffeomorphism, whose inverse is the parallel translation along the reversed part of c from $c(t)$ to $c(0)$. From condition (2) for H it follows that $\tau_t(\alpha\cdot e) = \alpha\cdot\tau_t(e)$ for $\alpha \neq 0$. This is true even for $\alpha = 0$, i.e., τ_t takes the zero vector at $c(0)$ to the zero vector at $c(t)$; this follows from the fact that $H_{s(p)} = s_*(M_p)$ when s is the zero section. It now follows that τ_t is a vector space isomorphism, because of the following

Clever Observation: If $f\colon \mathbb{R}^n \longrightarrow \mathbb{R}^m$ is differentiable at 0, and $f(\alpha\cdot v) = \alpha f(v)$ for all $v \in \mathbb{R}^n$ and $\alpha \in \mathbb{R}$, then f is linear.

Proof. Let $T = Df(0)\colon \mathbb{R}^n \longrightarrow \mathbb{R}^m$. Then

$$\begin{aligned} T(v) &= \lim_{\alpha\to 0} \frac{1}{\alpha}[f(\alpha v) - f(0)] = \lim_{\alpha\to 0} \frac{1}{\alpha} f(\alpha v) \\ &= \lim_{\alpha\to 0} f(v) = f(v) \quad . \ \blacksquare \end{aligned}$$

We now define the connection H in $F(E)$. Let $u \in \varpi^{-1}(p)$. For every C^∞ curve $c\colon [0,1] \longrightarrow M$ with $c(0) = p$, we define the curve $c^*\colon [0,1] \longrightarrow F(E)$ by

$$(c^*(t))_i = c^\dagger(t)(u_i), \qquad i = 1,\ldots,n,$$

the subscript i denoting the ith component of the frame. Then we define H_u to be the set of all vectors $c^*(0)$ for all such curves c. The reader may verify that H is an Ehressman connection, and that it gives rise to the connection H on E (the fact that τ_t is a vector space isomorphism is used to prove that $R_{A*}H_u = H_{u\cdot A}$).

Given a connection H on E, we use the direct sum decomposition $E_e = H_e \oplus V_e$, to define the horizontal component hX and vertical component vX of any tangent vector $X \in E_e$. Condition (2) for H is equivalent to the equation $\bar{\alpha}_* hX = h(\bar{\alpha}_* X)$, or to $\bar{\alpha}_* vX = v(\bar{\alpha}_* X)$. Remember that the tangent space $\pi^{-1}(p)_e$ at e of the vector space $\pi^{-1}(p)$ can be identified with $\pi^{-1}(p)$ itself. Since $vX \in i_*(\pi^{-1}(p)_e)$, we therefore can, and henceforth will, regard vX as an element of $\pi^{-1}(p)$.

The equation $\bar{\alpha}_* vX = v(\bar{\alpha}_* X)$ then becomes

$$(*) \qquad v(\bar{\alpha}_* X) = \alpha \cdot vX \quad .$$

Now consider a section $s: M \to E$ of the bundle $\pi: E \to M$, and a vector $X_p \in M_p$. We define

$$\nabla_{X_p} s = vs_*(X_p) \in \pi^{-1}(p) \quad .$$

It is clear that $\nabla_{X_p} s$ is linear in X_p and in s. For a C^∞ function $f: M \to \mathbb{R}$, the analogue of Proposition 3 is the formula

$$(**) \qquad (f\cdot s)_*(X_p) = \overline{f(p)}_*(s_* X_p) + X_p(f)\cdot s(p) \quad ,$$

where $s(p) \in \pi^{-1}(p)$ is identified with a tangent vector in $\pi^{-1}(p)$ at $f(p)\cdot s(p)$; to prove this formula, we introduce a local trivialization, and observe that it becomes the product rule for the derivative. If we take the vertical component of both sides of (**) and use (*), we find that

$$\nabla_{X_p}(fs) = f(p)\nabla_{X_p} s + X_p(f)\cdot s(p) \quad .$$

Thus ∇ is a Koszul connection. If the connection H in E comes from the connection H in F(E), then this ∇ is the same as that determined by H; this is an easy consequence of Lemma 8 (although Lemma 8 is concerned only with $F(M) = F(TM)$, it is easy to see that it generalizes to any F(E)).

Finally, we point out that if we are given ∇, then H_e is simply $\{s_*(X_p)\colon p = \pi(e) \text{ and } \nabla_{X_p} s = 0\}$; it can also be defined as $s_*(M_p)$, where s is a section such that $\nabla s(p) = 0$.

Addendum 4. Flat Connections, and an Apology

Consider the trivial principal bundle $\pi: M \times G \longrightarrow M$, and let $\pi_2: M \times G \longrightarrow G$ be projection on the second factor. We can define the canonical flat connection H on $M \times G$ by letting $H_{(p,a)} = \ker \pi_{2*}: (M \times G)_{(p,a)} \longrightarrow G_a$. It is easy to see that the corresponding $\mathfrak{g}$-valued 1-form ω on $M \times G$ is given by $\omega = \pi_2^{\ *}(\omega')$, where ω' is the natural $\mathfrak{g}$-valued 1-form on G (defined on page I.10-49). Using the equations of structure of G, in the form given on page I.10-51, we now have, for all $Y_1, Y_2 \in (M \times G)_u$

$$\begin{aligned}
d\omega(Y_1,Y_2) &= d\pi_2^{\ *}(\omega')(Y_1,Y_2) \\
&= \pi_2^{\ *}(d\omega')(Y_1,Y_2) \\
&= d\omega'(\pi_{2*}Y_1,\pi_{2*}Y_2) \\
&= -[\omega'(\pi_{2*}Y_1),\omega'(\pi_{2*}Y_2)] \\
&= -[\omega(Y_1),\omega(Y_2)] \quad .
\end{aligned}$$

Comparing with the structural equations of the principal bundle $M \times G$ (Theorem 16), we see that for this connection ω we have $\Omega = 0$.

Conversely, suppose $\pi: P \longrightarrow M$ is a principal bundle with group G, and a connection ω such that $\Omega = 0$. By Proposition 18, the distribution H corresponding to ω is integrable. From this it is easy to see that around any point $p \in M$ there is a neighborhood U and a diffeomorphism $t: \pi^{-1}(U) \longrightarrow U \times G$ of the form $t(u) = (\pi(u),\varphi(u))$, where $\varphi(u\cdot a) = \varphi(u)\cdot a$, such that $t_*(H_u)$ is the horizontal subspace at $t(u)$ for the canonical flat connection in $U \times G$. (One part of our final proof of the Test Case essentially used this fact.) The connection on $F(S^1)$ which was constructed in Addendum 2 shows that we may not be able to choose U as all of M. However, it can be shown that if M is simply connected, then we can choose $U = M$.

Despite the promise made in Volume 1 regarding the important role to be played by the equations of structure of a Lie group, the reader has probably already realized that they have not been used once in this volume, except in this Addendum. The second structural equation of Euclidean space (Proposition 7-1), with which we began our whole investigation of moving frames, could have been deduced from the equations of structure of $GL(n,\mathbb{R})$ (by a process essentially equivalent to the deduction just given for the equation $d\omega(Y_1,Y_2) = -[\omega(Y_1),\omega(Y_2)]$). However, it seems to me that this would have just complicated everything unnecessarily, by bringing in the bundle of frames before we needed it. The first form of the equations of structure, on page I.10-50, and the notation $[\omega \wedge \omega]$ are never needed at all.

The reader might also wonder how such an approach would have yielded the first structural equation of Euclidean space. To obtain this equation, we would have had to consider the equations of structure for the group of affine motions [an "affine motion" is a translation followed by an element of $GL(n,\mathbb{R})$]. In general, for a connection ω on $F(M)$, the torsion form and the first structural equation for ω can be interpreted in terms of a connection in the bundle $A(M)$ of affine frames of M, where an "affine frame" of M_p is a pair $(v,u_1,\ldots,u_n)$, for $v \in M_p$ and $(u_1,\ldots,u_n)$ a frame for M_p. For this interpretation, the reader is referred to Kobayashi and Nomizu, Foundations of Differential Geometry, Vol. 1, pp. 125-130.

NOTATION INDEX

Chapter 8 (cont.)

(cont.)

Chapter 7

(cont.)

Chapter 8

INDEX

CONTENTS OF VOLUME I

[Although the Chapters are not divided into sections, the list for each Chapter gives some indication which topics are treated, and on which pages. The numbers in brackets after the PROBLEMS refer to Problems which, for one reason or another, I felt should be brought to the attention of the reader.]

CONTENTS OF VOLUME I

CONTENTS OF VOLUME I

CONTENTS OF VOLUME I

Errata for Volume I

page 1-23, line 3 from bottom
change a "closed manifold" to "non-bounded"; a non-bounded compact manifold is called a "closed manifold"

page 2-7, line 4
change a function to a C^∞ function

page 2-8, line 6
change for Problem . to for Problem 14.

page 2-16, line 3
change $\frac{\partial x^j}{\partial y}(p)$ to $\frac{\partial x^j}{\partial y^i}(p)$.

page 2-30, line 9 from bottom
change rank k on $f^{-1}(y)$ to rank k on a neighborhood of $f^{-1}(y)$.
add to the statement of the Proposition In particular, if y is a regular value of $f\colon M^n \longrightarrow N^m$, then $f^{-1}(y)$ is an $(n-m)$-dimensional submanifold of M (or is empty).

page 2-44, line 10 from bottom
change rank k on to rank k on a neighborhood of

page 3-23, footnote
change category of C^∞ maps to category of C^∞ manifolds and C^∞ maps

page 3-50, Problem 19
It must also be assumed that $a\cdot(1,0,\dots 0) = a$ for all a.

page 4-39, line 2 from bottom
change Problem 8-12 to Problem 7-12.

page 5-19, line 6
change $\beta(t)$ for $|t| < \varepsilon$ to $\beta(t) = \alpha(t,\alpha(s,x))$ for $|t| < \varepsilon$

page 7-22, line 7
change PROPOSITION to PROPOSITION (FROBENIUS INTEGRABILITY THEOREM; SECOND VERSION)

page 8-16, line 7 from bottom
change $(c_{(i,\alpha)})_{j,\beta}$ to $(c_{(i,\alpha)})_{(j,\beta)}$

page 8-36, line 3
change Clearly to Thus (Problem 15)

page 9-76, line 4 from bottom

change $+\sum_{\ell=1}^{n} \frac{\partial^2 x^\ell}{\partial x'^\alpha \partial x'^\beta}$. to $+\sum_{\ell=1}^{n} \frac{\partial^2 x^\ell}{\partial x'^\alpha \partial x'^\beta} \frac{\partial x'^\gamma}{\partial x^\ell}$.

Errata for Volume I

page 9-78, line 3 from bottom

change $\int^{t} e^{M(s)}ds,$ to $\int_{c}^{t} e^{M(s)}ds,$

page 9-85, line 4

change $t_0,$ to $t_1,$

page 10-1, line 6 from bottom

change $x,y \longrightarrow xy^{-1}$ to $(x,y) \longmapsto xy^{-1}$

page 10-50, line 3 from bottom

change $\sum_{i=1}^{n} \sum_{j=1}^{n} C_{ij}^{k} \omega^{i} \wedge \omega^{j} \cdot X_k$ to $\sum_{k=1}^{n} \left(\sum_{i=1}^{n} \sum_{j=1}^{n} C_{ij}^{k} \omega^{i} \wedge \omega^{j} \cdot X_k \right)$

page 10-53, line 2 from bottom

change (n) to $\mathcal{E}(n)$

page 10-60, line 2 from bottom

change sufficiently close to 0, and ... to sufficiently close to 0.

add (c) If $[X_i, X_j] = \sum_k C_{ij}^{k} X_k$ for certain constants C_{ij}^{k}, show that this multiplication is associative for points sufficiently close to 0.

page 10-61, line 1

change (c) to (d)

page 10-65, line 7

change matrix multiplication. to matrix multiplication, and P^{-1} denotes the map $A \longrightarrow A^{-1}$ on G.

page 10-65, Problem 24(e)

change to (e) Using $dP = P \cdot \omega$, show that $0 = dP \cdot \omega + P \cdot d\omega$, where the matrix of 2-forms $dP \cdot \omega$ is computed by formally multiplying the matrices of 1-forms dP and ω. Deduce that ...

page 11-25, line 4

change $0 \leq u < p(x)$ to $0 \leq u \leq \rho(x)$

page 11-37, lines 3 and 5

change f_x to f_X

page 11-42, line 6

change THEOREM to THEOREM (POINCARÉ-HOPF)